A FIRST COURSE IN

Geophysical Exploration and Interpretation

ROBERT E. SHERIFF
University of Houston

INTERNATIONAL HUMAN RESOURCES DEVELOPMENT CORPORATION
Boston

ISBN: 0-934634-02-5

Library of Congress Catalog Card Number: 78-70766

Printed in the United States of America

TABLE OF CONTENTS

PREFACE

This book was originally written as the textbook for the videotape program of the same name, produced and distributed by International Human Resources Development Corporation. As a summary of the course, however, it is a text which stands on its own and can be used by anyone interested in the principles of geophysical exploration and interpretation.

The course includes problems and exercises in several chapters (indicated in the Table of Contents). The exercises are discussed following the statements of them, and it is intended that they be worked at the points where they are stated before reading the discussion.

The goal of this course is to aid geologists in understanding geophysics more clearly so that they can make better geological interpretation of geophysical data. Geologists are using more and more geophysics and most geologists spend a great deal of their time doing geophysical interpretation. It is necessary to understand geophysical principles and how they affect interpretation in order to use geophysical data to best advantage.

I thank the organizations who provided the illustrations shown in this text and on the tapes. Abbreviations in the lower right corner of figures in this text refer to the sources, which are identified in the back of this book. I especially thank Seiscom Delta for their cooperation, Phillips Petroleum Company for making the videotapes, Scott Carlberg, the director of the videotaping sessions, Susan Secord of IHRDC, my wife Margaret for counsel and help in many ways, including the writing of this text, and Lynda Allen for typing the text.

NOTE TO VIDEOTAPE USERS

This text follows the videotapes in subject order but in abbreviated form. Some of the videotape figures and discussions are not included in the text or are expressed differently, but the text is intended as a summary and review of the videotapes rather than a text to be followed while watching the tapes.

Each chapter has been numbered to correspond with the videotape session of the same number. It is suggested that each chapter be read after viewing the appropriate tape.

CHAPTER 1

INTRODUCTION: THE TOOLS OF A GEOPHYSICIST

A geophysicist needs many tools to solve problems of different kinds. The different geophysical tools have in common the fact that they involve measurements of physical properties of rocks. Let us consider some of these tools.

From magnetics, we can learn:

Limits of basins,
Depth to basement,
Basement lineations.

Chapter 6 of this course deals with magnetics.

Electrical exploration is not used a great deal in the United States, although the French and the Soviets use it extensively. We may be missing opportunities by not doing more with electrical methods, but you probably will not work with electrical data. Electrical exploration can give us information about:

Depth of thick high resistivity members,
Resistivity vs. depth,
Prospect leads.

Gravity data are much used in looking for hydrocarbons. Gravity, along with magnetics, is one of the principle tools used in the early phases of exploration. Gravity data can tell us something about:

Basin shape and extent,
Structural trends,
Fault locations,
Prospect leads.

The main use of gravity is to develop leads for prospects. However, gravity also has considerable use in later phases of exploration as a tool for checking the validity of interpretations. An interpretation must be compatible with all information that is available, including gravity. Chapters 2 through 5 deal with gravity interpretation.

The earliest seismic exploration was refraction. Refraction tells us something about:

Presence of high velocity members,
Maps on high velocity members,
Velocity gradients.

Refraction work was especially useful in finding salt domes in Tertiary sediments because of the large velocity contrast compared with the surrounding sediments. While refraction is little used today, there are situations where it can be very useful.

The main tool used by geophysicists is seismic reflection, which can be used for both regional and detailed work. Seismic reflection can tell:

Types of structures,
Structural character,
Relation of different features,
Sediment velocities,
Sediments which flow,
Unconformities,
Direction of sediment source,
Elements of geological history,
Inferences from reflection character,
Inferences of depositional environment,
Inferences of age,
Problems in mapping,
Prospect leads and prospect definition,
Leads as to gas accumulations.

Another important tool for a geophysicist is information from stratigraphic holes. From such holes we can learn about:

Nature of sediments,
Age of sediments,
Possible source/reservoirs.

All types of information must be tied together in order to make a worthwhile interpretation. An interpretation should include not only a given set of geophysical data but also everything else known about the section, including information from surface geology, concepts of the regional tectonics, gravity and magnetic data, and information from boreholes. Worthwhile geophysical interpretation can not be done in a vacuum.

All exploration methods involve signal and noise. Signal is what we wish to measure. In gravity work signal is the field of the mass distribution produced by some feature, and in seismic work it is reflections from interfaces in the subsurface. Noise is anything which obscures signal; noise is always present in greater or lesser extent. Sometimes noise is predominant. A discriminant is needed to separate signal from noise; discriminant is an aspect in which signal differs from noise. Signal compared with noise can be improved in one of two ways: by strengthening the signal or by attenuating the noise.

Signal and noise are always mixed together. However they mix in different ways as the method of making measurements varies. We can use this property of mixing differently as a discriminant to aid in separating signal from noise. Common-depth-point seismic work measures reflections from the same point in a number of different ways and then combines the measurements to improve signal-to-noise ratio.

A general characteristic of measurement is that they become more definitive if the measurements can be made closer to the cause. Anomalies become broader and signal-to-noise ratio becomes poorer as distance to the cause of the anomaly becomes greater; this can be seen in gravity analysis and in seismic work. Wave equation migration is a method of showing features more sharply by effectively planting the geophones closer to the features.

Exploration has to rely on inferential reasoning. A typical line of reasoning is:

a is often related to b,
b is often related to c,
c is often related to d, etc.

Therefore, if a is observed, perhaps d can be inferred. This is not very rigorous logic, yet we often have to follow such a line of reasoning for want of anything better. It must be realized that the "answers" we come up with are hypotheses and much of our work is spent testing hypotheses against other criteria to determine if they are wrong.

As an example of inferential reasoning, we might observe magnetic measurements which we interpret to indicate features in basement rocks. The magnetic features might suggest a fault in the basement. A fault in the basement might be a zone of weakness which might be reactivated during subsequent periods. Therefore, the trend of this basement fault might define hingelines, changes in the depositional environment, or other features in sedimentary rocks above it. Thus we use the magnetic evidence of a basement fault which we can see, to infer the location of sedimentary features which we cannot see.

The problem with inferential reasoning is that a whole set of alternative explanations are usually available. Perhaps the result of a set of measurements is that the answer is a, or b, or c, or d, or some other answer. The multiple possible answers are eliminated by introducing additional constraints. The line of reasoning is diagrammed in Figure 1.1.

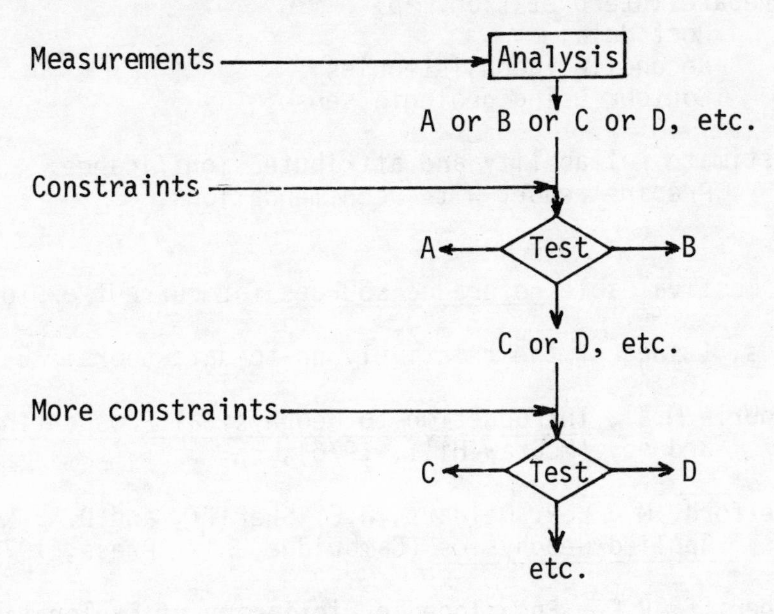

Figure 1.1

Data reduction is the conversion of a set of measurements into another set which is more useful. The steps of data reduction usually involve the following:

Examine for bad data:
- Failure of instruments,
- Variations beyond allowable tolerance,
- Disturbances from outside;

Resolve location:
- Locate measurement points,
- Adjust to tie,
- Prepare base map;

Collate various measurements:
- Apply appropriate corrections,
- Mark line intersections,
- Prepare profile plots;

Prepare map:
- Post data on map,
- Contour with geologic sense;

Select anomalies for study:
- Locate anomalies,
- Select portions of data free of interference,
- Make appropriate measurements,
- Compare measurements of same features,
- Attribute significance to anomalies,
- Assign grading factors;

Prepare interpretation map:
Post data,
Reconcile inconsistencies,
Contour using geologic sense;

Estimate reliability and attribute significance:
Prepare report with recommendations.

As the most valuable reference sources for current exploration geophysics, I suggest the relatively up-to-date books:

Dobrin, M.B., Introduction to Geophysical Prospecting, 3rd ed. (McGraw-Hill, 1976);

Telford, M., L.P. Geldart, R.E. Sheriff, and D.A. Keys, Applied Geophysics (Cambridge Univ. Press, 1976);

Sheriff, R.E., Encyclopedic Dictionary of Exploration Geophysics (Society of Exploration Geophysicists, 1973).

More specialized references include the journals:

Geophysics (Society of Exploration Geophysicists);

Geophysical Prospecting (European Association of Exploration Geophysicists);

and several specialized textbooks:

Anstey, N.A., Seismic Prospecting Instruments, Vol. 1 (Gebruder Borntraeger, Berlin, 1970);

Fitch, A.A., Seismic Reflection Interpretation (Gebruder Borntraeger, Berlin, 1976);

Kanasewich, E.R., Time Sequence Analysis in Geophysics (Univ. Alberta Press, 1973);

Nettleton, L.L., Gravity and Magnetics in Oil Prospecting (McGraw-Hill, 1976);

Tucker, P.M., and H.J. Yorsten, Pitfalls in Seismic Interpretation (Society of Exploration Geophycists, 1973).

A number of more specialized texts are also available.

Geophysics, especially seismic exploration, has been a very fast-moving discipline. New developments occur rapidly, and in this course, we'll discuss both new techniques and established basics.

Geophysical Activity Analysis

Source of Data

Statistics of geophysical activity are compiled on an annual basis by the Society of Exploration Geophysicists and published in the journal "Geophysics." These activity summaries cover most of the geophysical exploration work which is done in the world outside of work in Communist countries, especially Russia and China. The last year for which such data are available is 1977, which was published in "Geophysics", Volume 43, Pages 1277-91, October, 1978.

In addition to these annual summaries, monthly surveys of work in the United States have been carried out by the International Association of Geophysical Contractors and the SEG since May, 1974; these figures are published in "Geophysics".

Activity Summary

Worldwide expenditures for geophysical exploration in 1977 exceeded 1.1 billion dollars, an increase of 2% over 1976. Of the reported expenditures, 43% were spent for exploration in the United States. The geographical distribution of expenditures is shown in Figure 1.2. Of the total expenditures, 94% involve petroleum exploration and 92% was for seismic work. Almost all of the petroleum seismic work was reflection work. Of the petroleum seismic expenditures, 79% were for exploration on land, up 11% over 1976, whereas marine work was down 23%; land work includes work in marsh areas and shallow water where marine vessels cannot operate.

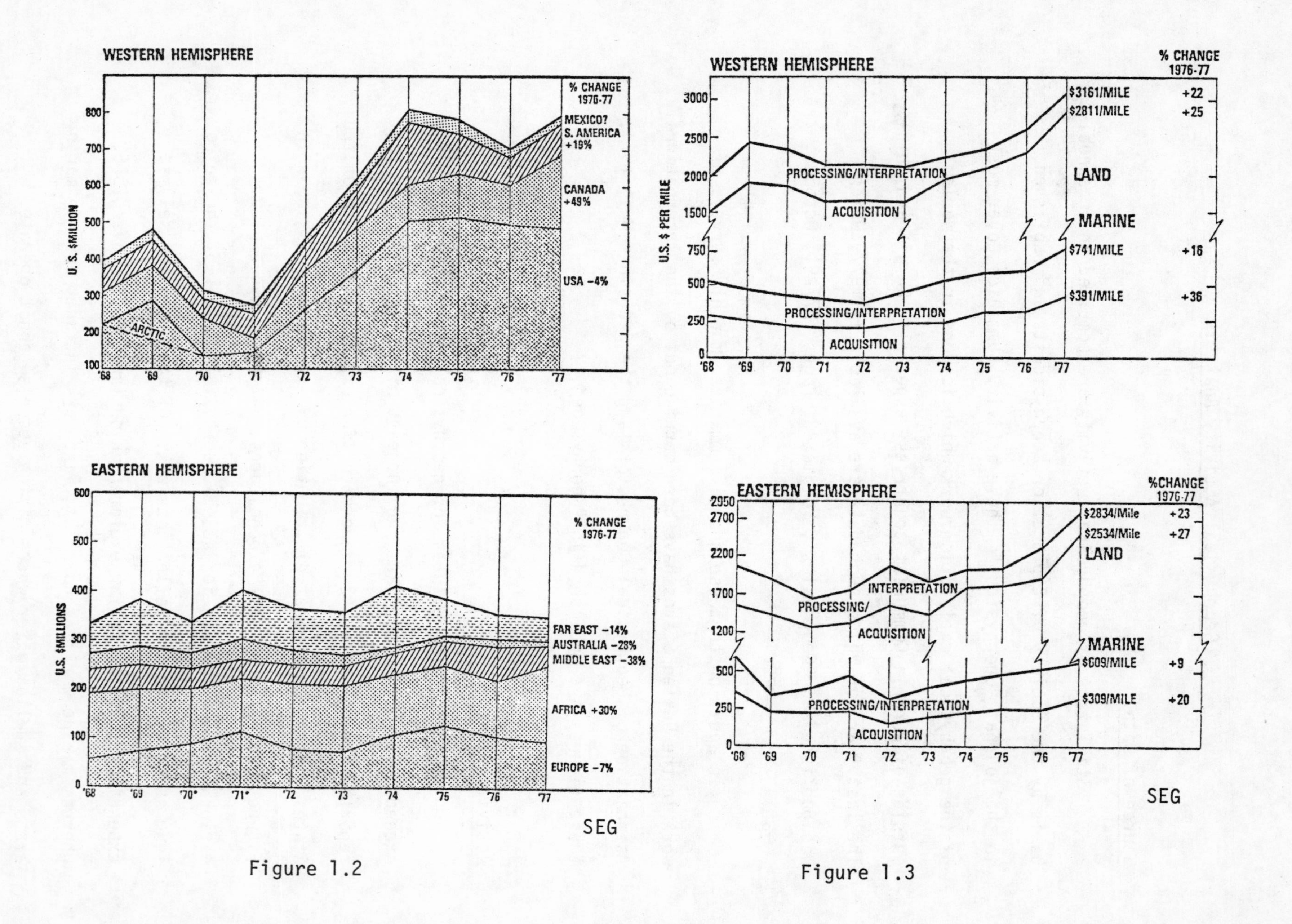

Figure 1.2

Figure 1.3

Cost per mile for seismic work is shown in Figure 1.3. The data are gross averages and costs for individual surveys may vary considerably, not uncommonly by factors of 5 or 6, depending upon local difficulties of data acquisition. In getting data economically, logistics, terrain, access and legal restrictions are of first order importance. Costs in remote areas or areas which are difficult to access may be much greater than in easier and more accessable areas.

Annual figures for crew months and for line miles is shown in Fig. 1.4 and 1.5. The number of miles of land work world-wide using various energy sources was tabulated as follows

dynamite	57%,	up 7% compared to 1976;
vibroseis	31%,	down 1%;
airgun	2%;	down 4%;
mechanical sources	5%,	down 1%;
gas exploder	1%;	down 3%;
solid chemical	4%;	up 2%;
other sources	1%;	

For marine work the breakdown by energy source is:

airguns	76%;	up 16% compared to 1976;
gas exploder	3%;	down 17%;
implosive	19%;	up 12%;
solid chemical	1%;	down 6%;
electrical	2%;	down 4%.

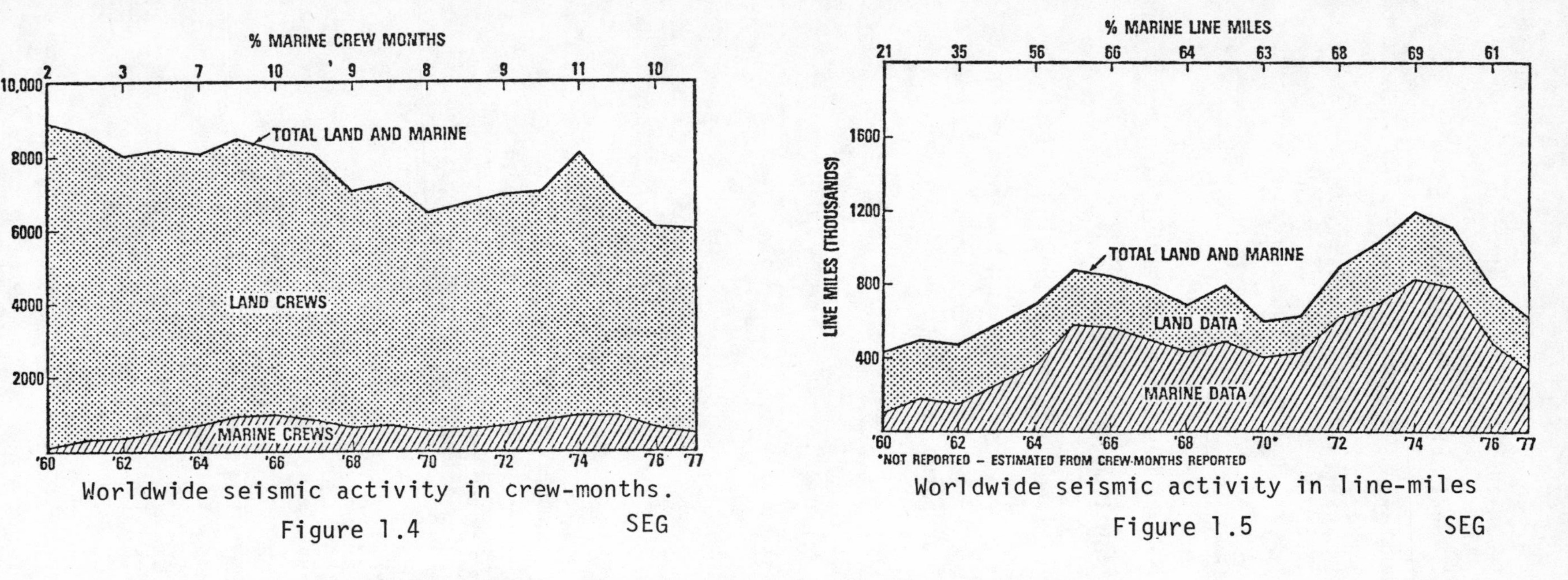

Worldwide seismic activity in crew-months.

Figure 1.4 SEG

Worldwide seismic activity in line-miles

Figure 1.5 SEG

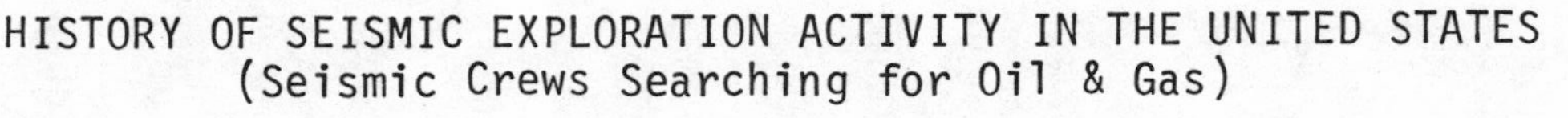

HISTORY OF SEISMIC EXPLORATION ACTIVITY IN THE UNITED STATES
(Seismic Crews Searching for Oil & Gas)

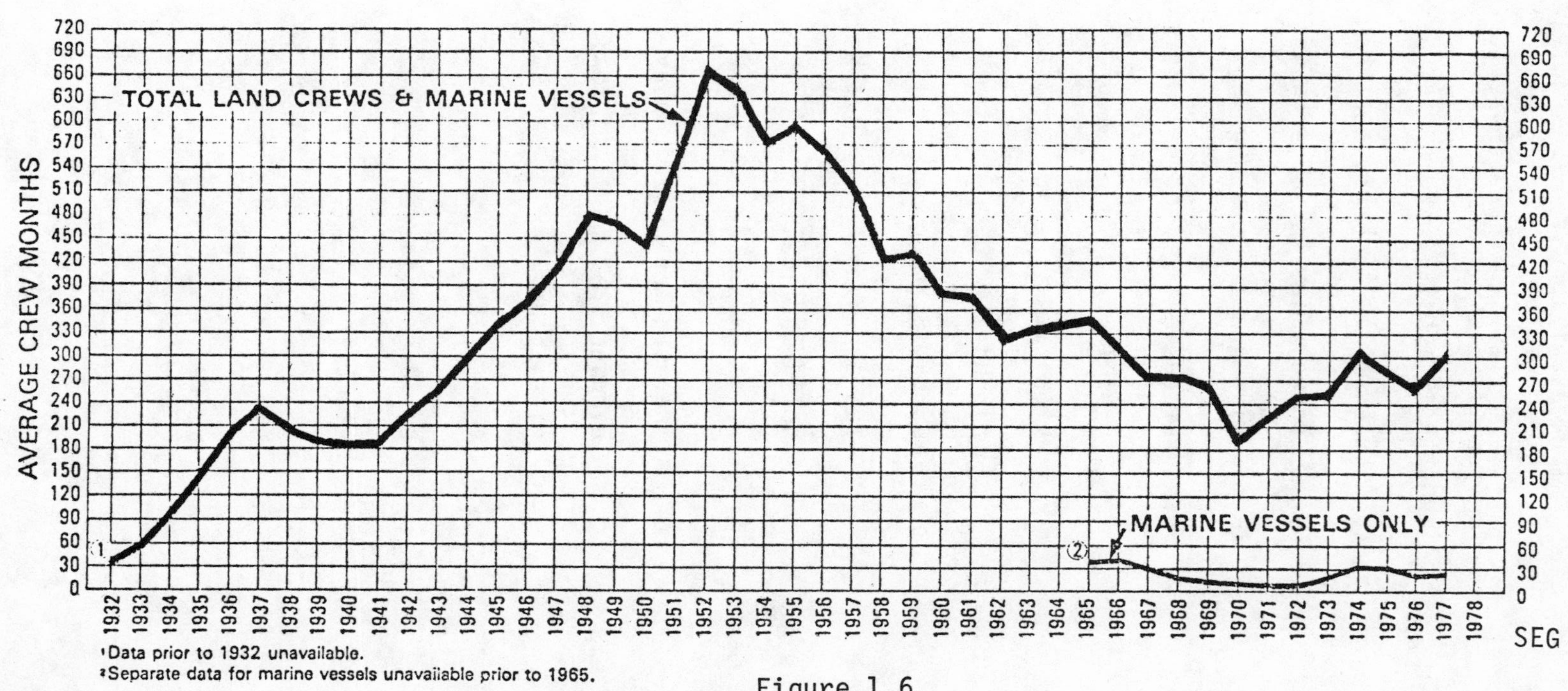

[1]Data prior to 1932 unavailable.
[2]Separate data for marine vessels unavailable prior to 1965.

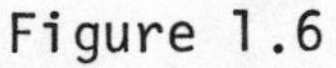

Figure 1.6

The data used to be broken out according to analog vs. digital and single coverage vs. common-depth-point, but such breakouts are no longer made because virtually all data are now recorded in digital form and virtually all data are now common-depth-point recording.

Emphasis in data recording is on maintaining constancy along the seismic line; one needs to be confident that changes which are seen result from changes in the geology rather than changes in recording techniques. Routine practices in data acquisition thus favor both economy in data acquisition and also the objective of being consistent along the line.

Historical trends are shown in Figure 1.6 over the long term, and for recent work in the United States in Figure 1.7.

The curve of wildcat well drilling generally lags the curve of seismic activity by 2-3 years and the level of seismic activity is generally taken as a leading indicator of hydrocarbon discoveries.

The amount and reliability of geologic conclusions drawn from seismic work have been considerably improved in the last few years and seismic improvements have been cited as one of the factors responsible for improvement in the success ratio of wells drilled over the last few years, as shown in Figure 1.8.

SEISMIC CREW COUNT

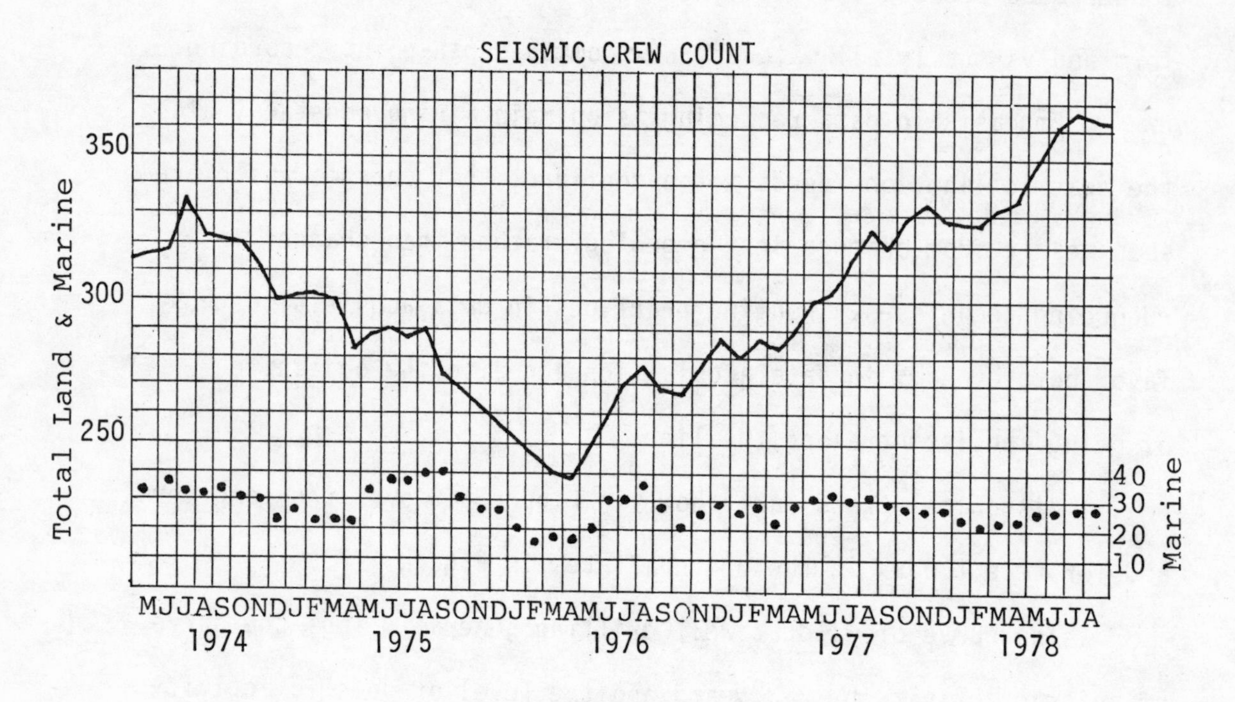

Figure 1.7

EXPLORATORY HOLES COMPLETED AS PRODUCERS

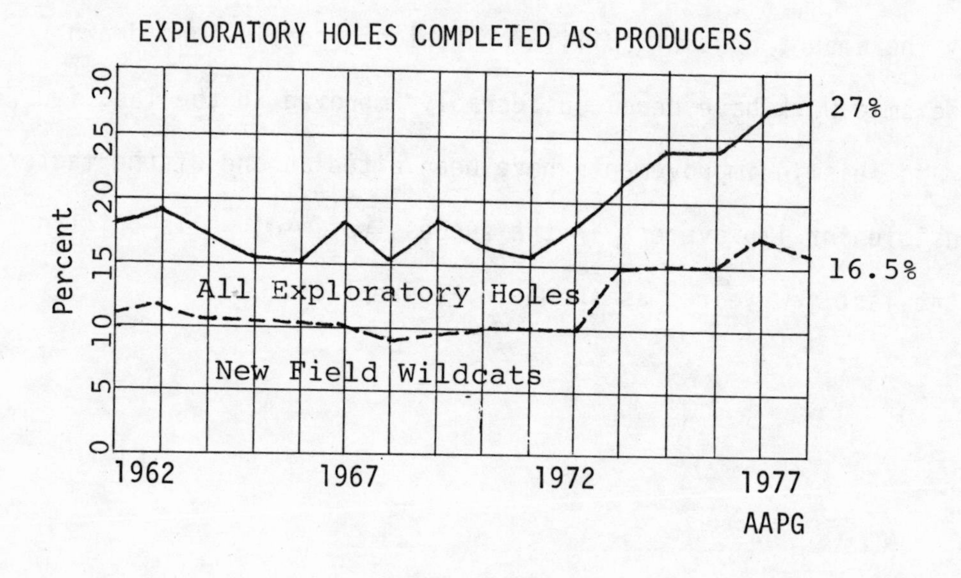

Figure 1.8

CHAPTER 2

GRAVITY I: MEASURMENT AND DATA CORRECTION

Gravity exploration is based on the familiar Newton's Law of Universal Gravitation. It very simply states that:

$$\begin{pmatrix}\text{Gravitational force}\\ \text{between two masses}\end{pmatrix} = \begin{pmatrix}\text{universal gravi-}\\ \text{tational constant}\end{pmatrix} \times \left(\frac{\text{product of masses}}{(\text{distance between masses})^2}\right)$$

In gravity exploration, we ordinarily measure the force on one mass, a "proof" mass within the gravity meter; this is called the acceleration of gravity.

Of course in the earth we have many masses, all of which attract the mass in the gravity meter. Acceleration and force are vector quantities and the effects of the different masses add vectorially. With gravity we have a big advantage in adding vectors, because we know that the net result is in the vertical direction; thus we only need to add the vertical components, knowing that the horizontal components will add to zero. Hence we only need scaler addition and we don't have to use vector addition at all. While the earth is a continuity of masses, we usually think of it as broken into many small cells and the net effect is obtained by summing the vertical effects of each of these cells.

Actually, gravity meters measure differences in acceleration of gravity. The measuring procedure is to level the meter and balance the meter by changing the tension on a spring so that the aggregate gravitational forces on the proof mass is balanced by the force of the spring; then we move the meter to another location, level and balance again, measuring the change in the spring tension required to bring the meter back into balance at the new location. The effect, therefore, is one of measuring the change in the acceleration of gravity between the two locations.

Gravity values are usually referenced to some starting point. If the absolute force of gravity is known at some point, we may use that as the starting point. However, if the absolute value is not known, we can still carry out a meaningful survey with reference to any arbitrary starting point because only differences between locations is of interest in interpretation.

The usual unit of gravity measurement is the milligal, abbreviated mgal, an acceleration of 0.001 cm/sec/sec or 0.001 dyne per gram. Sometimes a gravity unit is used instead; it is 0.1 of a milligal. A milligal is named for Galileo.

As an example of how land gravity surveys might be carried out, Fig. 2.1 shows meter readings made during one day's work. The dots represent the meter readings at various gravity stations.

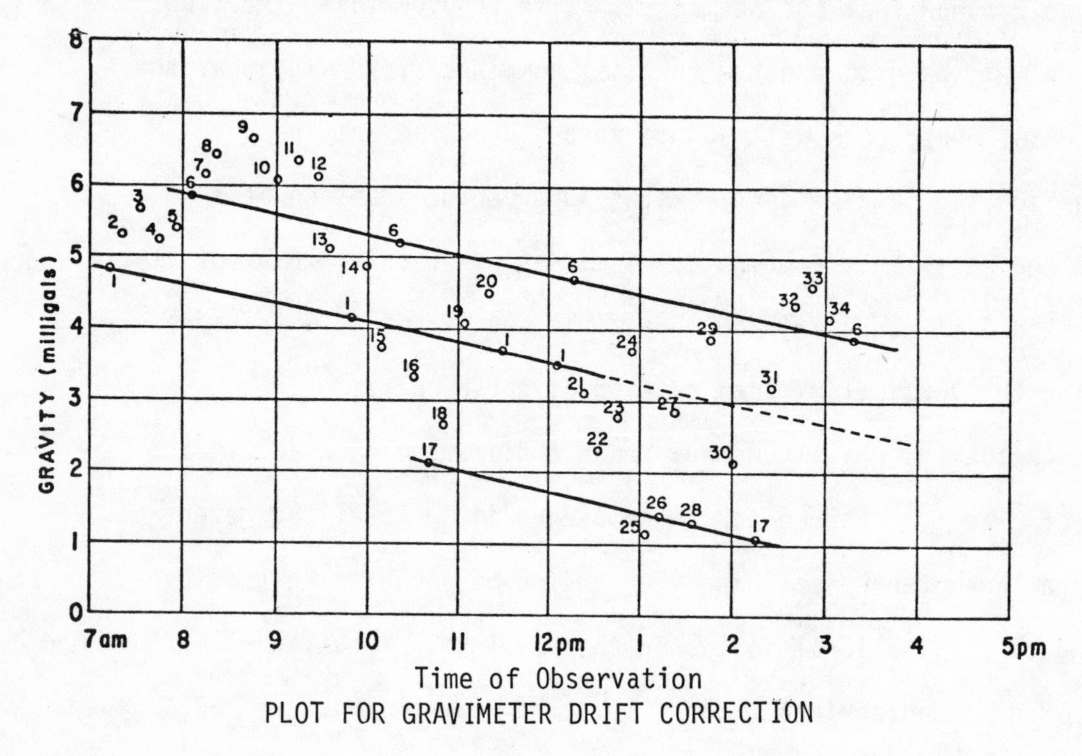

Time of Observation
PLOT FOR GRAVIMETER DRIFT CORRECTION

Figure 2.1

Note the crew started early; the first reading was just after 7 a.m. at station number 1, probably a base station close to the crew camp. Thereafter they measured 3 to 4 points/hour in an accessable area and the time between readings allowed them to drive about ½ kilometer to each new location, set up the meter and make the reading.

Occasionally they repeat measurements at earlier locations. On the day shown in Fig. 2.1, the first station of the day was repeated about 9:45 and also several other times during the day. The gravity meter is an extremely sensitive device and its properties change gradually with time, that is, the meter drifts slightly. By periodically repeating measurements at the same location, we determine the amount of this drift and correct for it. Also during the course of the day the position of the sun and moon change with respect to the portion of the earth being surveyed, which changes the gravitational pull exerted by these bodies; such changes are called tidal effects and are sometimes lumped together with the corrections for meter drift. During a day's survey, readings at several stations will have been repeated, lines can be drawn through the values at these stations, and differences between values at other stations and these lines gives the difference between gravity values at those stations and the reference stations.

The attraction of the earth as a whole is about 980 Galileos, that is, 980,000 mgals. Ordinary gravity anomalies of interest are the order of magnitude of 1 mgal. Thus we are interested in variations in the earth's gravity field of about 1 part in a million. In proportion to the weight of a 200 pound man, the part of interest is of the order of 0.003 ounce. Gravity meters can be read to 0.01 mgal and gravity meter accuracy is usually not the limiting factor on survey accuracy.

Gravity measurements depend upon many things besides the distribution of masses of interest for petroleum exploration. We correct for these other effects, which are summarized in Fig. 2.2. We have already discussed the correction for meter drift and tidal effects. Tidal effects can be calculated from the time of observation and knowledge of the phase of the moon if they are not to be lumped with the meter drift correction. The accuracy with which the corrections can be made really determines the accuracy of the survey.

The centrifugal force caused by the rotation of the earth will have a component which will make the gravity meter read

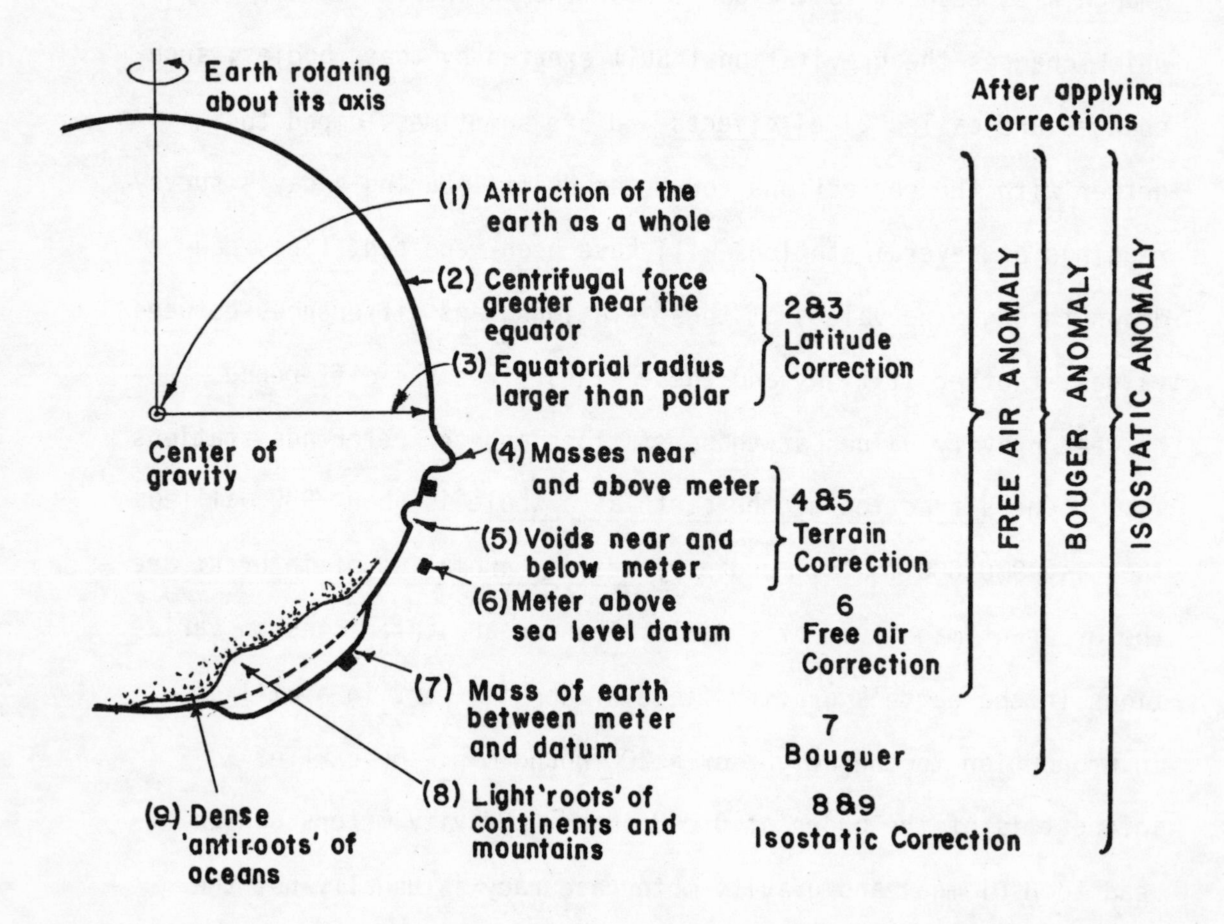

Figure 2.2

less than if the earth were not spinning. Centrifugal force depends upon the distance of a station from the axis of the earth; this and the component in the vertical direction both depend on the latitude. The earth is also not a sphere, the equatorial radius being larger than the polar radius, so that the distance to the center of gravity of the earth varies with latitude. Correction for this fact and for the variation of centrifugal force are often combined in a latitude correction.

If masses near the gravity meter protrude above the level of the gravity meter, they will exert an upward component on the meter and so make the meter read less than if they were not present. If there are voids near the meter and below it, the lack of mass in these voids will fail to pull down on the meter and hence will make the meter reading smaller than it would be in the absence of the void. Thus, a hill on one side of a gravity meter and a valley on the other side will cause effects which add rather than compensate each other. Corrections for nearby masses above the meter or voids below the meter are made in a terrain correction. The terrain correction is the most difficult correction to be made, because we usually do not have the very detailed information about the terrain required to make the correction properly. Inadequacies in the terrain correction are most often the limiting factor on the accuracy of gravity surveys, especially in areas of rough terrain.

Ordinarily gravity readings are referred to values at sea level, though other reference or datum elevations can be used. If the meter is at a different elevation than the reference elevation, then it will be at a different distance from the center of the earth, and therefore will read a different gravity value due to the attraction of the earth as a whole. A meter with air between it and the reference elevation would read a lower value and a correction would have to be added; such a correction is called a free air correction. Gravity maps are often made in which all of the foregoing corrections have been applied; such a map shows free air anomaly values. Free air anomalies indicate variations in gravity readings for reasons other than those for which the corrections have been applied.

If the meter is not at the reference elevation, it is probably because the ground elevation differs from the reference elevation, and hence there is mass between meter and reference elevations which will make the meter read larger than if such mass were not present. The attraction of this mass depends upon the density as well as the thickness of this mass. Correction for this mass is called the Bouguer correction. The free air corrections and Bouguer corrections both depend on the elevation of the gravity meter and are often combined into an elevation correction. Gravity data with the Bouguer and other foregoing corrections applied are called Bouguer anomalies. A Bouguer anomaly indicates gravity variations for reasons other than latitude, terrain, or elevation. Most of the exploration maps used in petroleum exploration are Bouguer maps.

With a shipboard gravity meter, an additional correction is required, the Eötvös correction for motion of the meter during the measurement. The speed of the gravity meter moving eastward adds to the rotational speed of the earth; hence the centrifugal force is more than allowed for in the latitude correction, and the meter reading will be too low. On the other hand, if the meter is moving westward, the meter reading will be too large. To illustrate the magnitude of this effect, a 200 pound man walking eastward at about 2 miles per hour weighs about 0.03 ounce less. The Eötvös corrections depends on the speed and direction during a measurement and these have to be known very accurately. Instantaneous ship speed is not normally known to the required accuracy and this usually limits the accuracy of marine gravity data.

The light roots of continents and mountains or the dense antiroots of ocean basins affect gravity readings; these are sometimes corrected for by an isostatic correction. However, isostatic corrections are not ordinarily made in gravity surveys for petroleum unless the survey is adjacent to some major change in the earth's crust.

To summarize the sensitivity of gravity data to errors of various types, an error of 0.1 mgal could be caused by

a) latitude error of 200 meters (at 45^{0} latitude),

b) elevation error of 0.5 meters,

c) east-west velocity error of 0.01 meters/sec (0.02 knots, at 45^{0} latitude),

d) north direction error of 0.1 degrees (at 45^{0} and 10 knot speed).

Land gravity data in areas of fairly flat terrain may be accurate to 0.1 and 0.2 mgal. In rough terrain the accuracy deteriorates because of terrain correction uncertainties, which may be 1 mgal in magnitude. Marine gravity data are usually no more accurate than 1 mgal because of uncertainties in Eötvös correction.

CHAPTER 3

GRAVITY II: ANOMALIES

The cause of gravity anomalies is the variation in density of rocks. Our geological objectives in locating and interpreting gravity anomalies is to relate the density variations to structural and stratigraphic features.

Data on the density of different rock types taken from the GSA Handbook of Physical Constants are shown in Fig. 3.1, adapted from Grant and West. This figure shows the range in density of various rock types. The ranges are "80% limits", that is, 10% of the samples measured fell below the values indicated and 10% above. The principal reason for such wide ranges of density within any given rock type is the varia-

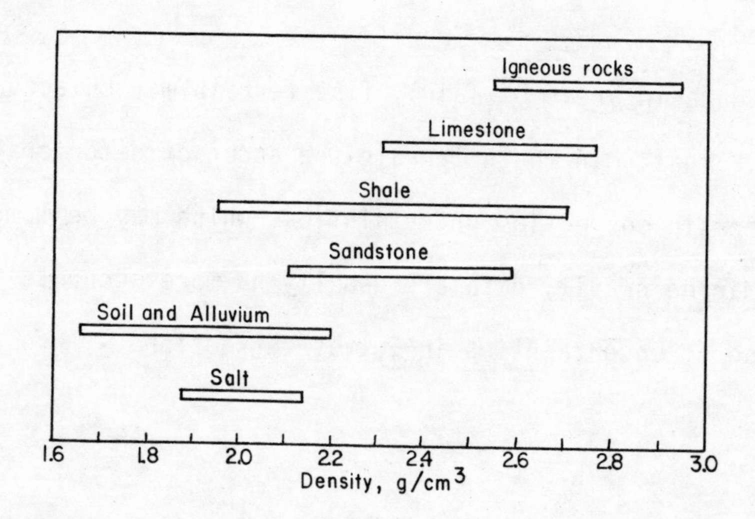

Grant and West

Figure 3.1

tion in porosity, void space in the structure of the rock. The porosity is usually filled with fluid of considerably less density than the mineral grains of the rock itself.

There is considerable overlap in the density of different rock types and density is not very diagnostic in distinguishing one rock type from another. Salt might be considered an exception; salt is lighter than sediments except for very shallow and very porous sediments. Basement rocks are heavier than most sediments except those without signficant porosity. Density contrasts encountered in the earth involving significant volumes of rock are rarely larger than about 0.3 g/cm^3.

Density ordinarily increases with depth because porosity decreases with depth. However, our gravity maps are not sensitive to the vertical variation of density provided it is the same under each gravity station. Our concern is with horizontal

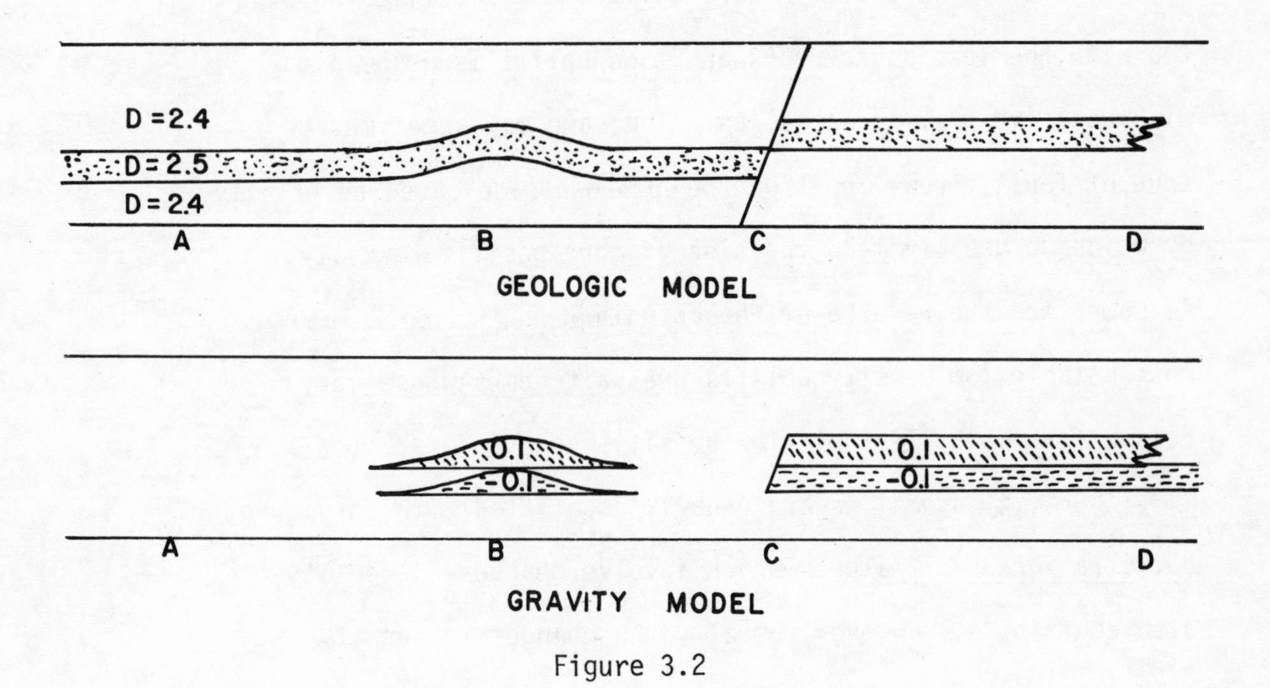

Figure 3.2

variations in density which occur as a result of structural features.

Several models of geologic structural features are shown in Fig. 3.2 along with the gravity models with which they might be associated. The gravity models involve only those portions of the geologic model where density changes in the lateral (horizontal) direction. In particular, uniform bedding (A) produces no gravity anomaly no matter what vertical variations of density may exist.

Because density usually increases with depth, a structural uplift usually involves bringing more dense rocks nearer the surface and laterally adjacent to light sediments. Hence structural uplifts are generally associated with positive gravity. Thus for a fault evidenced in gravity data, we usually call the side with more positive gravity the upthrown side. However, in actual nature uplifts are sometimes associated with the less dense rocks and then uplift is associated with negative gravity anomalies. The same may sometimes be true of fault throws and thus the upthrown side based on gravity, by which we usually mean the side of more positive gravity, is sometimes the reverse of the actual upthrown side. The most notable low density uplifts are salt domes whose very reason for rising is their low density.

Gravity anomalies are usually associated with structure, but stratigraphic features which involve changes in density also contribute. However, usually the changes in density associated with stratigraphic variations are relatively small, not much thickness is involved, and the changes are often gradational and therefore distributed over appreciable distance. Gravity data therefore are not especially amenable to stratigraphic interpretation.

The interpretation of gravity data possesses inherent ambiguity, that is, there are an infinite number of possible mass distributions which can account for the observed anomaly. However, in actual practial situations, many of these distributions are impossible or unlikely because they require densities outside the range of actual earth materials, or because they represent improbable geologic situations. It should be realized that ambiguity is not unique to gravity and almost all data are capable of a variety of interpretations; indeed this is the very nature of interpretation.

Nettleton illustrated gravity ambiguity with a "cone of sources" as shown in Fig. 3.3. Several mass distributions are illustrated, any one of which accounts for the gravity anomaly. The deepest of these possible mass distributions is a concentrated mass at the apex of the cone. One of the things that can be determined from gravity data is the <u>maximum depth</u> at which possible masses may lie. When we calculate the depth of a gravity feature, we are usually calculating the maximum

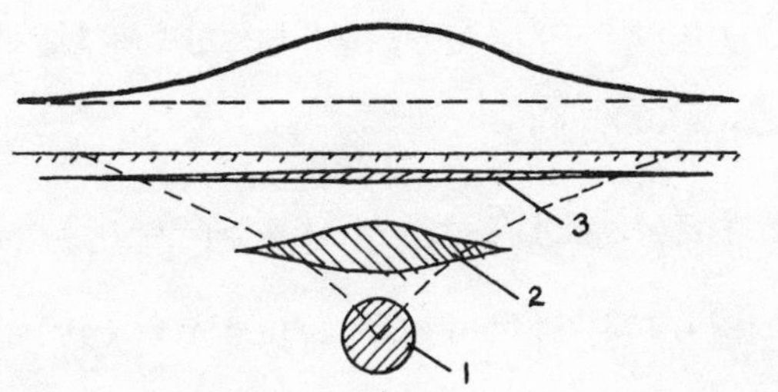

Nettleton

Figure 3.3

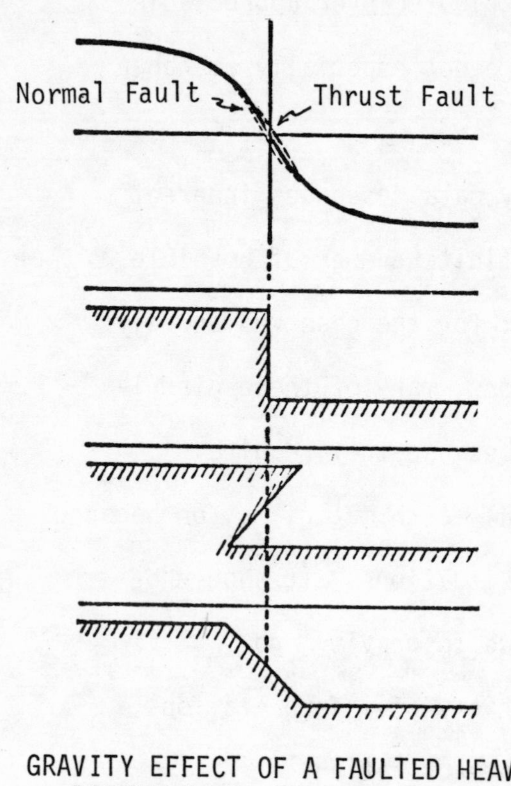

GRAVITY EFFECT OF A FAULTED HEAVY BASEMENT VS. TYPE OF FAULT

Figure 3.4a

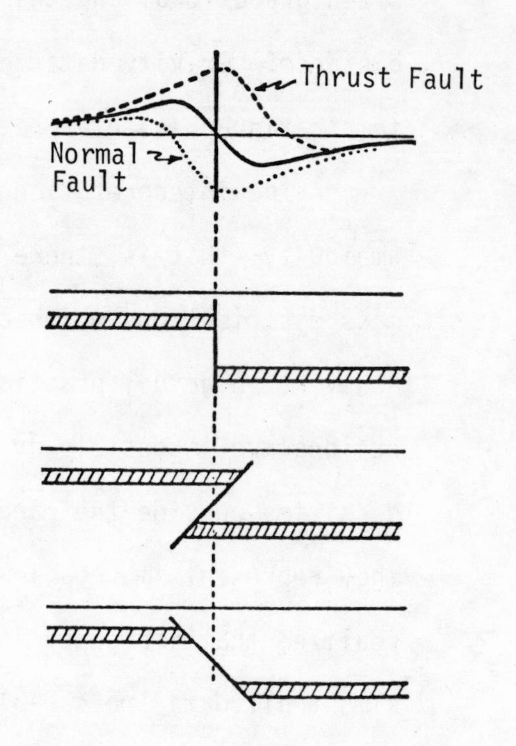

GRAVITY EFFECT OF A FAULTED HEAVY LAYER VS. TYPE OF FAULT

Figure 3.4b

depth at which it can lie and it is always possible that the cause lies more shallow and involves a more distributed anomalous mass.

All of these possible mass distributions contain the same total <u>anomalous mass</u>, which can be found by integrating the volume of the gravity anomaly. However anomalous mass is not usually related to hydrocarbons in any easily definable way and hence is not usually calculated in petroleum exploration. It is of importance in mineral exploration because it relates to the total volume of an orebody.

Gravity data are insensitive to shape of the anomalous body. In Fig. 3.4a, we show 3 different types of faults, each of which lies at the same depth. The gravity anomalies from these three features are nearly the same and it would be difficult to

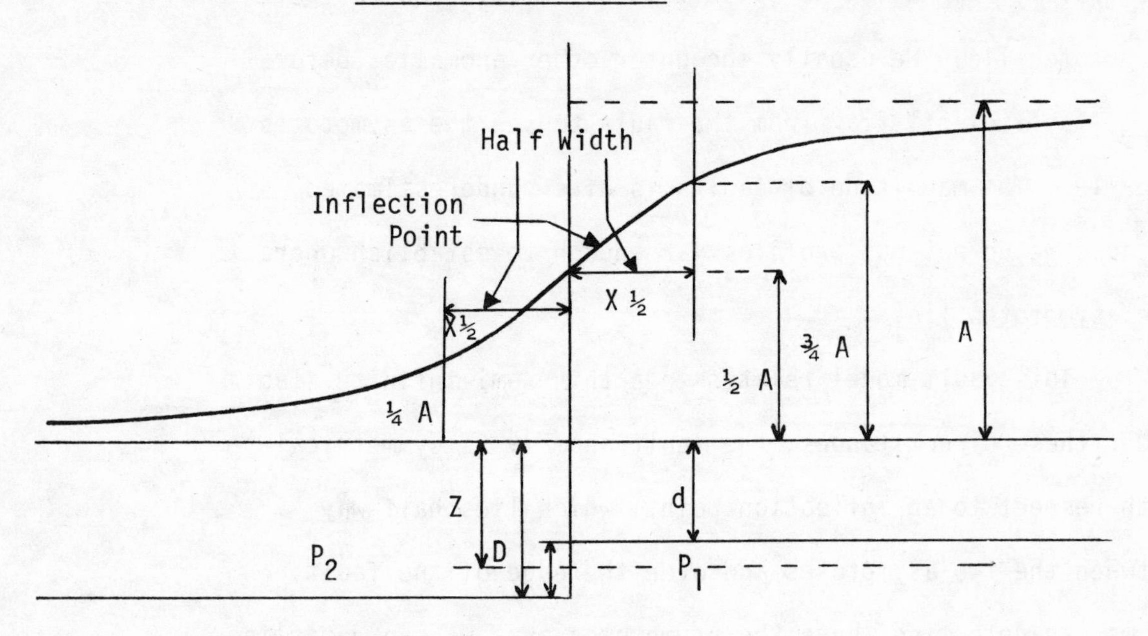

Figure 3.5

tell the shape of the faults from gravity profiles. However, if we modify the model slightly by making the offset bed of finite thickness, the gravity profiles are very different, as illustrated by Fig. 3.4B. The reverse fault involves a double thickness of the dense layer in the fault zone which results in a positive anomaly whereas the normal fault involves a zone where the bed is absent and hence a negative anomaly. This example should suggest caution about jumping too quickly to conclusions based on a model.

Fault evidences in gravity data are among the more useful anomalies to analyze. A profile across the edge of a thin semi-infinite slab is shown in Fig. 3.5; this is often taken as the model for a fault. The gravity levels out asymptotic to parallel lines as the distance from the edge becomes large. One of the more difficult aspects of analyzing the gravity expression of a fault is determining where the asymptotes lie. We usually encounter other anomalies before we get sufficiently far from the fault to see the asymptotes clearly. The magnitude of faults is often underestimated because we do not run profiles far enough to establish where the asymptotes lie.

This fault model relates to a thin semi-infinite slab. Under these circumstances, the fault anomaly is symmetrical with respect to an inflection point, which lies half-way between the two asymptotes and over the edge of the fault. If we can determine where the asymptotes are, we can determine the locations at which the anomaly has one-quarter, half, and three-quarters values. We can then measure the distance between 1/4 and 1/2 value points and between the 1/2 and 3/4 value points; these distances are called the half-widths. For the thin slab model, the half width equals the depth to the center of the slab.

Actual fault anomalies are apt to differ from this model in several regards. Usually the anomalous mass is distributed over appreciable vertical distance and the density contrast varies along the fault because the fault juxtaposes different materials at different depths. Often density contrasts are greater shallow than deep; this tends to make a fault anomaly asymmetric, with the upthrown side sharper than the downthrown side.

The gravity difference A because of a fault, the difference between the upper and lower asymptotes in Fig. 3.5, is related to the thickness of the anomalous mass and the anomalous density ρ

BURIED HORIZONTAL CYLINDERS

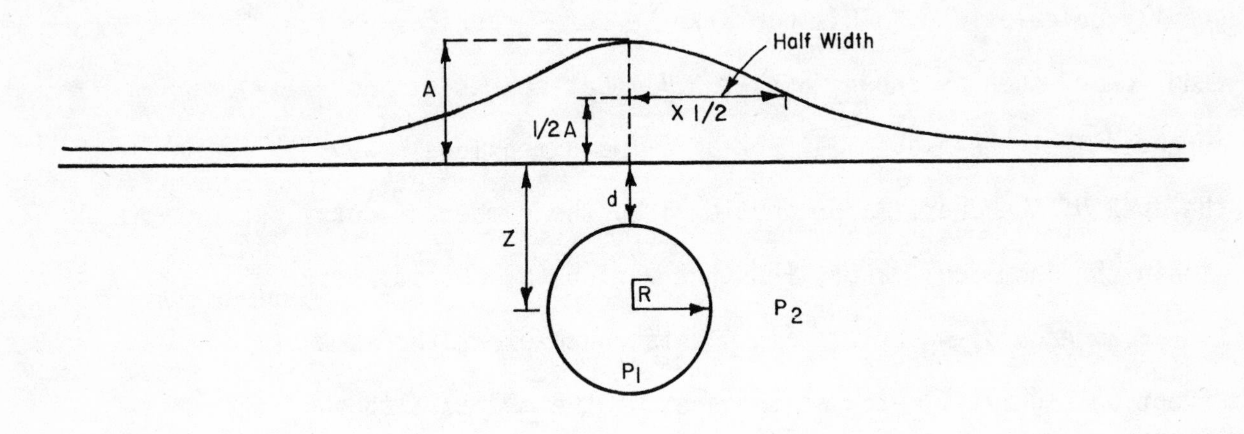

Figure 3.6

by the relationship,

$$A = 0.01276\ \rho h \quad \text{mgal} \quad \text{if } h \text{ is in feet,}$$

$$A = 0.04185\ \rho h \quad \text{mgal} \quad \text{if } h \text{ is in meters.}$$

Another model of considerable utility is the buried horizontal cylinder, illustrated in Fig. 3.6. Half of the width of the anomaly at values mid-way between the maximum anomaly value and the value of the asymptote is called the half-width; it equals the depth z to the center of mass of the horizontal cylinder. Because gravity anomalies are generally insensitive to shape of the anomalous mass, horizontal cylinders often provide useful approximations for anticlines. The half width technique can also be used for a buried sphere in which case the center of the sphere is 1.3 times the half-width.

The gravity anomaly for a body of simple shape can often be calculated by integration. However, the anomalies for bodies of irregular or more complicated shape are usually calculated by dividing the body into many small cells, calculating the effect of each and then summing the effects. This is the procedure usually used by digital computers in making gravity calculations. This can also be done graphically with a dot chart such as shown in Fig. 3.7. Dot charts are so arranged that the gravity effect of a two-dimensional body at the apex of the chart is proportional to the number of dots within the anomalous mass. Thus the various blocks shown in Fig. 3.7, each of which contain 4 dots, each exert the same effect on a gravity meter at the apex of the chart, although they differ considerably in size, shape and depth. The shape of the anomaly due to an arbitrary body can be calculated by drawing the body on an overlay of the chart and counting the number of dots within the body, which gives the value at the apex of the chart. The overlay is then moved successively with respect to the dot chart to determine the values at other locations.

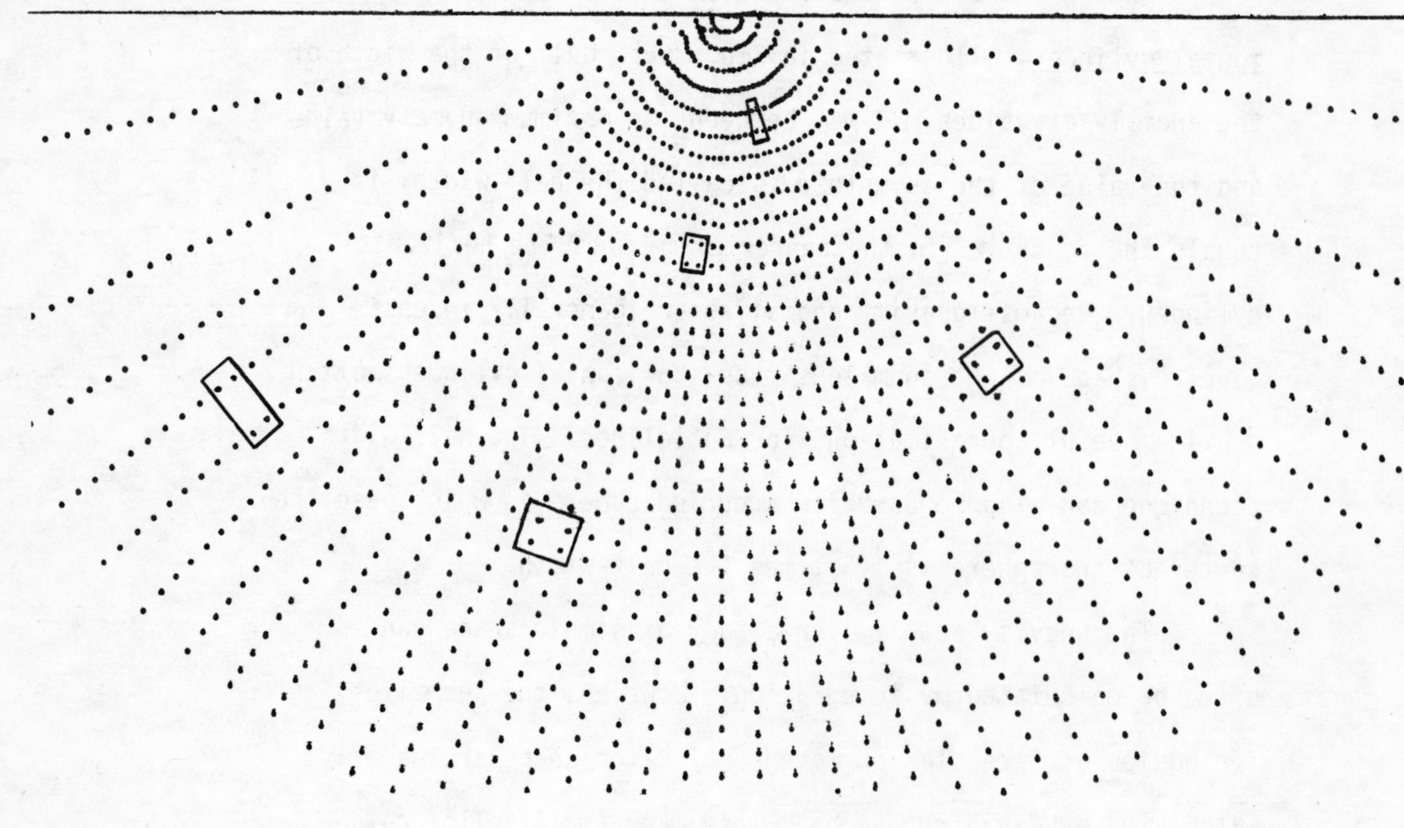

Figure 3.7

CHAPTER 4

GRAVITY III: ANOMALY ISOLATION (AND PROBLEMS)

We have been looking at the gravity effects of bodies which are isolated, that is, their gravity expression is not confused by overlap with the expressions of other bodies. In the interpretation of real gravity data, usually our <u>first order problem is isolating anomalies</u>, that is, finding out which areas are anomalous.

An anomaly is simply a variation from the normal. The gravity maps we usually interpret are Bouguer anomaly maps. The values contoured are Bouguer anomalies, differences between actual gravity values and what would be expected for a uniform earth, which has the same latitude, elevation and terrain; in other words, variations from the Bouguer normal.

We also use <u>gravity anomaly</u> to indicate variations which are of exploration interest. We want to find variations in Bouguer anomalies which represent structural features which may relate to hydrocarbon accumulations. These are usually local, but not too local, places where the Bouguer values depart from uniform.

We have used subjective terms to define a gravity anomaly: "of exploration interest" and "local, but not too local". We have different interests in analyzing data at various times and what constitutes an anomaly at one time may differ from another time. For example, when first looking at data we may be interested in determining basin shape and location, perhaps even in finding if there is a basin, so the deviations caused by the existence of the basin are our gravity anomaly. However, later we are probably interested in the location of local uplifts within the basin and what are their features; the effects of these then are the gravity anomalies.

The process of separating our anomalies is called residualizing. It involves the separation of the gravity field into two parts: gravity anomalies of interest, which we call the <u>residual</u>, and various larger features in which we are not interested, which we call <u>regional</u>. The regional is smoother than the residual.

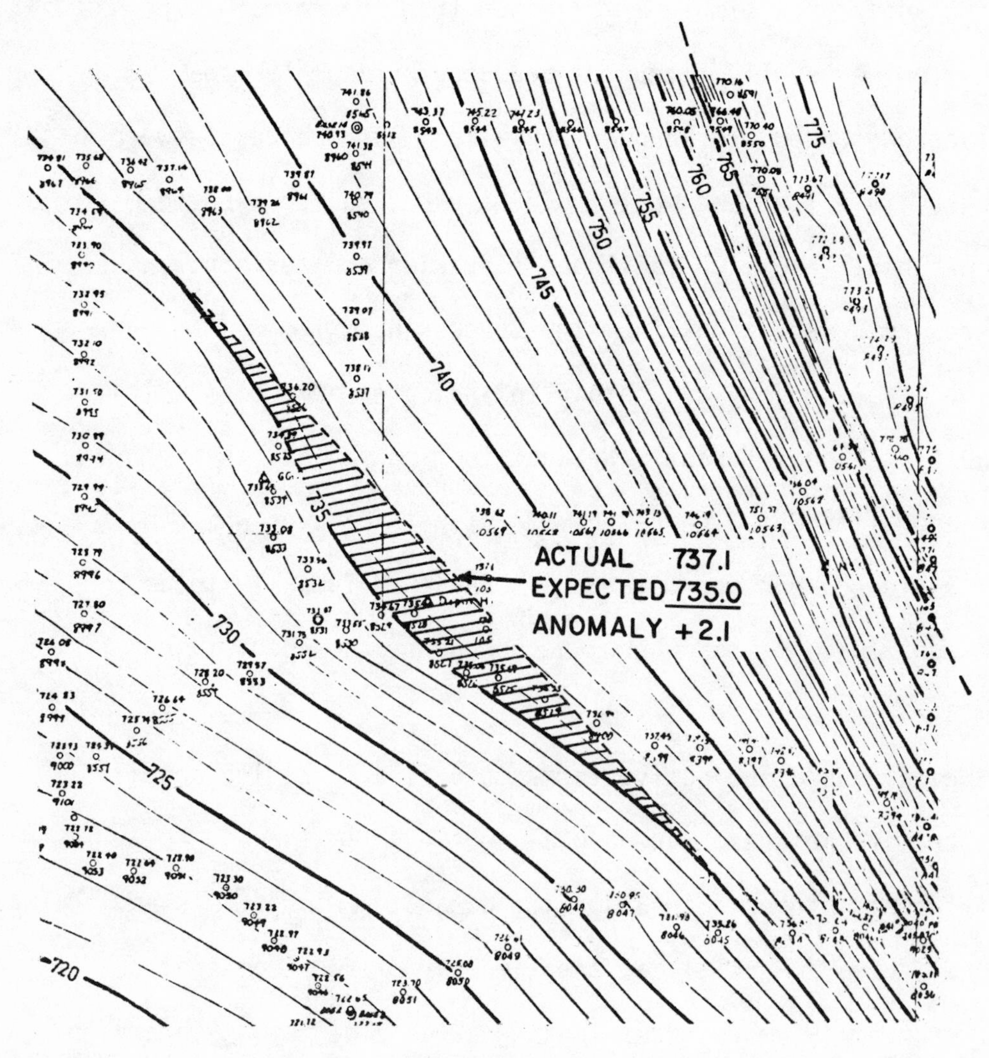

Figure 4.1

Once we have isolated anomalies, the second problem is determining the cause of these anomalies.

As example of the contour smoothing method of residualizing, consider the portion shown in Fig. 4.1. The contours over this region are generally fairly straight except in the middle of the map where they bow towards the southwest, towards an area of lesser gravity values. This bowing of the contours is a departure from uniformity and evidence of an anomaly. On this diagram one of the contours has been dashed in as we might expect it to be in the absence of this anomaly; the dashed-in contour represents the smooth regional. The region between the actual contour and the dashed contour has been shaded in to emphasize the anomaly. We can determine the magnitude of the anomaly at a point on the dashed line by subtracting the gravity we expect in the absence of the anomaly from the value actually found at that point; the difference is the residual, the amount of the anomaly. Here the difference is positive and this feature is a positive anomaly. The anomaly exists only over a limited area and we suspect it indicates a local uplift.

In the upper right corner of Fig. 4.1 the contours are spaced very closely together. This change in contour spacing is another departure from uniformity and another anomaly. In this particular case, the close spacing of contours is a linear feature attributable to a fault.

The best way of analyzing gravity anomalies is often on a profile. Figure 4.2 shows how one can easily make a profile from a gravity map by laying a piece of graph paper along the line of profile, marking where the contour crossings are, and then plotting these values above the contour crossings to give

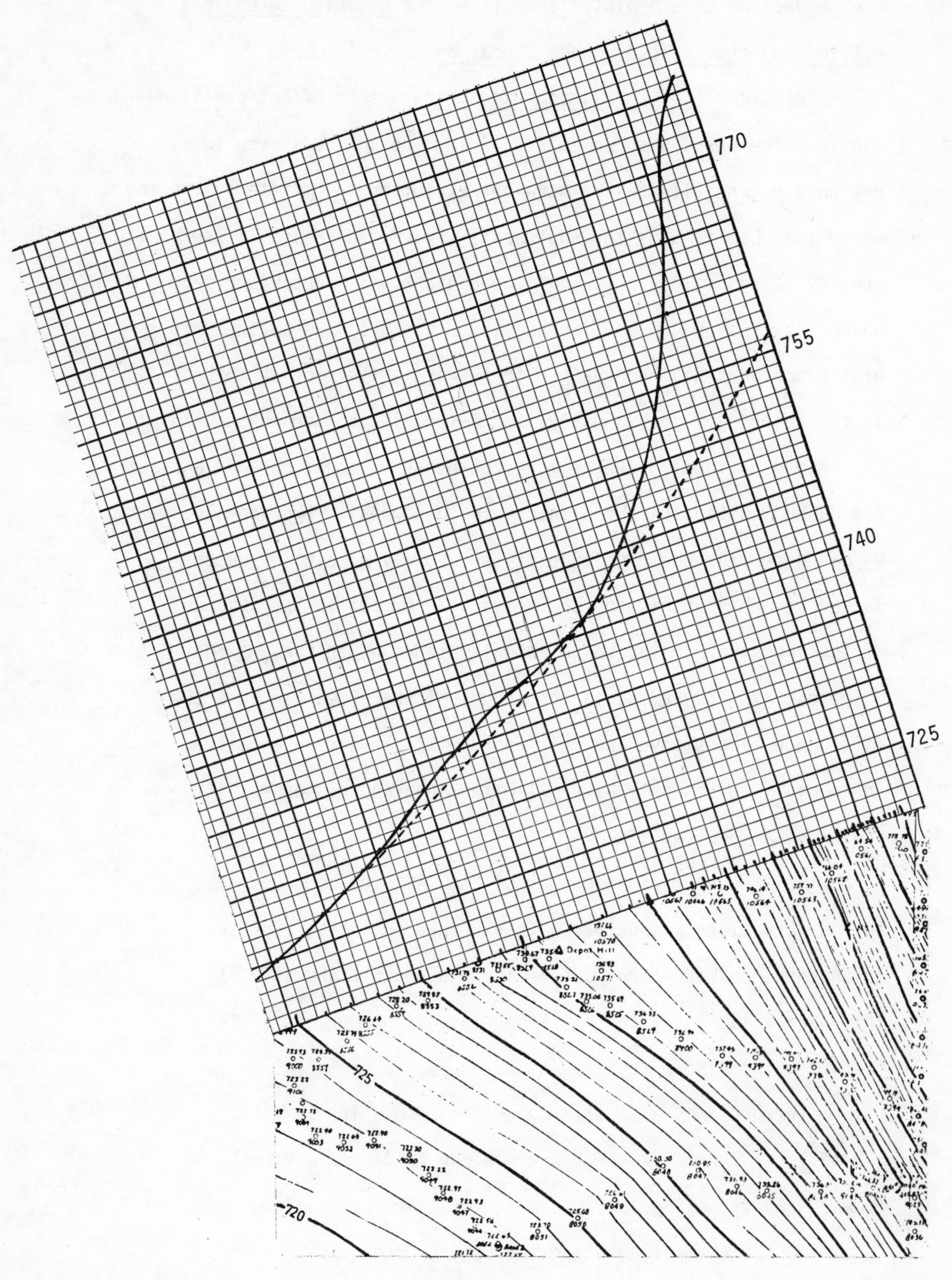

Figure 4.2

the profile. One advantage of such a profile is that it is at the same scale as the map so that once we have found a feature of interest on the profile, it is easy to lay the profile back on the map and relate the feature to its location on the map. We try to locate a profile roughly perpendicular to the structural axis, but we also consider the distribution of gravity control, so that the data plotted minimizes the interpretation involved in interpolating between lines of control.

The profile is residualized in the same way as the gravity map was residualized. We sketch a smooth curve which represents the regional, the expected field in the absence of local anomalies; the difference between the actual data points and the smooth regional is the residual. In making an actual gravity interpretation, we often draw many parallel profiles to be sure the smoothed regional curves are consistent over the entire region.

The same two anomalies seen as a result of the contour smoothing on the map show up by smoothing this profile. The localized bowing of contours shows as a small hump, and the fault shows up as a large departure from the regional curve with almost the classical gravity fault shape.

Grid methods are commonly used in residualizing. Grid methods determine the regional by averaging surrounding values. The departure of actual readings from the average gives the residual. Usually the gravity data are interpolated onto a regular grid and a template is used to select the values to be averaged. A simple template averages four values as shown in Fig. 4.3. This template results in the grid residual map of Fig. 4.4.

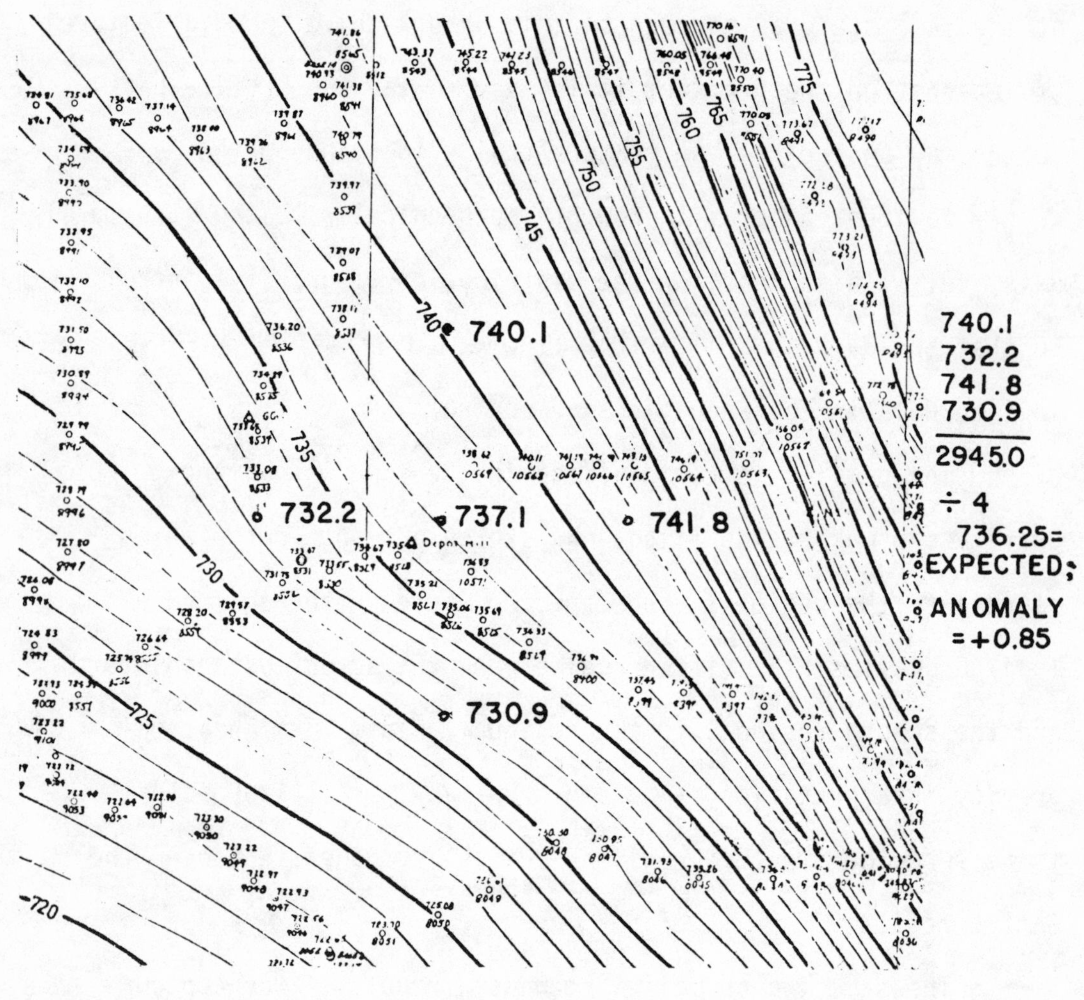

Figure 4.3

Various kinds of templates can be used, some averaging 6, 8 or 10 points rather than merely 4 points, and sometimes averages are formed of data at several distances surrounding the station in order to determine the amount of curvature of the regional field. Second derivitive maps are made in this way. Grid residualizing techniques are especially adaptable to computer processing. The template in Fig. 4.3 could be applied by multiplying each of the values surrounding the central station by -1/4, the central station by +1, and summing the results.

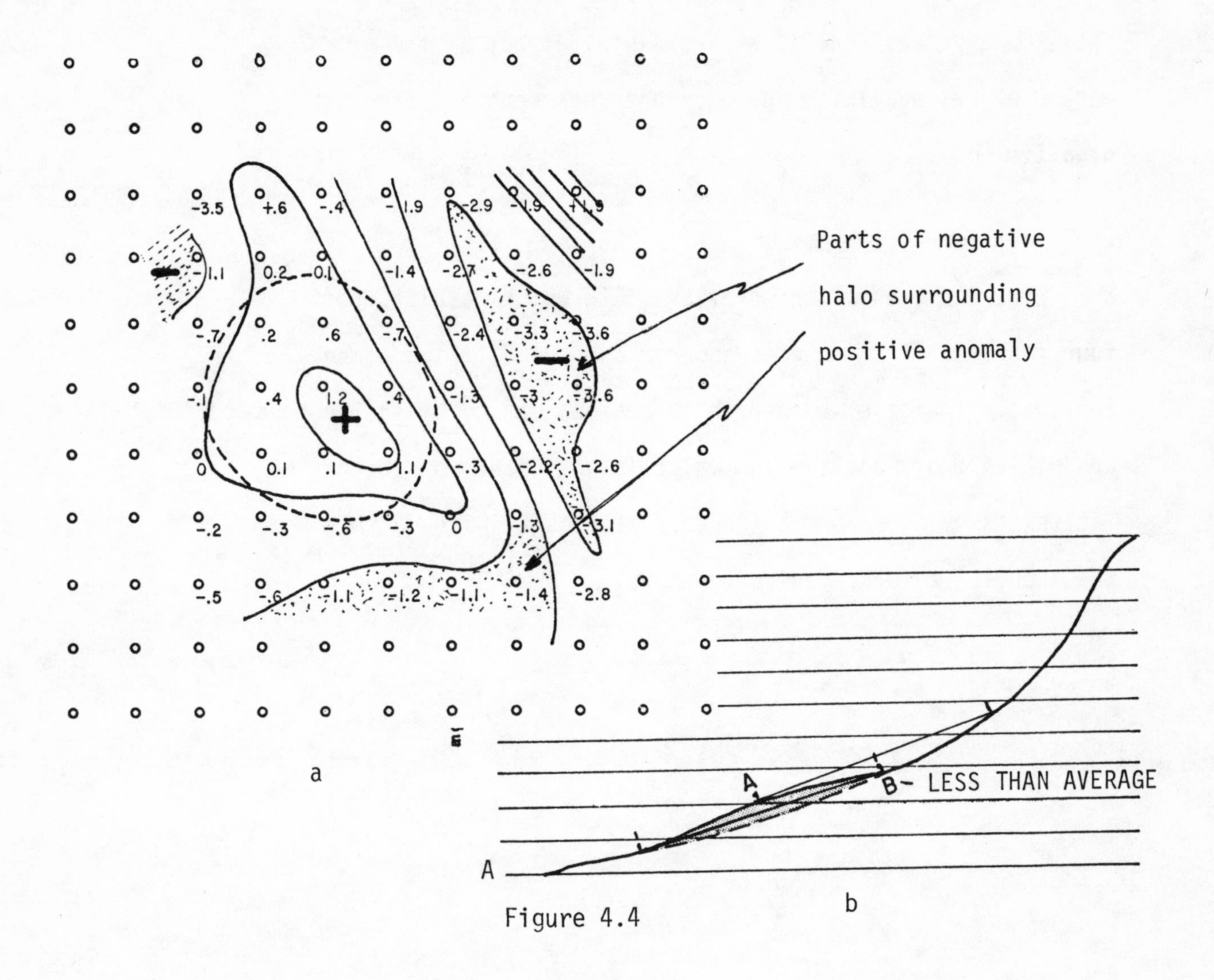

Figure 4.4

A consequence of grid residualizing is that it produces <u>halos</u> around anomalies. Examination of the profile in Fig. 4.4 illustrates this. In determining the expected value of the gravity by averaging points on either side of it, we include a distortion if some of the points being averaged are themselves anomalous. Over the central part of an anomaly, such as at point A, we average smaller values on either side and so produce a positive residual. When we get to the edge of the anomaly, as at the point B, the average values are biased by including anomalous values on the anomaly and we determine too large an expected value, tending to give a negative residual.

This negative residual is an incidental effect of the grid method of residualizing and does not represent a separate negative body.

Exercise:

Figure 4.5 shows the gravity field which results from four faults and two spheres superimposed on a tilted plane regional. Indicate whether the anomalies caused by the spheres are positive or negative and which is the upthrown side of the faults, assuming a normal increase of density with depth. Work the exercise before reading further.

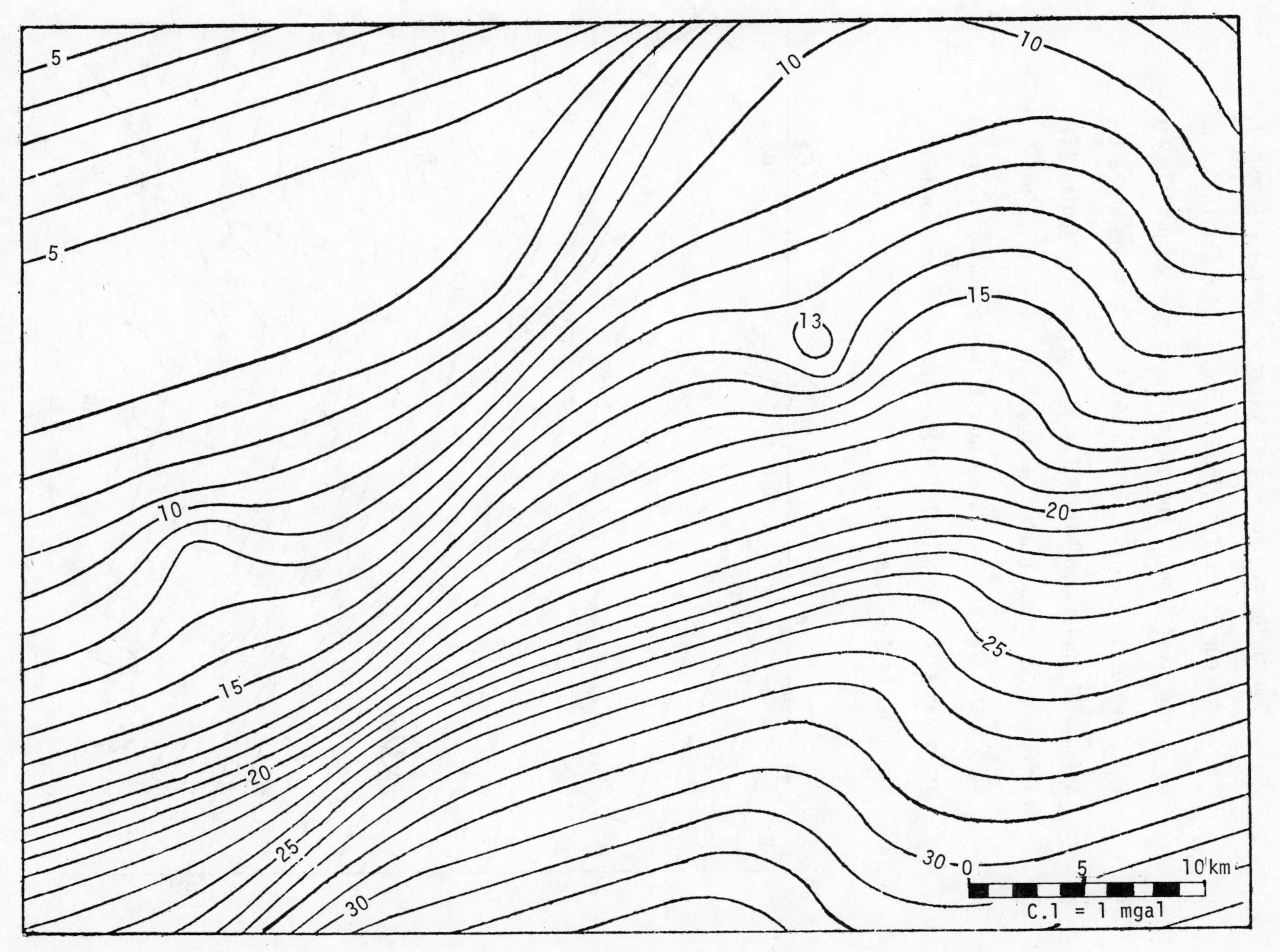

Figure 4.5

Discussion of exercise:

One of the two spheres is clearly located at about 1, the 12 contour on profile line AA' on Fig. 4.6 and the other at 2, the lower edge of the closed 13 contour on profile line CC'. Two of the faults strike about N 25^{0} E through the two systems of inflection points (3, 4). One of the faults (5) is evidenced by the tightly packed contours striking N 70^{0} E across the lower half of the map and the remaining fault (6) is evidenced by the parallel widely spaced contours across the upper half of the map.

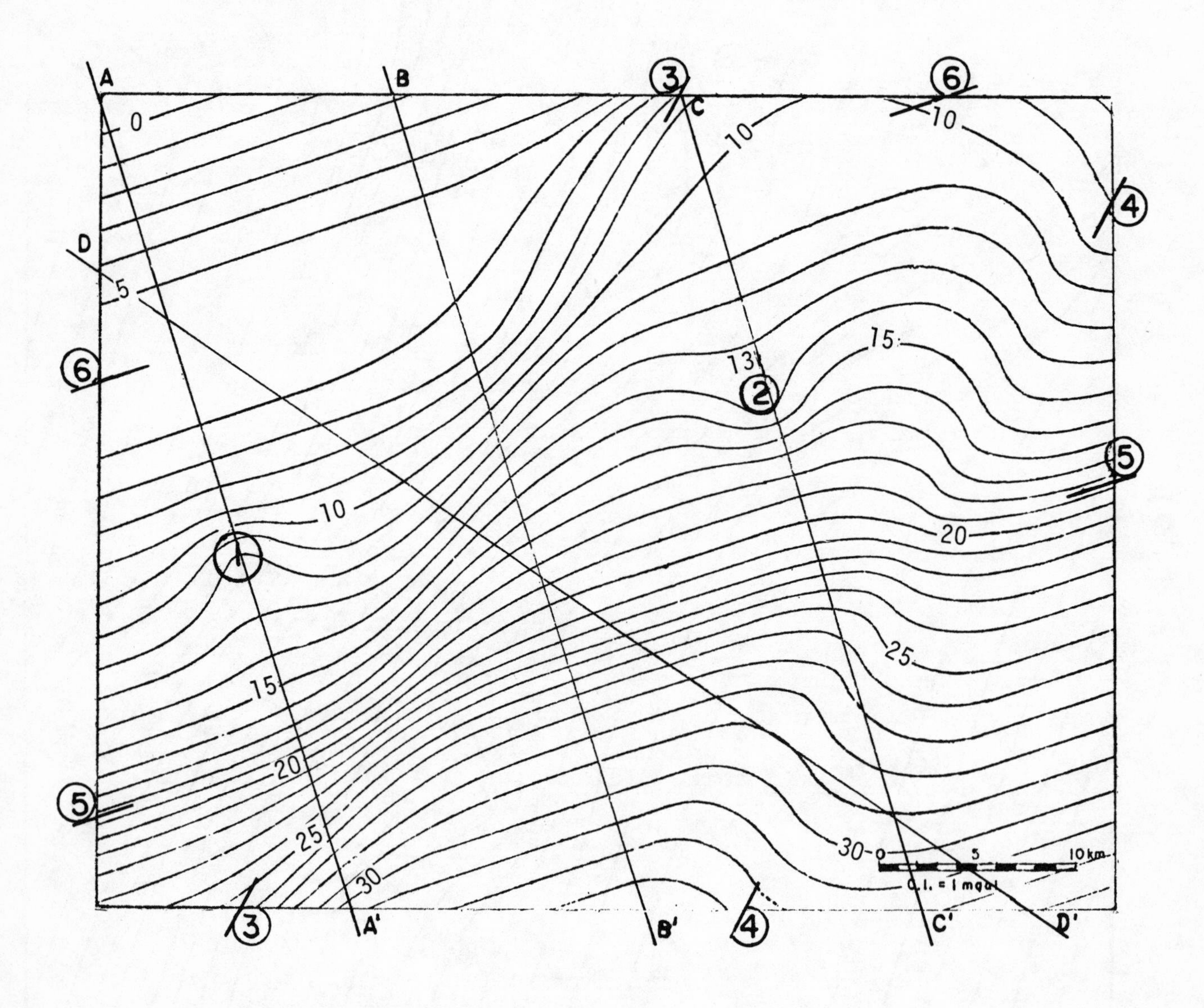

Figure 4.6

Exercise:

Using the profiles in Fig. 4.7a, b, c and d, drawn along the lines shown on Fig. 4.6, analyze the individual anomalies.

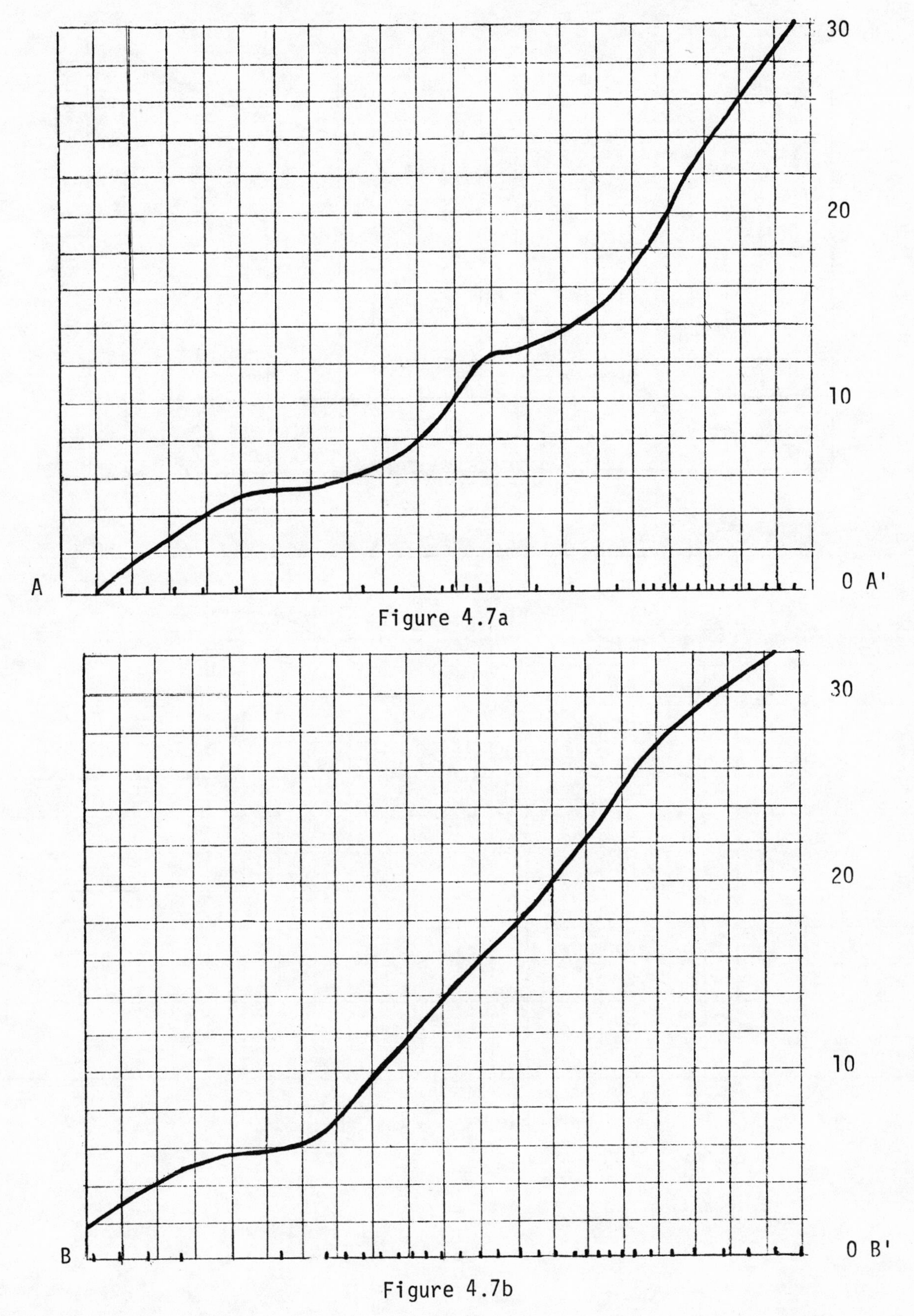

Figure 4.7a

Figure 4.7b

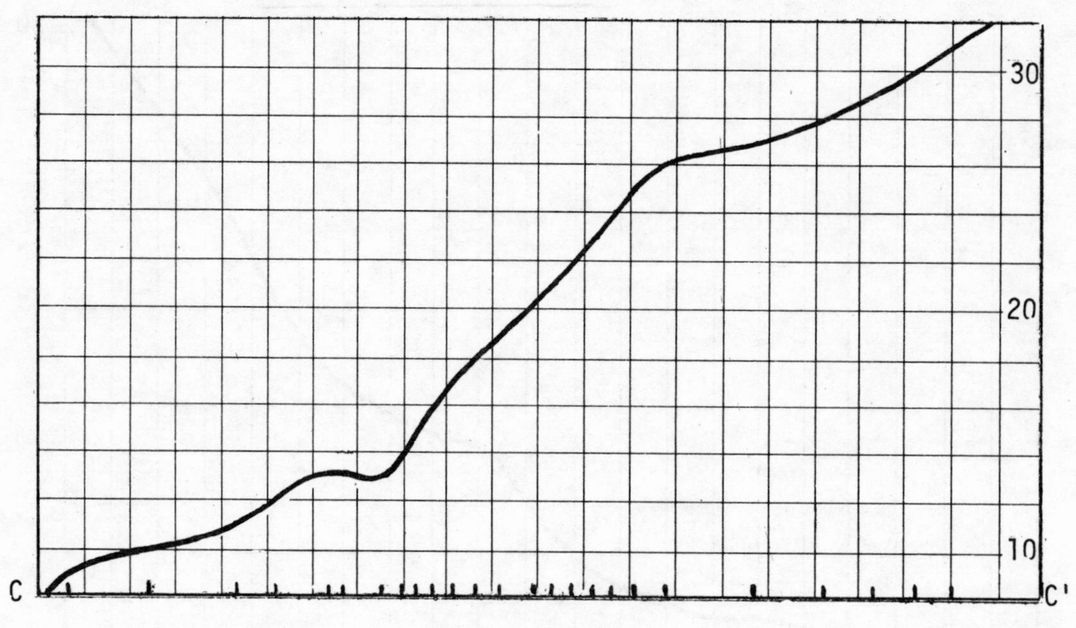

Figure 4.7c

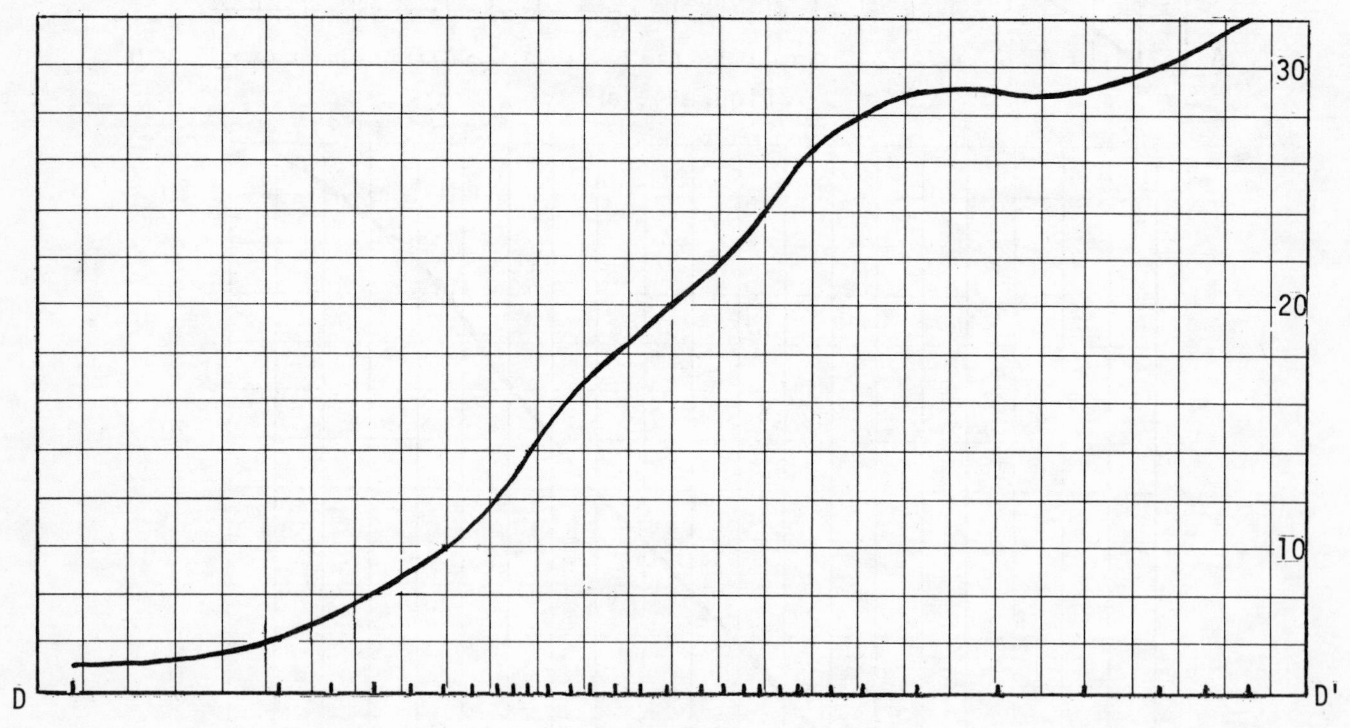

Figure 4.7d

Discussion of Exercise.

We note on Fig. 4.6 that the overall trend of the contours on the map is about perpendicular to the profile lines AA'; BB' and CC' and that these profiles (Fig. 4.7a,b,c) include relatively uniform portions which appear to be relatively free of disturbances from anomalies. We use this information to draw the regional slopes (Fig. 4.8a,b,c). We take special care that the regionals on all the profiles are consistent.

To analyze the fault on the left half of profile AA' (Fig. 4.8a), we draw a line parallel to the regional as the downthrown asymptote and other parallel lines to indicate the 1/4, 1/2 and 3/4 values. We then determine the half width h. The fault magnitude is 4 mgal.

The spherical anomaly on profile AA' has magnitude +2 mgal and half width j. The complicated shape and kink K in the right end of profile AA' tell us that this is not the

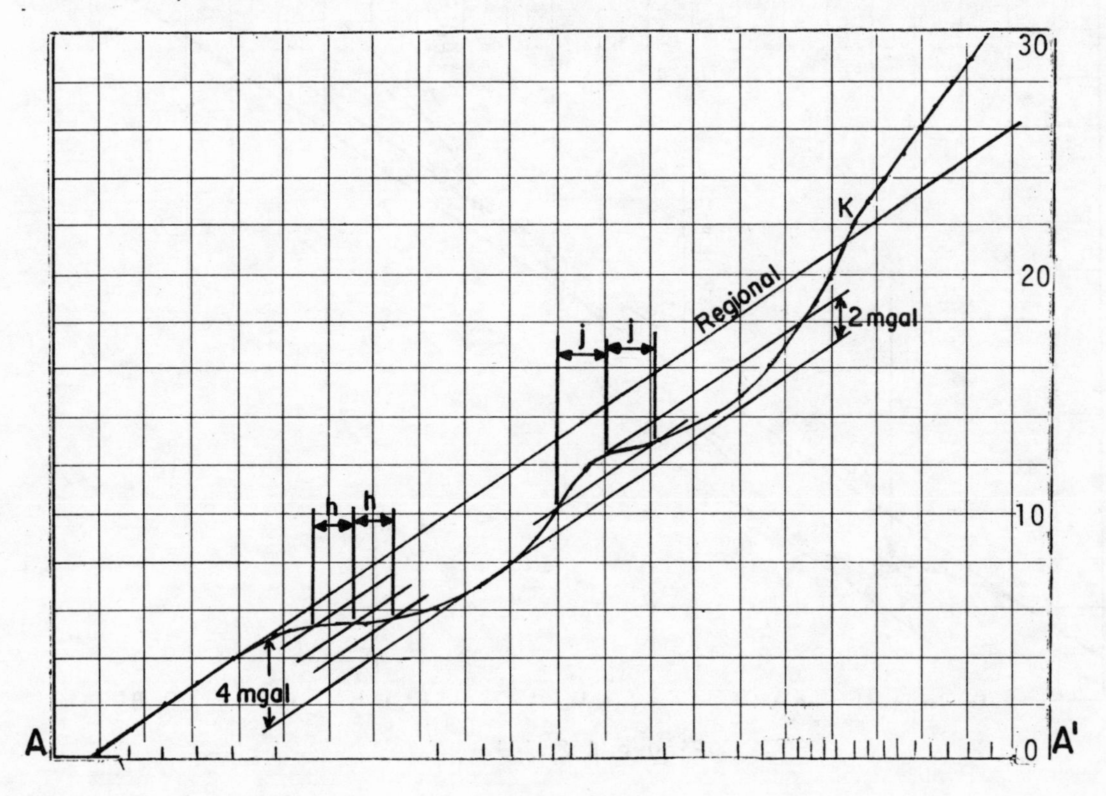

Figure 4.8a

simple anomaly of a single fault but that at least two anomalies are involved.

Figure 4.8b shows profile BB' again. The regional has been drawn so as to be consistent with the regional on the other profiles. The effects of the faults overlap slightly on this profile; the expressions of the individual faults have been dashed-in to make their individual effects clearer. Their respective midpoints then locate them. Faults 6 and 5 each have magnitude 4 mgal but fault 3 has 5 mgal magnitude. Most of the difficulties of gravity interpretation are consequences of anomalies being so close together that their expressions overlap.

If we were to do a half-width analysis of fault 3 on profile BB' we would obtain a value which is too large because

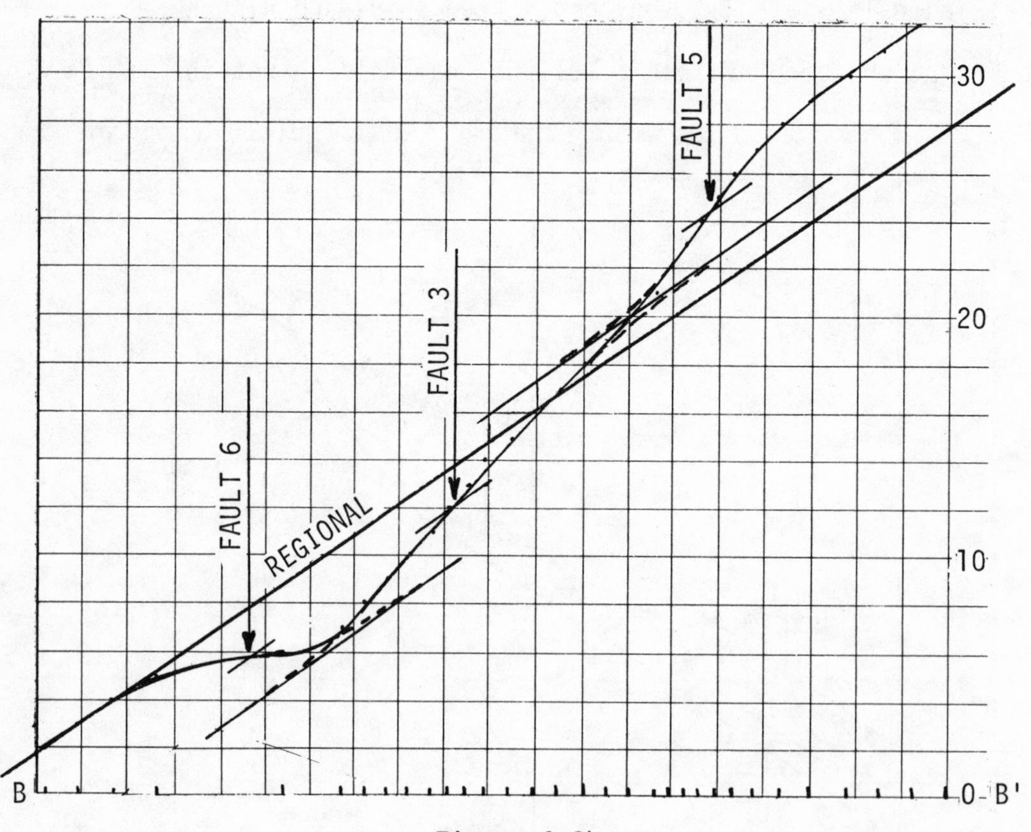

Figure 4.8b

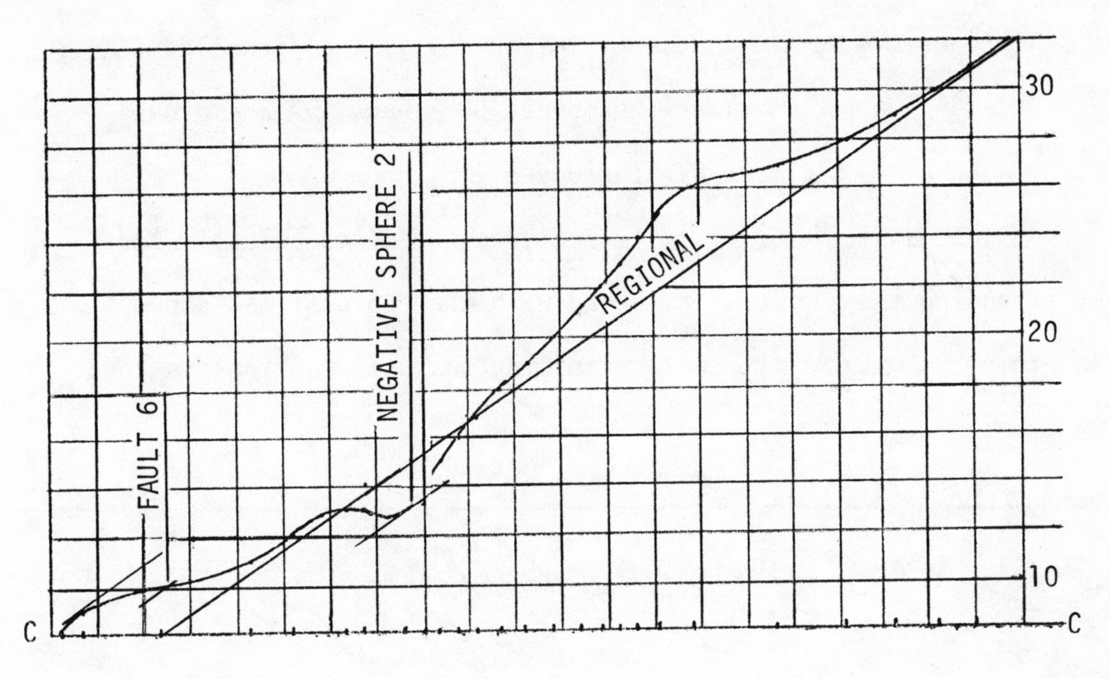

Figure 4.8c

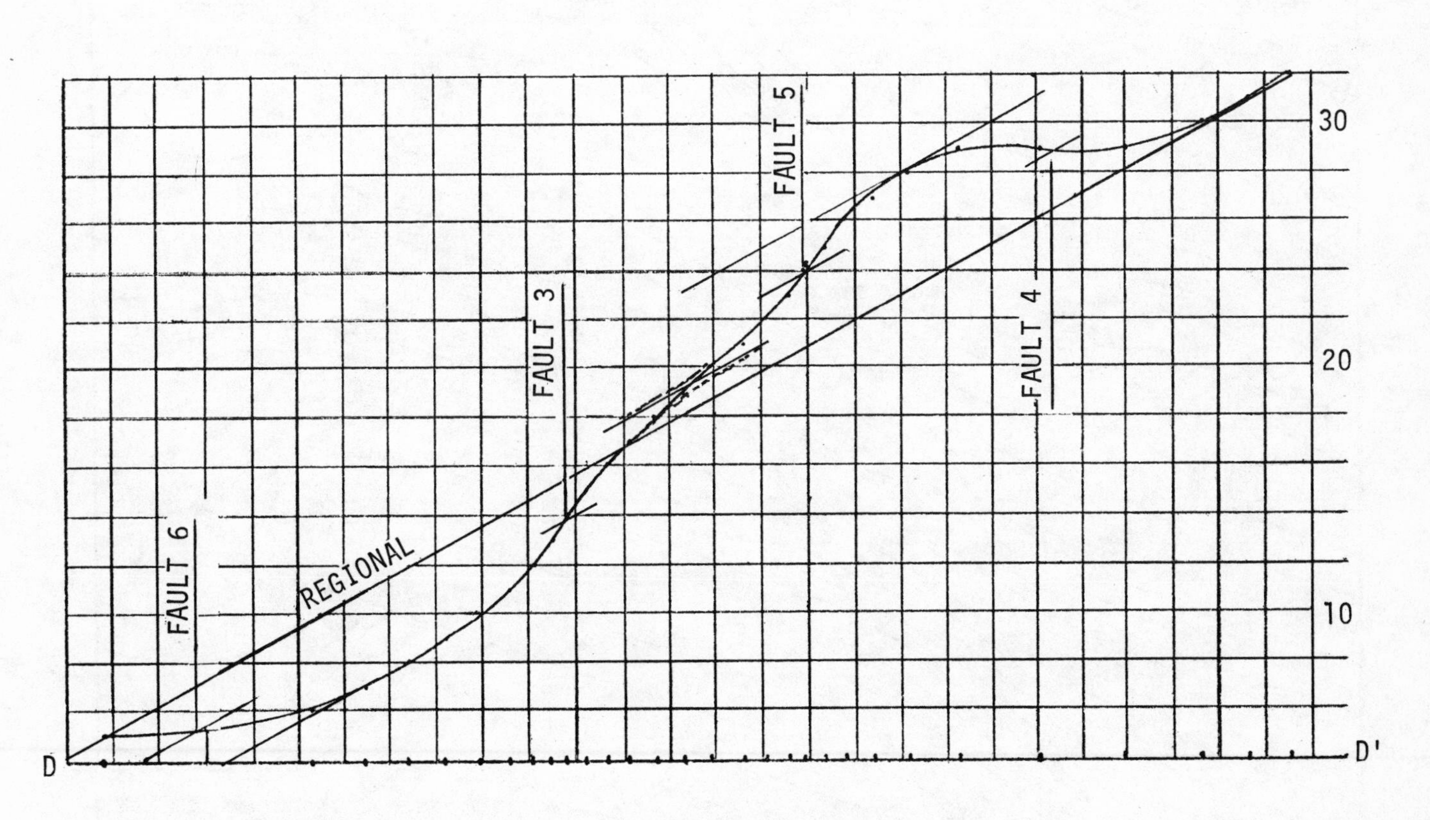

Figure 4.8d

the profile is not at right angles to the fault trace. Correct half-width is shown on profile DD'.

Figures 4.8c and 4.8d are similar analyses of profiles CC' and DD'. Anomaly 2 is clearly produced by a negative anomalous mass (such as a salt dome might produce).

The magnitudes of gravity features can often be estimated by extending the contours in accord with the regional, as shown in Fig. 4.9. At A the contour spacing has been continued across

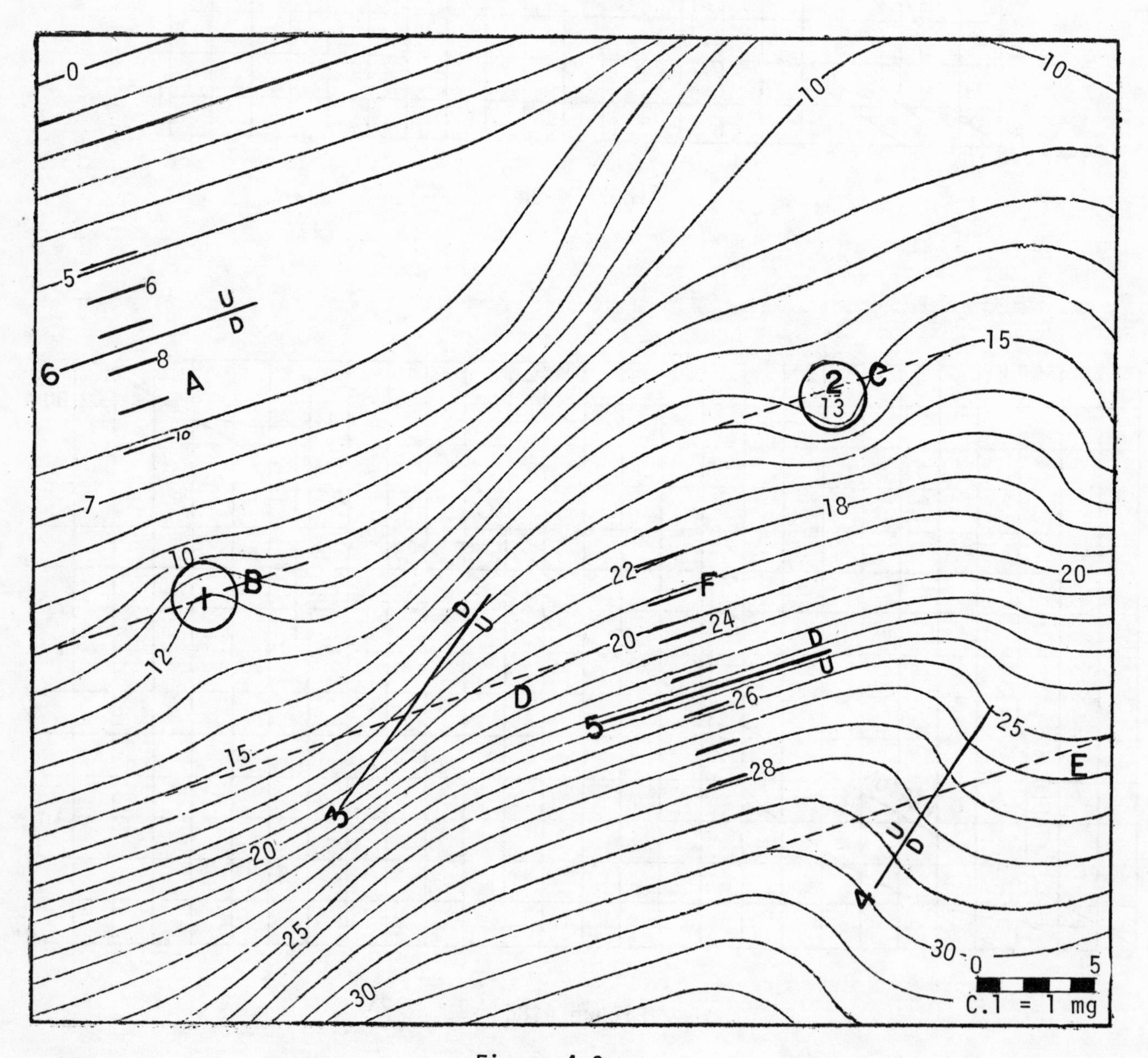

Figure 4.9

the anomaly produced by fault 6; the 11 contour obtained in this manner would roughly coincide with the 7 contour, indicating 4 mgal less gravity on the south side of the fault, which we would therefore probably call the down-thrown side because of the mass deficiency there. The 10 contour extended at B encounters a value 2 mgal higher as a result of spherical anomaly 1, indicating it is a positive anomaly of 2 mgal magnitude. Since the 15 contour at C finds a value slightly lower then 13, anomaly 2 is a negative anomaly of slightly more than 2 mgal magnitude. Extending the 15 contour across the anomaly of fault 3 (D) encounters the 20 contour; fault 3 is 5 mgal upthrown to the S.E. Likewise fault 4 is 5 mgal upthrown to the N.W. (e) and fault 5, 4 mgals upthrown to the south (F).

CHAPTER 5

GRAVITY IV: INTERPRETATION PROBLEM

Let us assume that it is our responsibility to explore a new concession area and that the first phase in exploration is a gravity survey. For efficiency in carrying out the survey, readings were made along traverses which were of the order of 4 X 6 miles in size, with station readings every half mile. Advantage was taken of easy lines of access such as roads and trails (mostly the crooked lines) and then additional cross-country traverses were run (the straight lines).

A map of the results of the gravity survey is shown in Fig. 5.1. The gravity data in this area proved to be extremely valuable in aiding the efficient exploration of the area, and at the conclusion of this exercise we will discuss planning the next exploration program in the light of the results of this gravity interpretation.

The area is essentially featureless, covered with sand without outcrops to give surface geologic information. Basement rocks outcrop just beyond the eastern extremity of the data. The gravity is very positive in both of these outcrop areas. The lowest gravity values are found on the south central portion of the map and gravity values generally increase towards the eastern edge of the surveyed area and also towards the northwestern corner. Most of the contours are

1 mgal intervals except half mgal dashed contours have been added in some of the very flat areas and the one mgal contours have been dropped in areas of extremely steep gradient where only the heavier 5 mgal contours are shown. There are over 100 mgal of gravity relief over this area; this is appreciably more than usually found in an area of this size. We therefore expect some appreciable structure.

Exercise: We usually begin a gravity interpretation with the features which are most obvious; examine the gravity map of Fig. 5.1 and interpret the most obvious features, before looking ahead at the continuing discussion of this map.

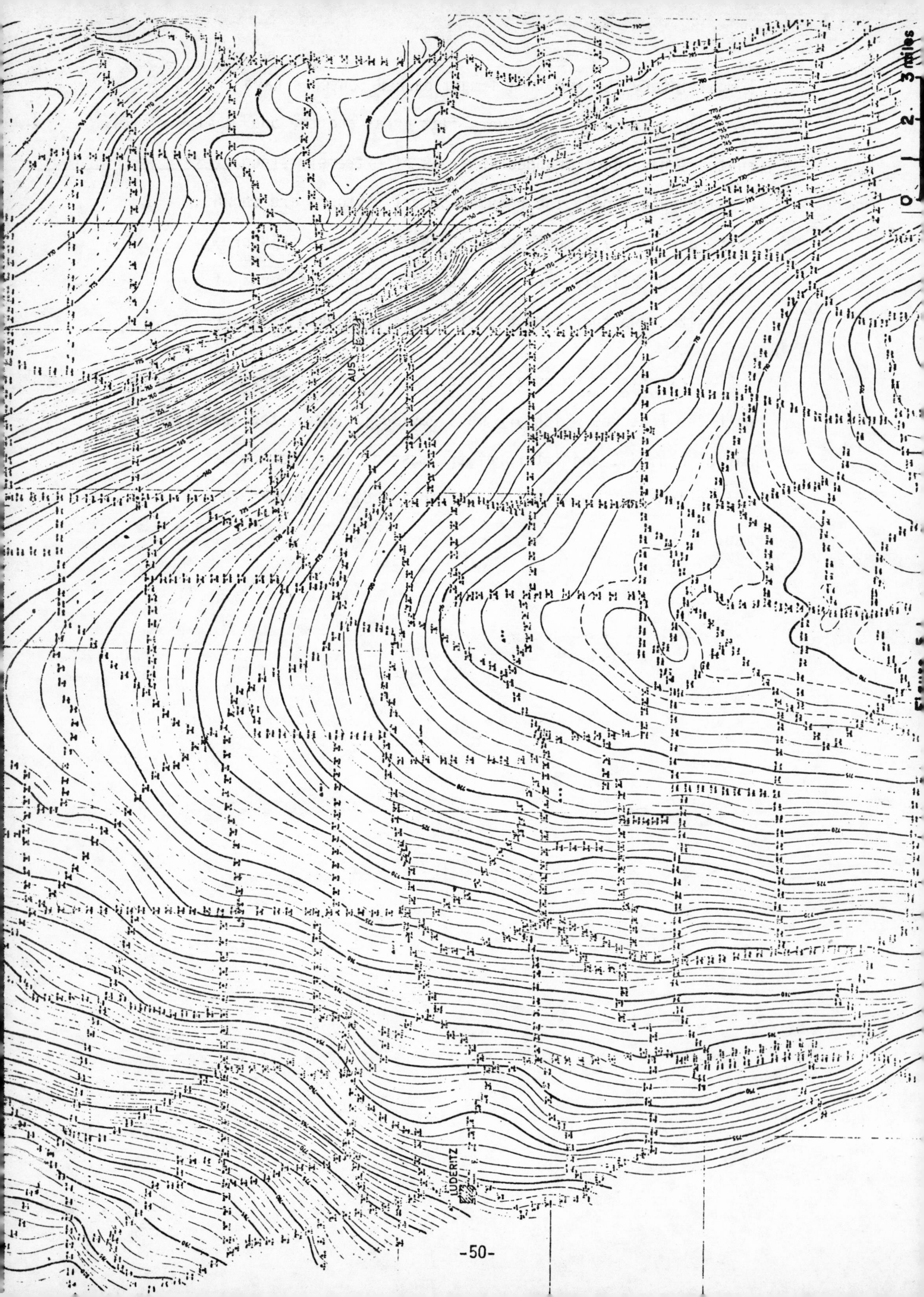
0 1 2 3 miles
AUS
LUDERITZ

Discussion of exercise:

The most obvious feature probably is the packing of contours on the right side of this map (Fig. 5.2) in a long linear fashion, which looks like evidence for a fault. The other major feature is the large region of negative gravity in the central portion of the map, indicating a large deficiency of mass and suggesting a sedimentary basin. The difference in gravity between the negative basinal areas and the positive basement outcrop areas is the order of 100 mgals. Using the equation $G = 0.0128\ \rho h$ (where G is the gravity difference, ρ is the density difference and h is the thickness in feet) we obtain about 8000 for the product ρh. If we assume a density difference of 0.3 between basement and sediments, we obtain of the order of 25,000 ft for the thickness of sediments. A density difference of 0.3 is, if anything, probably too large a value for an entire thickness of 25,000 ft, so that the actual thickness of sediments may be even larger. Certainly the area contains adequate thickness of sediments to be of interest.

To analyze the fault, we should draw several profiles across it, drawing the profiles roughly at right angles to the orientation of the fault, but locating the profiles as near as possible to control so that we minimize the interpretation involved in contouring. We draw several rather than only one profile to be sure that we have a representative profile to interpret which is not unduly distorted by other features, and also to get some idea of how the fault changes along its length.

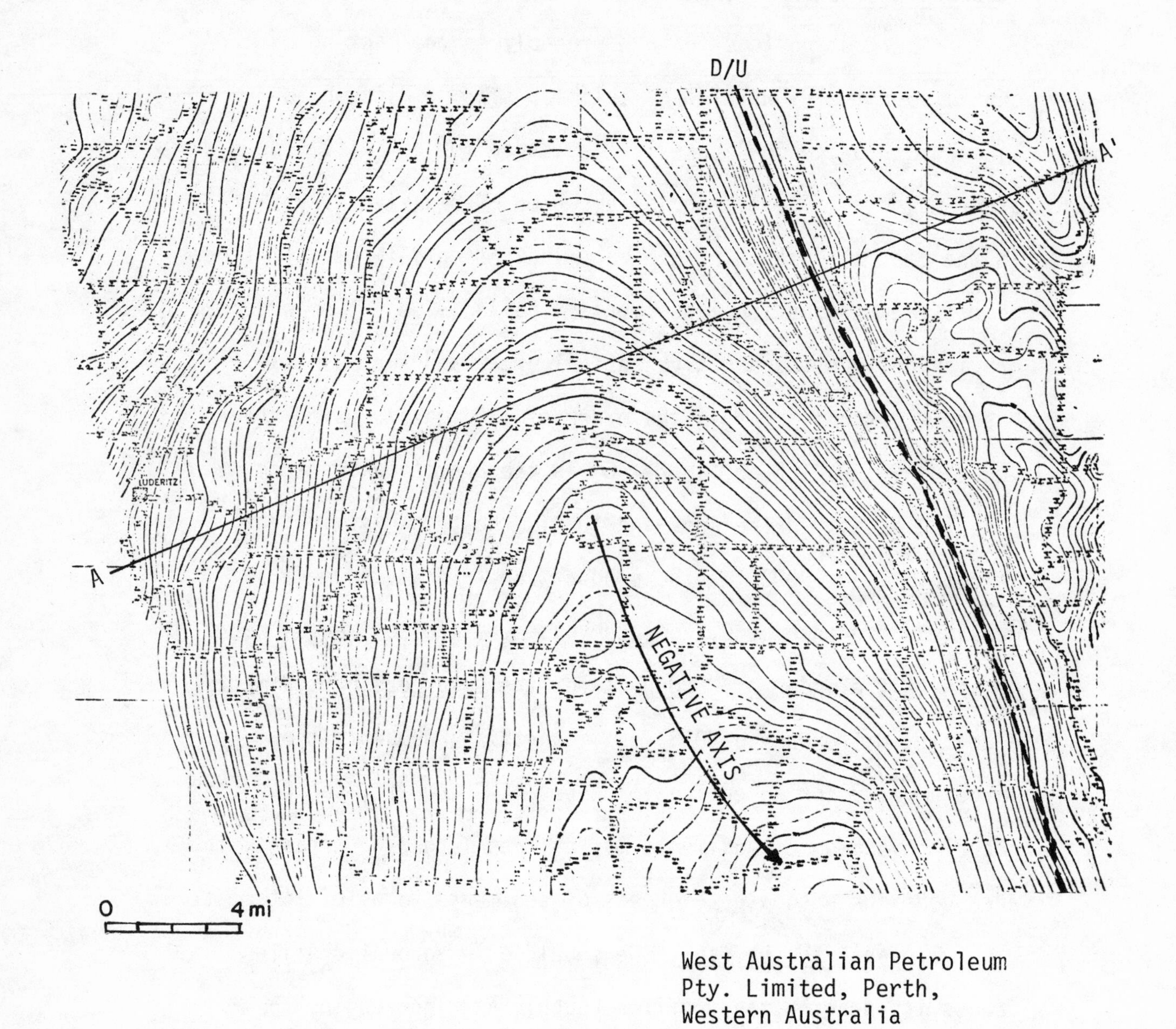

West Australian Petroleum
Pty. Limited, Perth,
Western Australia

Figure 5.2

Exercise:

Draw profiles across the fault and derive information about the fault from them before looking ahead at the continuing discussion. Include profile AA' in Fig. 5.2 as one of the profiles.

Discussion of exercise:

One profile, drawn between the points A and A' of Fig. 5.2, is shown in Fig. 5.3. The first problem after our profile is drawn is that of defining the regional. This is usually the most difficult problem in an interpretation. The regional represents gravity effects of mass distributions so deep and/or so large that we do not wish to consider them in our interpretation, and it is subjective. One usually looks at a number of profiles across the entire areas as a guide before deciding what to do about the regional. Let us assume that we end up concluding that we should use the regional shown in Fig. 5.3 by the dashed line. The ocean lies at the

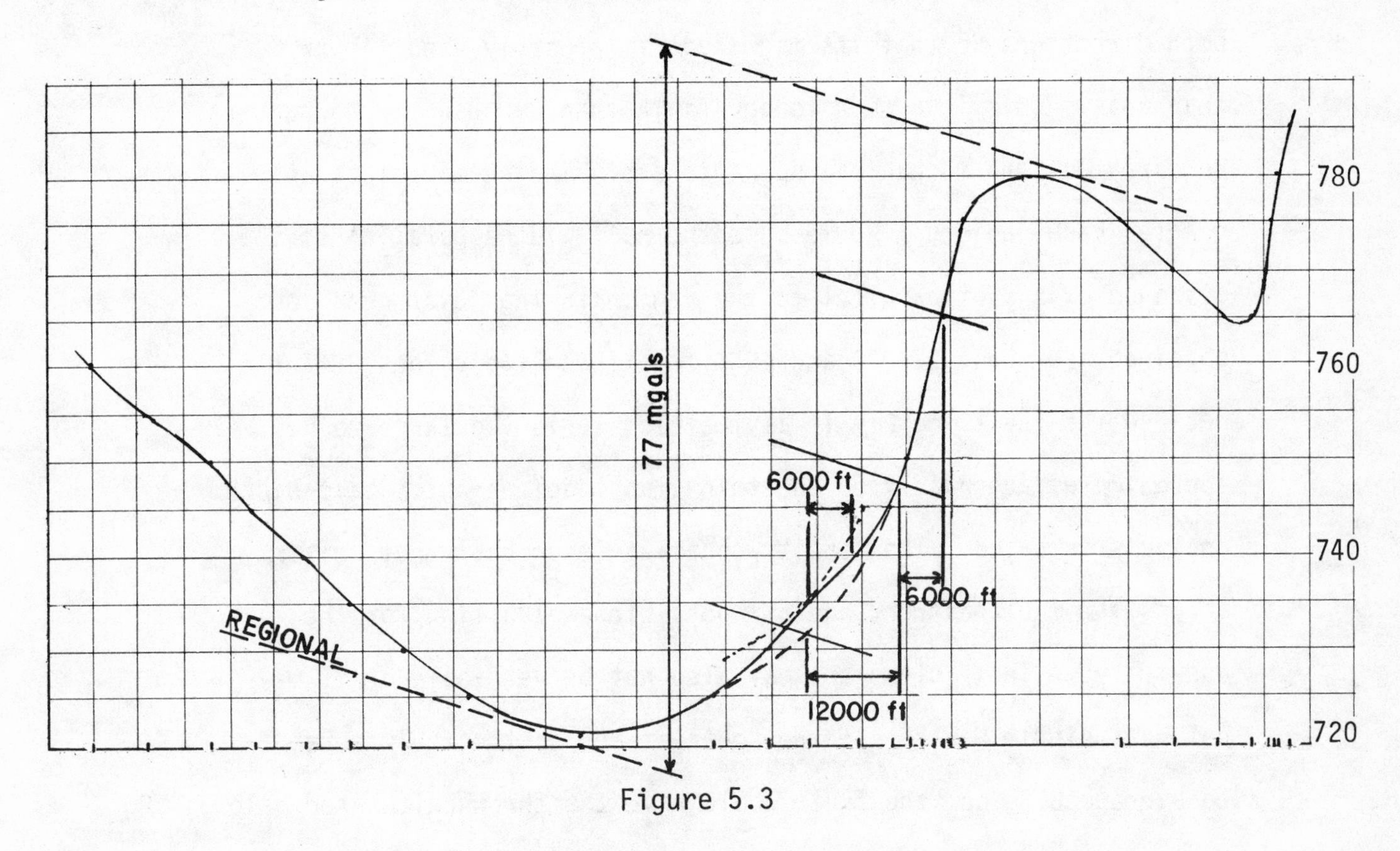

Figure 5.3

left edge of the map area and ocean basins are positive gravitationally with respect to continents, so we draw our regional with some slope to the right.

There is a bulge at A that is not part of the fault anomaly. On the map this bulge shows as a local nosing of the contours (see Fig. 5.2). We dash in the regional in this vicinity to illustrate what we expect is the part of the anomaly attributable to the fault. This results in a residual anomaly of about 3 mgal positive magnitude which we attribute to a local uplift. This anomaly has a half width of about 6000 ft so its depth if we use a spherical model is 1.3 times this or about 8000 ft. We should investigate this anomaly further for petroleum interest.

The fault has about 77 mgal expression as drawn. We have probably underestimated how far the fault effects extend in both directions so that its magnitude is probably even larger than this. This is a much larger fault than one usually encounters. We carry out the mechanical operations of finding midpoints and quarter value points. We measure the half widths parallel to the distance axis, not parallel to the regional; the distances so obtained are about 12,000 and 6000 ft. These two values differ because the fault anomaly is asymmetric; it is too large to be approximated adequately by the thin slab model on which half-width analysis is based. The density contrast is probably not uniform across the fault either, another possible explanation for the assymmetry. The fault plane may also not be vertical, or it may not be a single fault. If the contrast across the fault were to average 0.3g/cc, the fault would have the throw calculated below:

$$h = \frac{G}{0.0128\ \rho} = \frac{77}{0.0128(0.3)} = 20,000 \text{ ft.}$$

Combining this with the half width values suggests that the basement (assuming the contrast is sediments to basement) is nearly at the surface on the upthrown east side. This fault may be the main factor in the existence of the basin.

If we had drawn other profiles farther northward across this fault we would have seen slightly less expression and obtained values indicating slightly less depth. This fault is growing in magnitude toward the south.

If the basin is a half graben as sketched in Fig. 5.4, then the eastward thickening wedge would result in lower gravity values as we go eastward (AA'). Adding to this the anomaly because of the fault (BB') results in a minimum C, a negative gravity axis. This negative axis results from the superposition of these two anomalies and does not indicate the deepest part of the basin.

Exercise:

What local features can be found on Fig. 5.1?

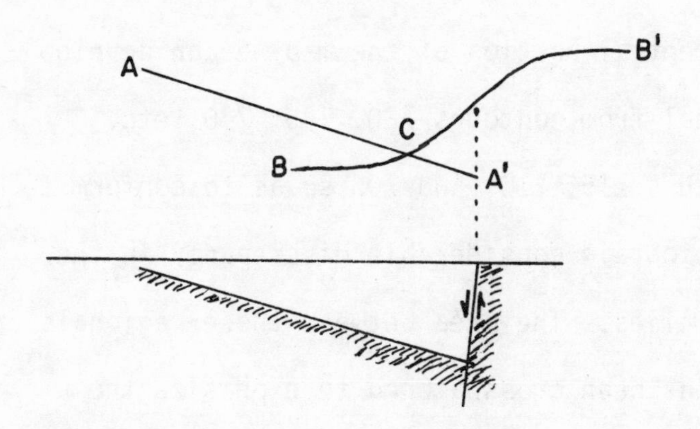

Figure 5.4

Discussion of exercise:

Contours appear to be packed more closely together at places marked F and G on Fig. 5.5, suggesting north-south trending faults as indicated. Note that these have been picked where controlled by the lines of the survey rather than in the intervening spaces where the contours are merely continued across in order to minimize errors because of excessive license in contouring. The faults marked F all appear to be upthrown to the west, those marked G upthrown to the east.

The contours east of Luderitz appear to be offset in steps as indicated by h1, h2, h3, etc. This offsetting indicates a system of east-west faults, J, which are upthrown to the north.

Just east of Luderitz a nosing of the contours P1, indicates a positive anomaly. Similar positive nosings are found elsewhere. Most of these indicate local uplifts worth investigating further. Additional similar features can also be found.

In the south-central portion of the map we can develop a feel for the regional from contours 720, 725, 730, etc. If we sketch in contours 715, 710, and 705 so as to conform to the regional shape we obtain considerable discrepancy in the form of positive anomalies. The area between these regionals and the actual contours has been crosshatched to exphasize the anomaly. These cross-hatched regions should, however, not be treated as outlining the anomalous area; the interpretation is not quite that simple. A central basin uplift area is

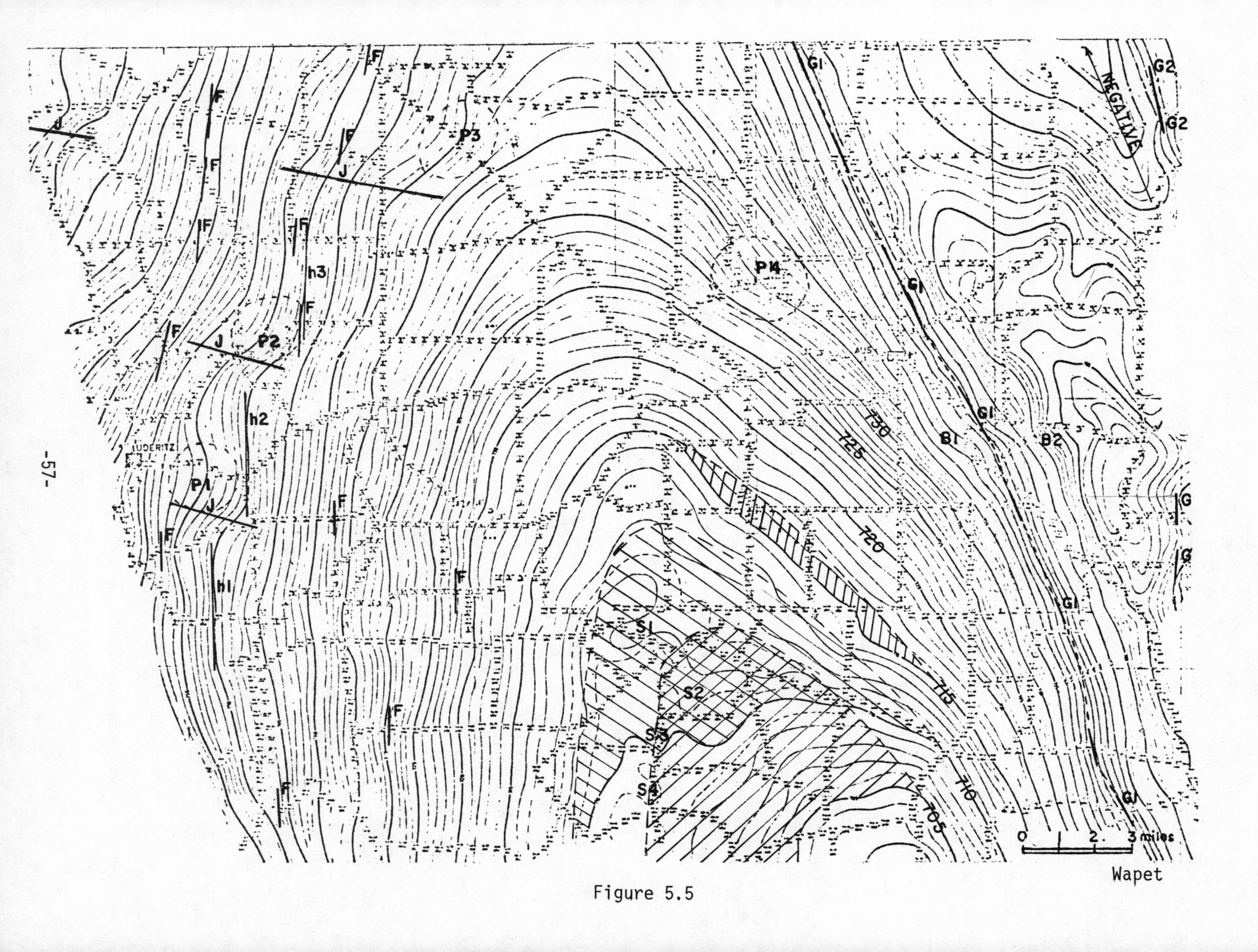
NEGATIVE
G2
G1
B1
B2
P1
P2
P3
P4
h1
h2
h3
730
725
720
715
710
705
S1
S2
S4
0
2
3 miles
Wapet
LUDERITZ

Figure 5.5

indicated and further analysis of this region would show fault-bounded north-south trending horsts.

The small contour undulations, S_1, S_2, etc., if analyzed for depth significance (as by drawing profiles across them) give very shallow depths - usually 1000 ft or so. Prospects in the center of the basin are thought to be much deeper than this and thus these features are largely ignored for structural significance although they are bona-fide gravity anomalies.

Some features of bowing (such as B1, B2) are ignored because they are not controlled but could be recontoured away without violating any of the data. Additional control should be run across them to verify their reality before much exploration effort is devoted to them.

To the east of the major fault G1 bounding the east side of the basin there appears to lie another half-graben basin with a negative axis plunging northward and bounded on the east by the up-to-the east fault G2.

Follow-up to the Gravity Survey

This gravity interpretation has suggested a thick section of sediments and a number of features of possible exploration interest. It was followed by a seismic survey to verify the features of the gravity interpretation and detail those features which appeared to have the best prospects. The first seismic lines actually run in the follow-up survey included a couple of east-west lines across the basin and across local anomalies; these lines verified that the features indicated structural anomalies and also helped show how the features were related to each other. Seismic work in the north-west corner where the north-south and east-west fault systems (F and J) lay suggested that different parts of the stratigraphic column subcropped under the surface sand in the different fault blocks and a series of shallow stratigraphic holes in these different fault blocks resulted in a picture of the nature of much of the stratigraphic column. Seismic detailing permitted the mapping of interesting anomalies on which deeper tests were subsequently run, leading to the discovery of oil and gas accumulations.

CHAPTER 6

MAGNETICS

Magnetic exploration is primarily used in the first phases of geophysical work in an area. Magnetic measurements can be made rapidly from the air and a large area can be surveyed quickly and cheaply.

Magnetic exploration is based on the fact that the earth acts as a large magnet; Figure 6.1 indicates the direction of the magnetic field in the space around the earth. The direction of the magnetic field at the surface of the earth depends on location with respect to the magnetic poles (rather than with respect to the geographical poles). How close a location is to the magnetic pole is measured by the inclination of the earth's field or the "magnetic latitude".

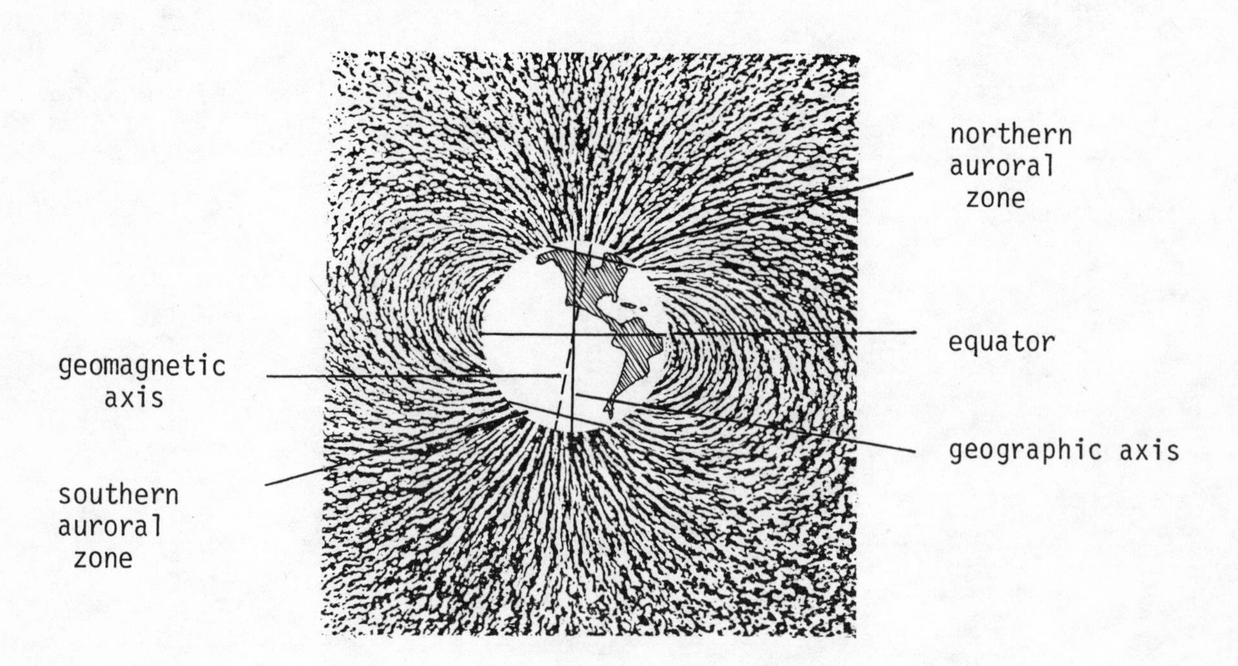

Figure 6.1

Rock types vary in magnetic susceptiblity, as shown in Fig. 6.2. The most important fact in magnetic exploration for petroleum is that sedimentary rocks are nearly non-magnetic, that is, have very small susceptiblity compared to basement rocks. Note that Fig. 6.2 shows susceptiblity in an exponential manner, successive vertical lines being a factor of 10 apart. The susceptibilities of non-sedimentary rock types are larger than those of sedimentary rocks by factors of 10-1,000 times. While some sediments are more magnetic than some basement rocks, this is the exception rather than the rule. This makes it possible to do petroleum exploration in a negative sense: <u>magnetic anomalies indicate the lack of sediments</u>. Thus, a magnetic survey is used to rule out areas which lack interest for petroleum exploration so that efforts can be concentrated elsewhere.

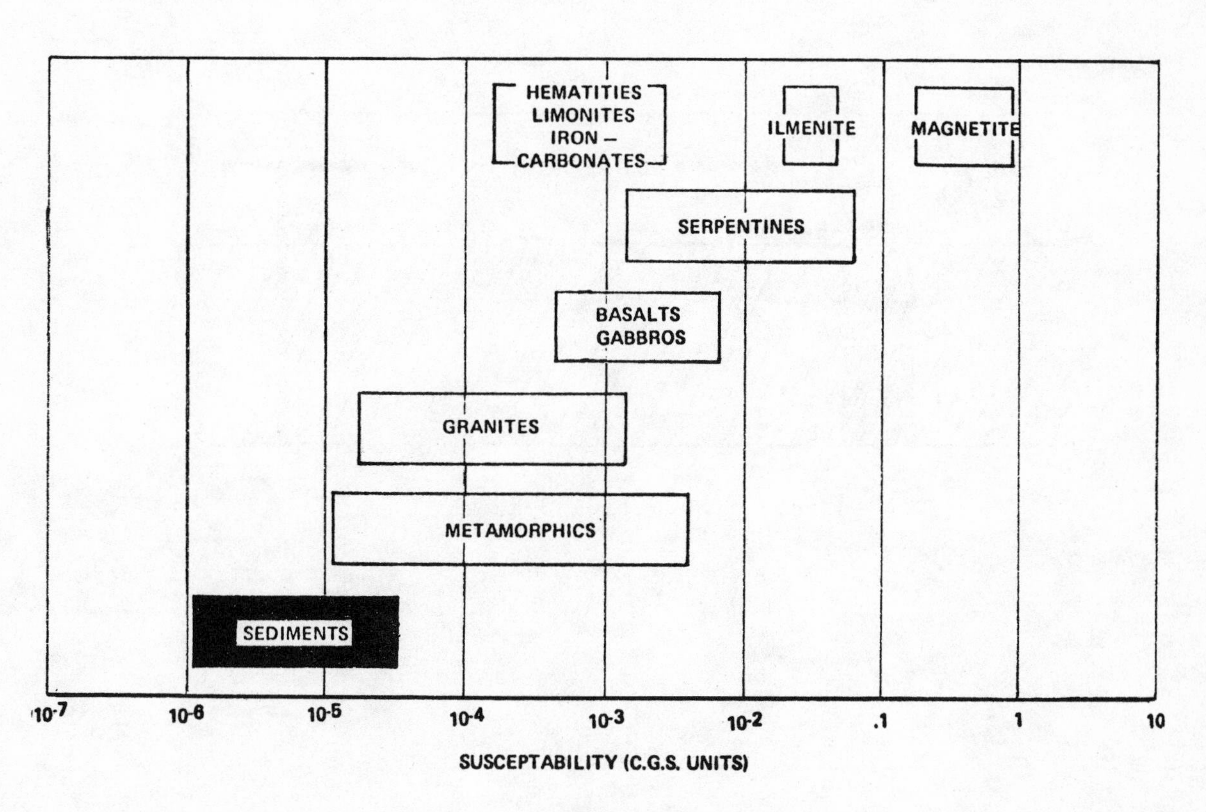

Figure 6.2

The chart in Fig. 6.3 suggests several <u>types of magnetic bodies</u> whose effects might be recorded. The most important feature of magnetic bodies is whether their vertical extent is small or large. Magnetically susceptible rocks in the presence of the earth's magnetic field become secondary magnets. The vertical component of the earth's field will effectively accumulate north or south poles on their upper and lower surfaces; which it will be depends on whether they are located in the northern or southern magnetic hemispheres. If a body is thin, the poles on its upper and lower surfaces are close together and their magnetic effects nearly cancel, resulting in only a small net effect. Only where bodies are sufficiently thick that the poles on the upper and lower surfaces are appreciable distances apart will magnetic effects be large. Thus of the features shown in Fig. 6.3, the

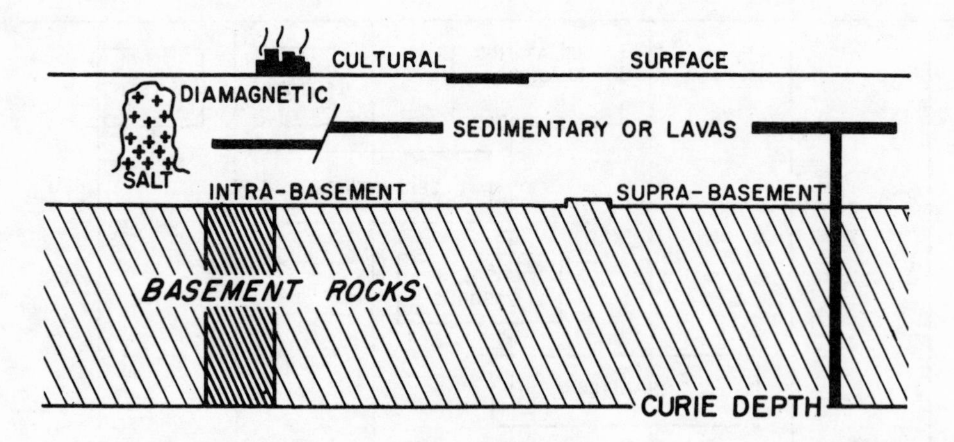

TYPES OF MAGNETIC SOURCES

Figure 6.3

only ones which will cause sizeable magnetic anomalies will be the large intrabasement blocks or features like volcanic necks which have large vertical extent.

Salt presents a rather special case. It is diamagnetic and tends to oppose the field that is polarizing it, resulting in a magnetic low over a salt dome. The diamagnetic effect is, however, of very small magnitude and anomalies due to salt show up only where other magnetic features are very remote and so do not produce obscuring effects. One could locate salt domes by magnetic means in deep basins, but other techniques such as gravity or seismic refraction are better suited.

A magnetometer needs to be as remote as possible from distorting magnetic fields. The magnetometer in an airborne survey may be located in a tailstinger which protrudes from the rear of the aircraft or in a bird suspended below the aircraft. Mounting in a bird locates it farther away from the magnetic field of the aircraft, but makes it more difficult to know its exact location.

Magnetic measurements are often made as an add-on to marine seismic surveys, a magnetometer being towed behind a seismic ship to provide magnetic measurements at little incremental cost. Such magnetic data might help in seismic interpretation, for example, in telling whether a diapir seen in the seismic data represents salt or a volcanic neck.

Magnetic measurements made close to magnetic bodies would see sharp anomalies whereas the anomalies would be smaller, broader, and smoother if the magnetic bodies were distant. We can determine the distance to magnetic bodies from the sharpness of the relief of magnetic profiles. Sharp magnetic anomalies indicate basement

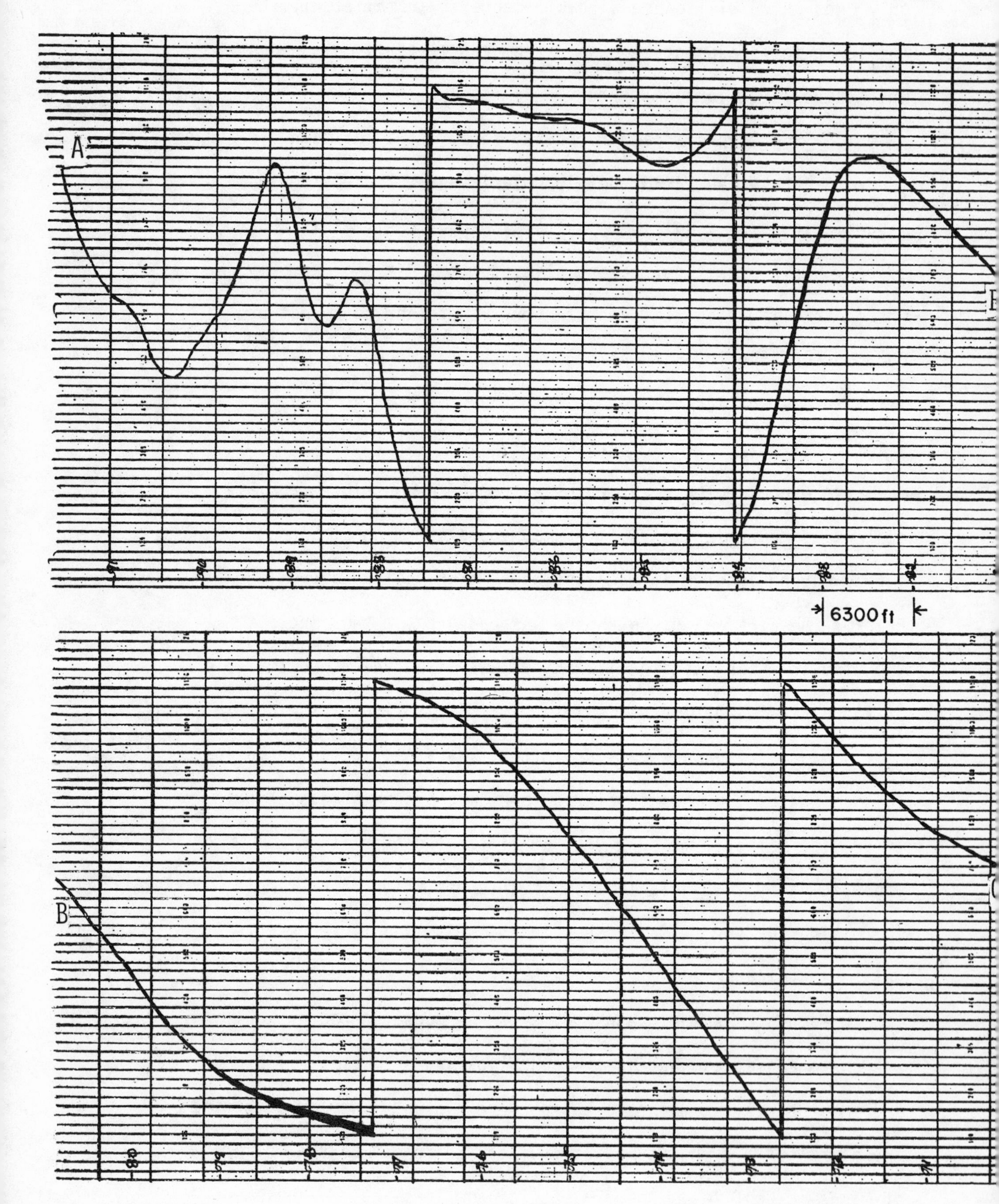

Figure 6.4

rocks not far beneath the surface, but where magnetic features are broad magnetic basement may be deep and we infer that the difference between the basement depth and the surface is filled with sediments.

Magnetic data are usually plotted in profile form such as Fig. 6.4. This figure shows how the total magnetic field intensity varies as an aircraft flies along a straight line, ABC. Magnetic surveys with petroleum objectives would usually be flown along parallel lines 1½ and 4 kilometers apart with occassional tie lines at right angles. Such a spacing would define the field adequately where basement is deeper than the line spacing below the aircraft; 1½ km is about the minimum basement depth of interest in petroleum exploration. A magnetic map is subsequently constructed from a series of profiles.

The shape of a magnetic anomaly varies somewhat with magnetic latitude and also with the shape of the magnetic body its attitude, its dimensions, its orientation with respect to magnetic north, and especially the depth of the body. In petroleum surveys, we are interested mainly in the last of these, the depth. Quick analysis techniques determine the degree of sharpness of the magnetic field, and hence a rough estimate of the distance to the magnetic body. Figure 6.5 shows two measurements which can be made on anomalies; the distance to the magnetic body which causes the anomaly can be estimated from these measurements. A straight line whose slope is the maximum slope of the magnetic anomaly provides a reasonable approximation to the anomaly curve for a certain horizontal distance, labeled "S" on Fig. 6.5. This distance is divided by an "index value" of 0.7 to 1.0 to give the distance to the magnetic body within about 25%.

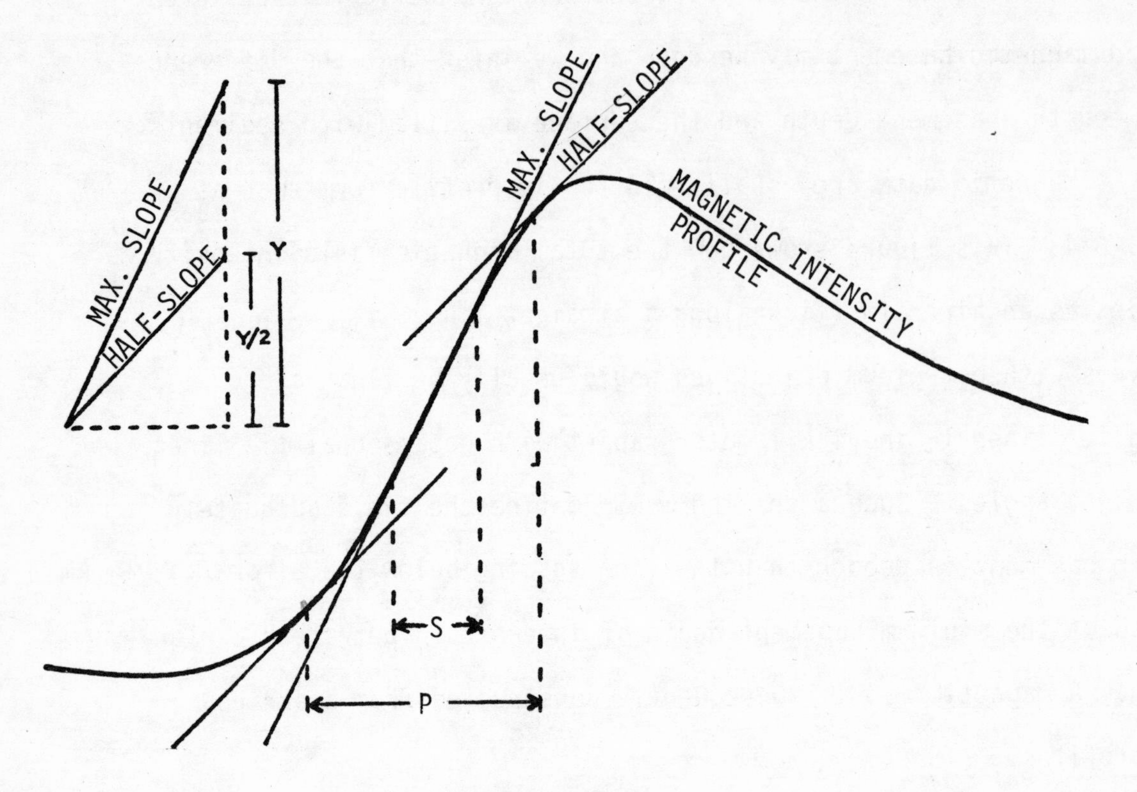

Figure 6.5

The Peters' half-slope method is less subjective than the maximum slope method. The points at which lines with half the maximum slope are tangent to the magnetic profile are determined. The horizontal distance between these points is labeled "P" on Fig. 6.5; this value is divided by 1.2 to 2.0 to give the depth to the body. In both the maximum slope and half slope methods the index values increase with the width to depth ratios for the body causing the anomaly, and the depth of a body is the distance to its upper surface. A number of other methods can also be used; all involve measuring some shape factor which is proportional to the depth of the body which causes the anomaly.

These analyses can be made from magnetic maps as well as from magnetic profiles. The maximum slope is indicated on the map by the contours being most closely spaced. The distance over which closely spaced contours are rather uniform is determined and multiplied by the appropriate index factor. Half-slope points on a contour map are where the contour spacing is double the closest spacing.

Magnetic maps of many areas are available in the literature. Magnetic surveys are extensively done by governmental organizations. While such maps often cannot be analyzed with the precision possible on actual profiles, much useful information often can be obtained from them.

Other geologic features besides basement depth can often be inferred from magnetic maps. Trends in magnetic features often have related trends in the overlying sediments. A magnetic map of the folded Appalachian region of Pennsylvania gives many basement depths all of the same magnitude, suggesting that basement is roughly the same over the area despite tremendous relief in the surface rocks, and hence the sediments must be detached from the basement. Systematic offset of magnetic anomalies may indicate strike-slip faults which have displaced basement rocks and possibly affected the sedimentary section.

Exercise

Interpret the magnetic profile shown in Fig. 6.4. The top and bottom are part of a continuous traverse, ABC. The fiducial marks at the bottom are a fixed time interval apart and during this time the aircraft flies 6300 ft. These marks help register the various documents from a survey (magnetometer, altimeter, navigation records) and they give the horizontal scale. The aircraft elevation was about 1500 ft above ground level. Interpret this exercise before reading further.

Discussion of exercise:

Let us use both the "length of maximum slope" and "half-slope" methods at three points, near fiducial marks 89, 83-84 and 73-75. The result is shown in Fig. 6.6 and the table below.

Fiducial	Measurement	Index	Depth	re ground level
89	S = 1300'	0.85	1500	0
	P = 1900'	1.4	1400	0
83-84	S = 4600'	0.85	5400	-3900'
	P = 6500'	1.4	4600	-3100'
73-75	S = 11,100'	0.85	13,000	-11,500'
	P = 31,100'	1.4	22,200	-20,500'

The S index value selected is the mid-range value. The measurements indicate basement rocks very near the surface near fiducial 89, 3-4000 ft of sediments near 83-84, and considerable thickness of sediments near 73-75. The discrepancy between the S and P results for the last instance suggests that something is disturbing the measurements, perhaps the anomalous magnetic body does not have the simple homogeneous, rectangular shape which the model assumes. On the basis of other measurements and overall fit we'd subsequently throw out values which don't fit an overall pattern.

This is an east-west profile in central Texas. The Llano uplift with outcropping basement nearby is at the left end of the line. Several down-to-the east faults are then crossed and sediment thicknesses of about 3500' and 12000' are believed to be approximately correct.

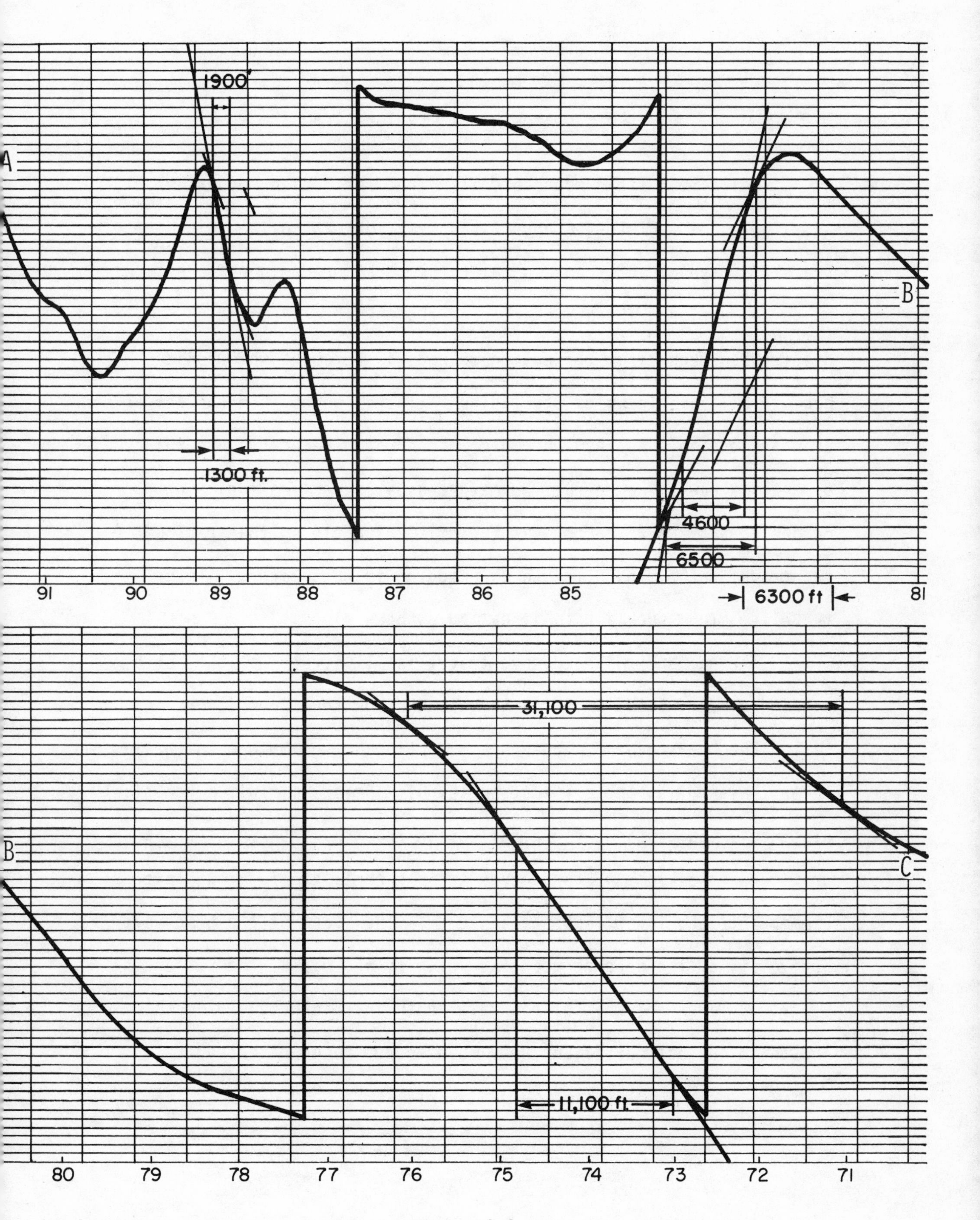
1900'
A
1300 ft.
4600
6500
B
91
90
89
88
87
86
85
6300 ft
81
31,100
B
C
11,100 ft.
80
79
78
77
76
75
74
73
72
71

Figure 6.6

Refraction Seismic Exploration

Refraction field methods differ from reflection methods mainly in using large shot-to-geophone distances, so that the seismic wave travel is predominantly horizontal rather than predominantly vertical as in reflection. The association of the word "refraction" to distinguish these from reflection methods is somewhat unfortunate because all seismic waves (including reflected waves) undergo refraction or bending when they encounter a change in velocity, according to Snell's Law. The refraction method usually implies impinging on one layer, the refractor, at the critical angle so that part of the travel path is parallel to the bedding in that layer. A wave traveling along such a path is called a "head wave".

The geometry of head wave travel is illustrated at the bottom of Figure 6.7. Seismic energy travels through the upper layer, called the "overburden", in such a direction as to impinge on the lower layer, layer 2, at the critical angle, θ_c. Beyond a minimum "critical distance", the distance for a reflection from the interface at the critical angle, some of the energy travels horizontally along layer 2, reemerging into layer 1 at the angle θ_c. The slope of the head-wave arrival-time line depends upon the velocity and dip of the refracting layer. A layer about 5% of the wavelength in thickness may be thick enough to conduct sufficient head wave energy to be detected.

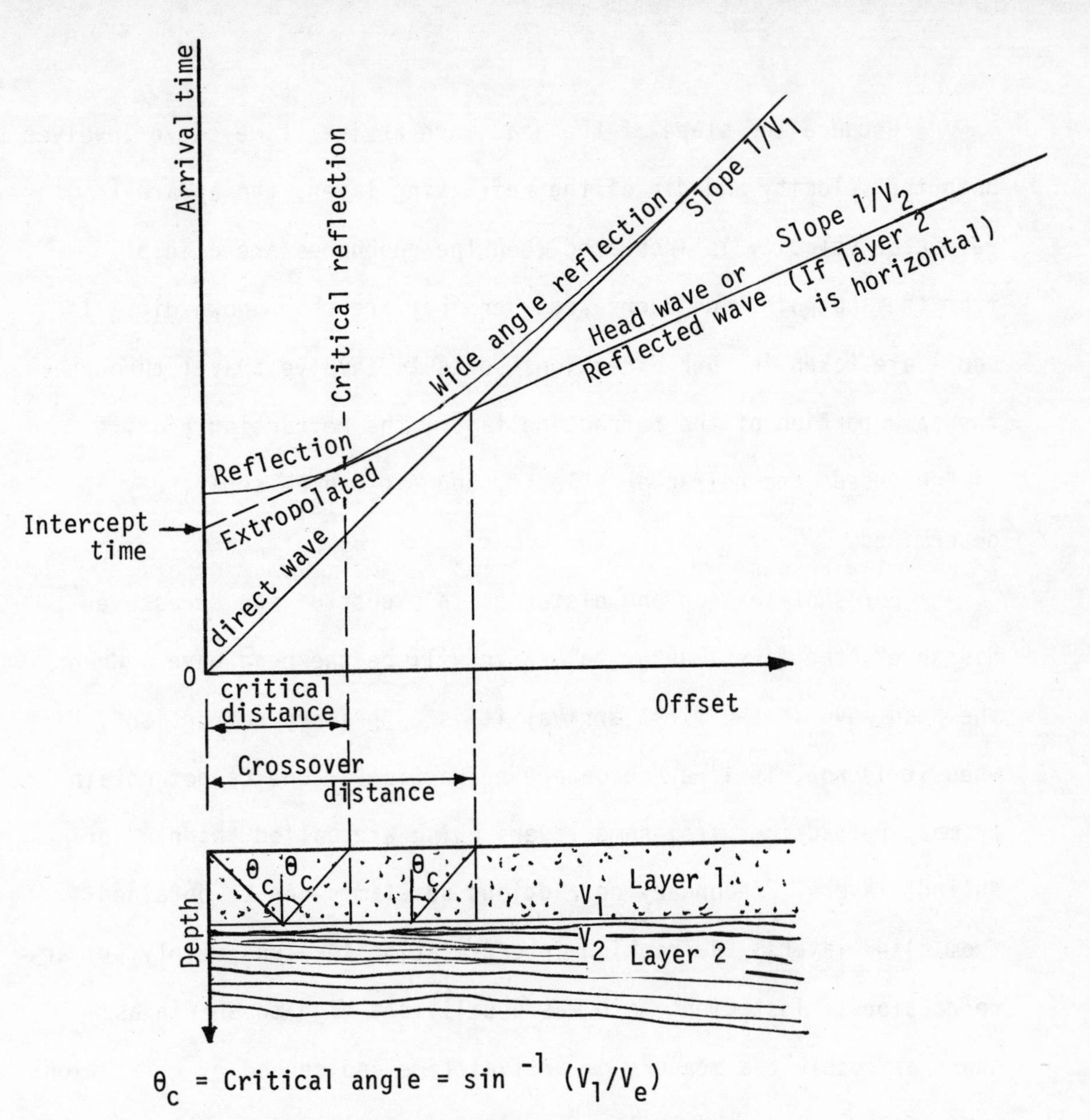

θ_c = Critical angle = $\sin^{-1}(V_1/V_e)$

NOTE: Reflection curvature is hyperbolic.
Reflection amplitude may be exceptionally large in vicinity of "critical reflection."

Figure 6.7

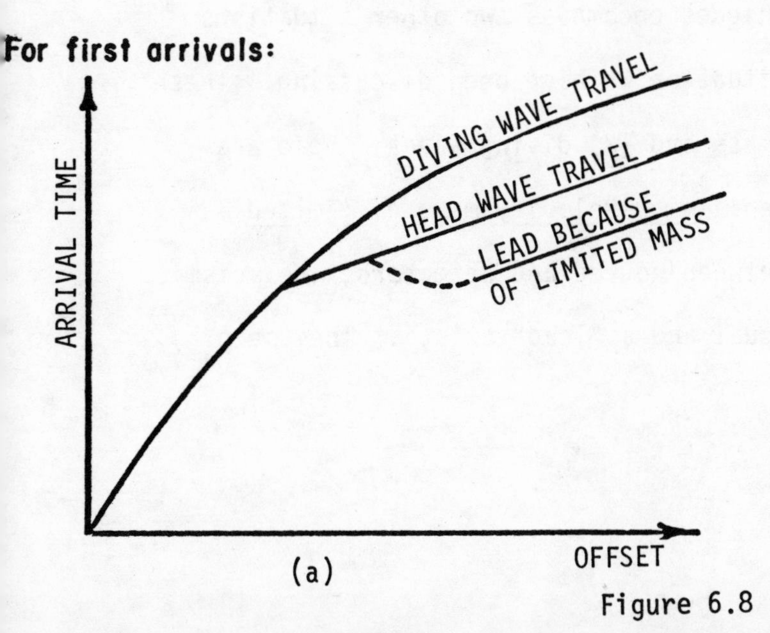

(a)

DIVING WAVES

Velocity increasing downward ⟶ upward raypath curvature

Greater offset gives greater penetration
Slope of T-D curve gives velocity of bottom of path

(b)

Figure 6.8

Because the slope of the head wave arrival-time curve involves both the velocity and dip of the refracting layer, the apparent refractor velocity is increased when the geophones are up-dip from the shotpoint and decreased when they are shown down-dip. If shots are taken in both directions so as to involve travel through the same portion of the refracting layer, the refraction profile is "reversed" and refractor velocity and dip can be separately determined.

For shot-to-geophone distances in excess of the "crossover distance", the first energy to arrive will be the head wave. When the head wave is the first arrival it is a "primary refraction"; when it is not, it is a "secondary refraction". One cannot obtain primary refractions from some layers which are called "hidden" or "blind" layers. Secondary refractions sometimes can be obtained from blind layers. Refraction interpretation involves mainly primary refractions. Instrument gain was usually set high to obtain as sharp as possible a measure of arrival time and secondary refractions were often lost in the tails of earlier arrivals. The ability to record data on magnetic tape and subsequently vary gain in processing has removed these restrictions.

Seismic refraction techniques encompass two other situations besides the simple head wave situation we have been discussing. These are (1) a local high-velocity mass and (2) diving waves; these are illustrated in Figure 6.8. When a high velocity mass of limited horizontal extent interposes between source and detectors, the seismic arrival time is earlier than usual and a "lead" exists. The use of seismic refraction to search for leads caused by salt domes was the first practical application of seismic methods in hydrocarbon exploration. Today this method is used only rarely.

Where distinctive high-velocity layers are not present, the first arrival shows successively greater penetration into the earth as the shot-to-geophone distance increases. This is the diving wave situation. The slope of the arrival-time curve then continually changes, the slope being the horizontal velocity at the maximum depth of penetration. The velocity-versus-depth relationship can be derived from the time-distance plot of diving-wave curves.

The reliability of refractor depth estimates depends fundamentally upon the velocity of the overburden. Good refraction interpretation usually needs information as to overburden velocity beyond what can be obtained from the time-distance curve itself.

Refraction interpretation involves identifying portions of the travel time curve which have a reversed relationship, that is, identifying segment A of time distance curve 1 as involving the same refractor as segment B of refraction curve 2. Making such identifications is the crucial element in refraction interpretation. Where more than 3 or 4 separate refractors exist, the identification process often is ambiguous and the reliability of the refraction interpretation decreases. To get a clear distinction between segments of a time-distance curve, refractors have to have appreciable velocity contrast, hopefully a factor of 1.5 in comparison with other layers.

Ambiguous refraction interpretation is apt to result (1) where there is insufficient velocity contrast between the refracting medium and overlying layers, (2) where the refractors are not distinctive, that is, where sometimes one layer and sometimes another layer is involved, and (3) where too many refractors, more than 4, are present.

Reliable refraction requires reversed control. The shot-to-geophone distance is sometimes critical for mapping a particular refractor and separate refraction programs may be required for each refractor which is to be mapped. Directional charges, an array of several charges detonated sequentially so as to match the horizontal velocity to the refractor velocity, improve the efficiency of refraction shooting. Occasionally extending a reflection profile to sufficient offsets that are recorded may allow tying refraction and reflection events.

The sonabuoy technique is used for marine refraction. A sonabuoy is an expendable hydrophone which radios the hydrophone information back to the recording ship. It is simply thrown overboard as a seismic ship does conventional reflection shooting. A refraction profile is obtained along with the reflection line at little incremental cost.

In plotting refraction data, time delays proportional to the offset distance may be introduced, the delays being approximately equal to the travel time of the head wave so that the head wave event appears roughly horizontal on the plotted profile, called a "reduced" profile. This makes it easier to pick refractions reliably.

Uses of Refraction Methods

While refraction techniques are not used often, under proper circumstances, they provide answers not obtainable otherwise. Refraction is sometimes used as a depth probe in a reconnaisance survey to determine the depth and velocity of high-velocity members, such as carbonate or evaporite layers or basement rocks. Depth probes are sometimes useful in the negative sense, to establish that no high-velocity layers exist and that there is an appreciable thickness of sand and shale sediments.

Where a distinctive refractor is present and not too deep and where a map on the refractor surface provides useful geological information, detailed mapping of the refractor may be practical. Refraction may also be useful in establishing correlations across faults.

The most frequently used application of refraction techniques is to map the base of the seismic weathering or low velocity layer. This is especially useful where surface seismic sources are used so that weathering layer information is not obtainable from the reflection work. Special small refraction crews sometimes accompany reflection crews for this purpose.

Refraction is also sometimes used in mapping salt dome limits, especially where a deep borehole is available so that geophones can be placed at various depths to record seismic energy from shot points on the surface. This method provides a way of defining the flanks of features such as salt domes with high precision than otherwise possible.

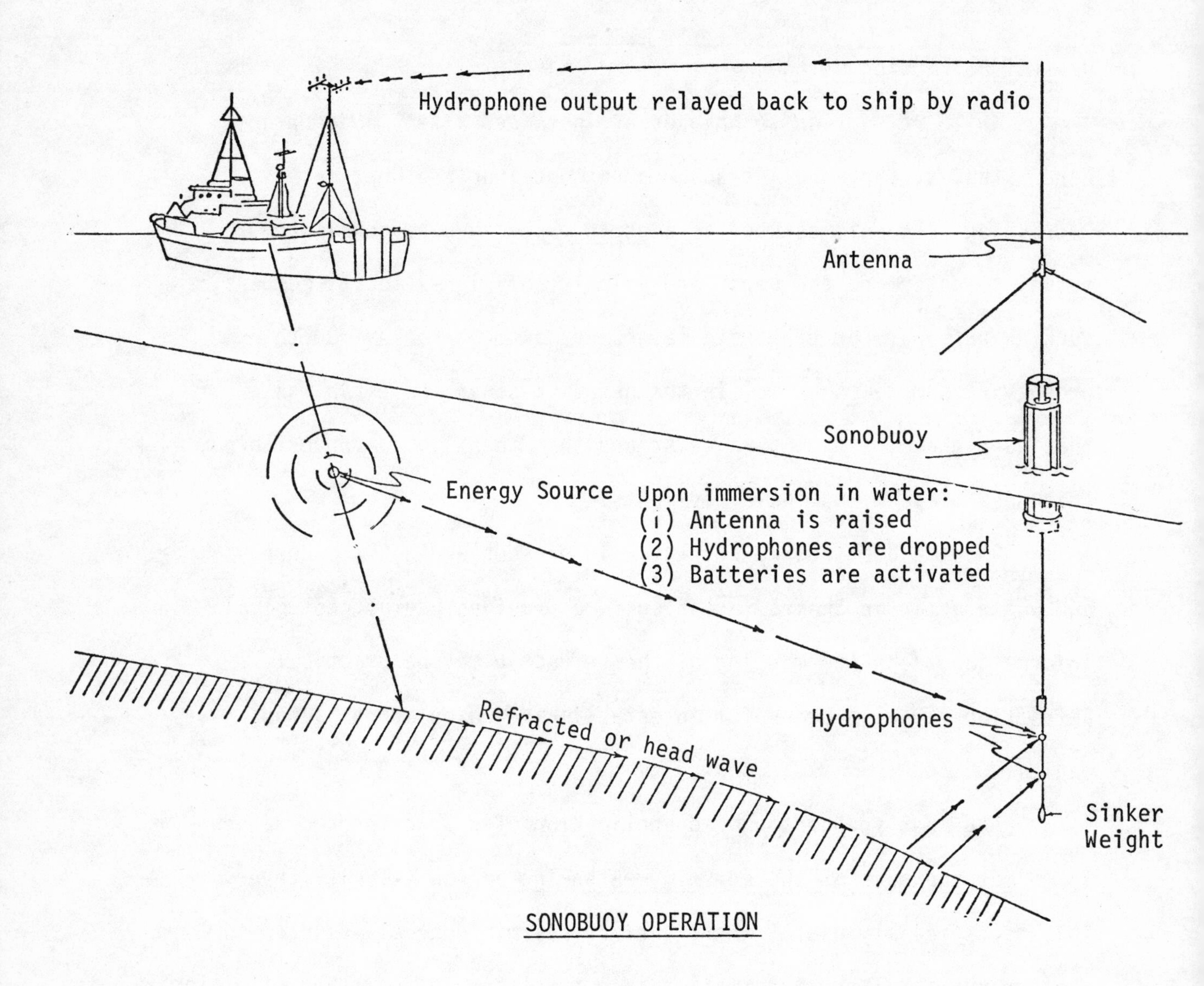

SONOBUOY OPERATION

CHAPTER 7

BASIC SEISMICS AND STATICS CORRECTIONS

The fundamental idea of seismic work is simple. Energy is generated at a source and transmitted into the earth as seismic waves; these waves bounce back from reflecting interfaces within the earth and are detected by geophones spread out on the surface of the earth.

An understanding of the properties of seismic waves and basic physical principles involved in seismic wave travel is needed in order to discuss seismic exploration.

Compressional wave: A type of seismic wave in which particle motion is in the same direction as the wave is moving. This is the type wave primarily involved in seismic exploration. Also called P-wave, primary wave, longitudinal wave.

Shear wave: A type of seismic wave in which particle motion is at right angles to the direction in which the wave is moving. Also called S-wave, secondary wave, transverse wave.

Wavefront: The surface on which particles are moving together at a given moment in time as a result of a seismic disturbance.

Raypath: An imaginary line perpendicular to the wavefronts along which it is often assumed that seismic energy travels. It is the shortest path between source and observing station which involves reflection.

Amplitude: The maximum displacement from equilibrium.

Period: The time of repetition of a periodic seismic wave. The time for a wavecrest to travel one wavelength of distance or the time for two successive wavecrests to pass a fixed point.

Frequency: The repetition rate for a periodic wave. The reciprocal of period.

Wavelength: The distance between successive similar points on a wave measured perpendicular to the wavefront.

Velocity: The distance traveled by a wavefront divided by the time to travel this distance. Equals the product of frequency and wavelength.

Apparent wavelength: The distance between successive similar points on a wave measured at an angle to the wavefront, for example, the distance between geophones on the surface of the earth which see successive wavecrests. The relationship between

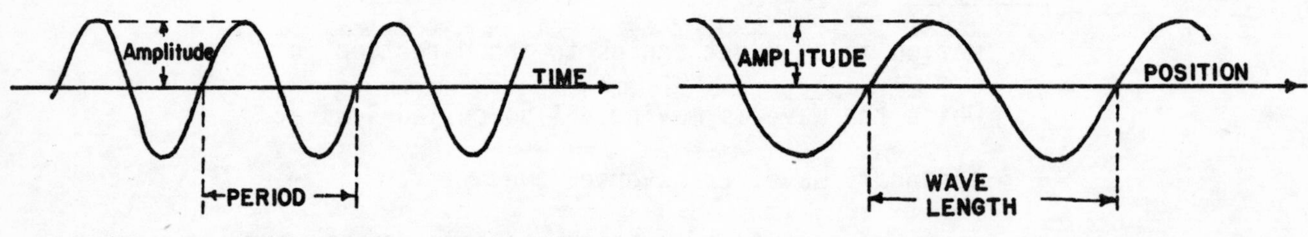

How one point varies with time   How various points look at same time

If wavefront approaches at an angle

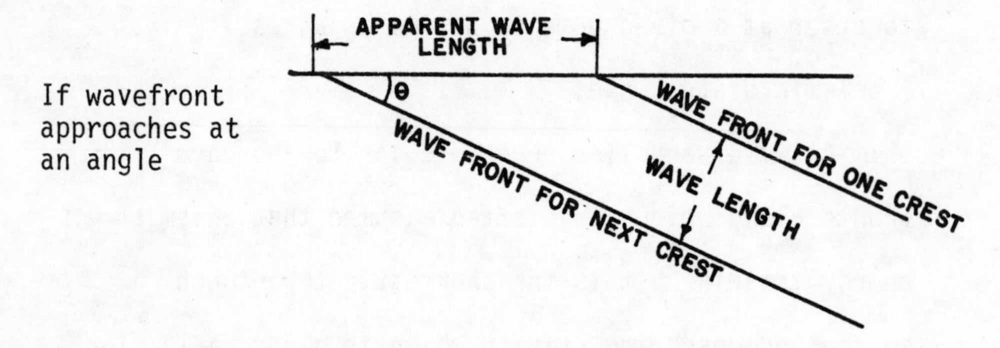

Figure 7.1

apparent wavelength and the true wavelength depends on the angle between the wave and the line of observation. This is illustrated in Fig. 7.1.

<u>Apparent velocity</u>: The product of frequency and apparent wavelength.

<u>Law of reflection</u>: The angle of incidence i equals the angle of reflection (Fig. 7.2).

<u>Snell's Law</u>: The sine of the angle of incidence i is to the sine of the angle of refraction r as the respective velocities (Fig. 7.2).

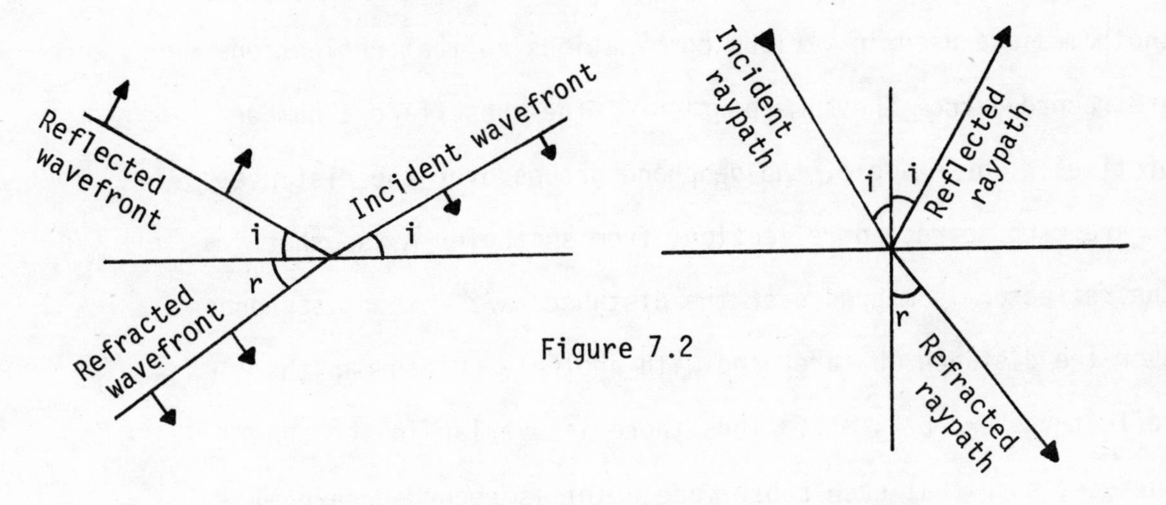

Figure 7.2

The classic method of recording seismic data is continuous coverage, single-fold shooting. Reflections from a single shot are recorded by groups of geophones laid out in a line on the surface of the earth. The geophones and shotpoints can have a variety of arrangements. One common arrangement is called a split-dip spread; it is a symmetrical arrangement with the source in the center of the geophone groups (Fig. 7.3a). The portion of the reflector involved is half the distance over which the geophone groups are spread (2d). The adjacent portion of the reflector is mapped by moving the geophones and shotpoint half the distance covered by the geophone groups. Another arrangement is the end-on spread (Fig. 7.3b); the shothole is at one end of the groups of geophones. Again, the portion of the reflector involved is half the length of the spread. To map the other half of the reflector, another shothole is used at the opposite end of the spread.

Most seismic shooting today is multiple-fold or <u>common-depth-point shooting</u> (Fig. 7.4). Geophone groups and shotholes are used in various combinations so that reflections are recorded from the same portion of the subsurface a number of times. For example, the geophone groups over the distance a are used to record reflections from shothole A so that the reflector is mapped over the distance a/2. Then geophones over the distance b are used with shothole B to map the reflector over b/2, etc. Thus there is overlap in the subsurface, such that each subsurface point is recorded several times. This is indicated by the raypaths in Fig. 7.4, all of which involve the same reflecting point. This example is of four-fold shooting.

Recording each point a number of times allows improvement of the signal-to-noise ratio. Changing the distance between shot and geophone changes the mixture of energy of different types and the differences can be used to attenuate some of the wave types and thus discriminate against noise.

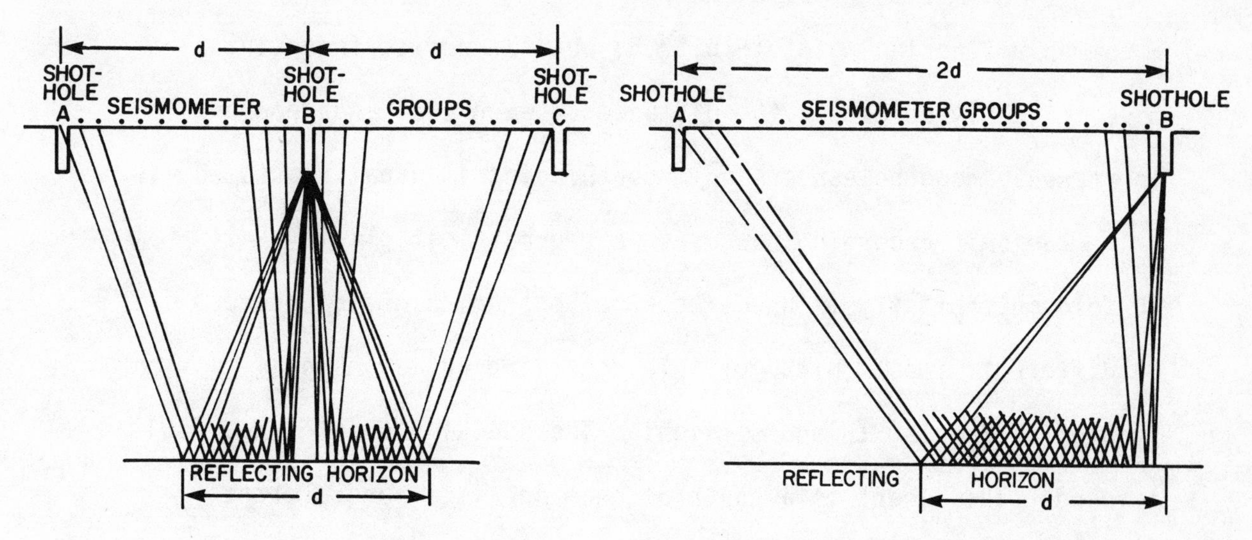

Vertical Section of 24-Group
Split Spread
Figure 7.3a

Vertical Section of 24-Group
Off-End Spread
Figure 7.3b

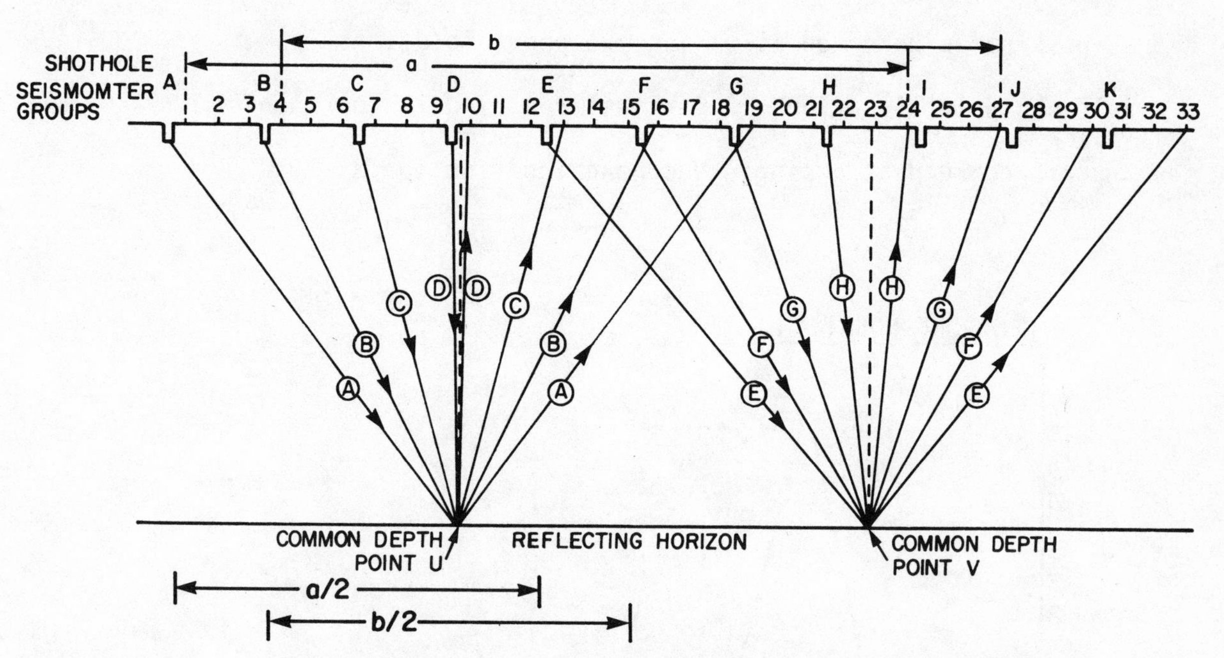

Vertical section of 4-fold common depth point shooting

Figure 7.4

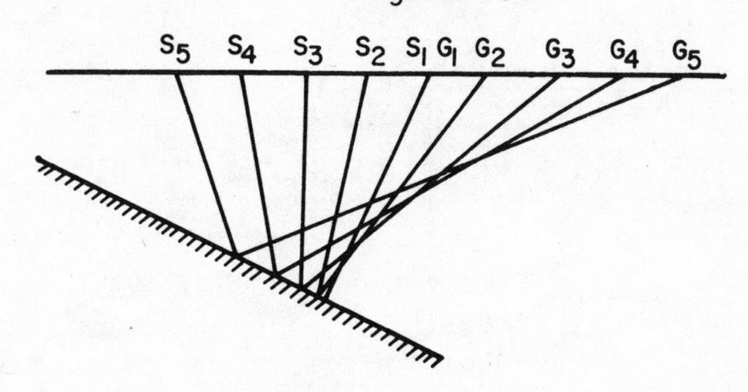

Figure 7.5

If the reflector is dipping, the data don't quite have a common reflecting point (Fig. 7.5) but the reflecting point moves slightly up-dip as the distance between shot and geophone increases. Nonetheless a single depth point is usually assumed.

Seismic recording now uses very great multiplicity. If 48-fold multiplicity is used, each reflecting point is recorded 48 different times. With multiple recording we may have a problem of keeping things straight. The stacking chart (Fig. 7.6) is used. This chart is a graph of geophone locations plotted horizontally against shotpoint location vertically, each dot representing one seismic trace. A horizontal row of dots represents traces recorded at various geophone locations from a single shotpoint location. A vertical row of dots represents the traces recorded at a single geophone location for a

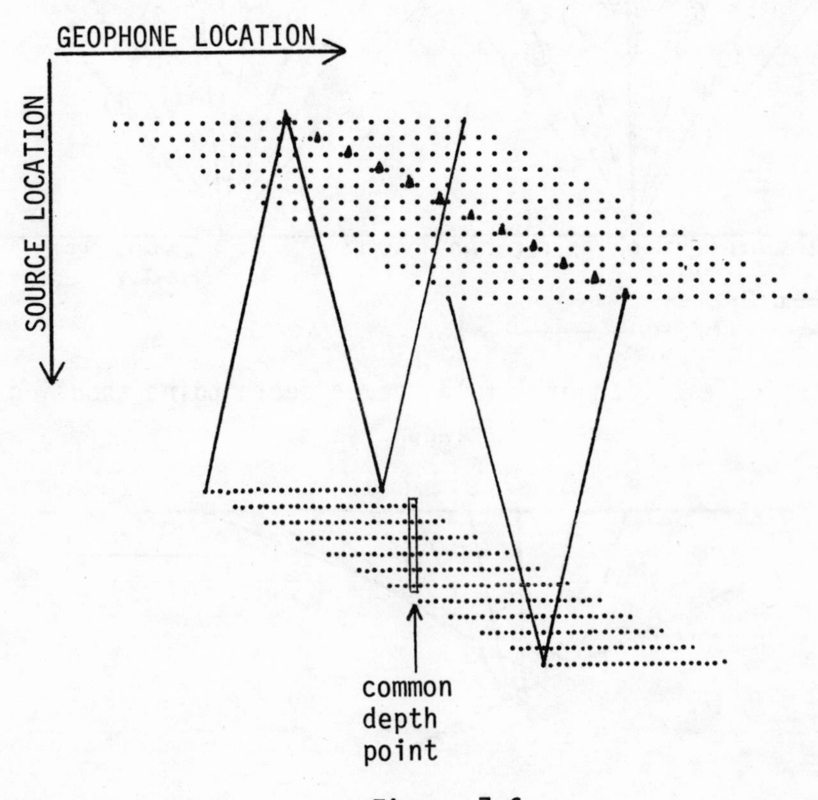

Figure 7.6

succession of shotpoints. A diagonal row of dots in one direction represents traces recorded from different combinations of geophones and shotpoints for all of which the offset or shot-to-geophone distance is the same. A diagonal row in the opposite direction contains the traces which have common reflecting points. Thus the stacking chart contains the information required to sort data with various aspects in common. Stacking charts can be prepared to indicate surface locations of shots and geophone, or subsurface locations. Reflecting points in the subsurface are half as far apart as the corresponding points on the surface. If we collect the traces with a common offset distance we have a common-offset gather or if we collect the traces reflecting from the same point we have a common-depth-point gather.

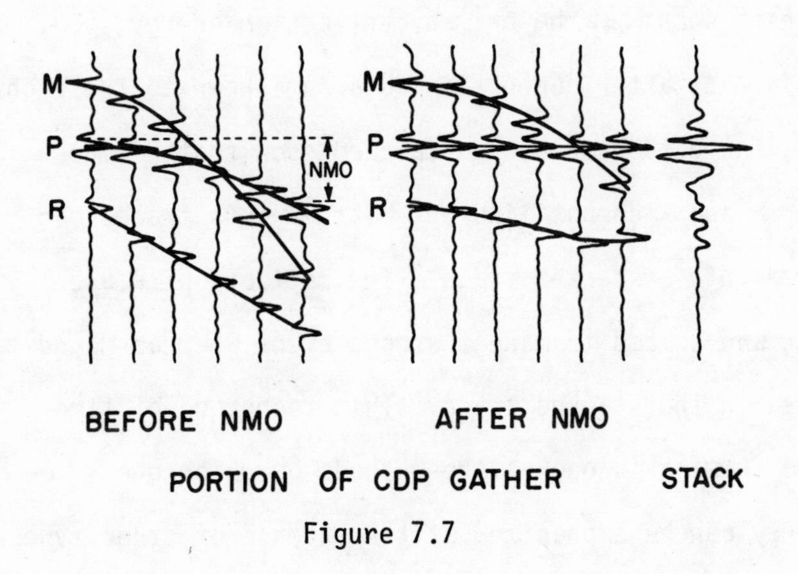

Figure 7.7

A few traces of a common-depth-point gather are shown on the left of Fig. 7.7. The distance from shot to geophone is gradually increasing for successive traces of the gather so that the reflected energy takes a bit longer for successive traces. The delay for reflections from a flat reflecting interface is called normal moveout, often abbreviated "NMO". Multiple energy, which has bounced around the shallow part of the section, will have traveled at a lower velocity than primary reflections usually and will have more moveout than the normal moveout of the primary reflection. The multiple is indicated in Fig. 7.7 by the letter "M" and energy which travels in a straight line along this spread by the letter "R". In normal moveout removal in data processing we displace each trace by a time shift such that the primary reflection energy, "P", lines up. In the "after correction" diagram shown to the right in Fig. 7.7, the data have been displaced such that P lines up. However M and R do not line up. The various traces can now be summed into a stacked trace. The reflections P are all in phase and so add to make a strong event whereas M and R are smeared in the stacked trace. This technique of time shifting and summing is one of the methods by which one type of energy can be emphasized at the expense of other types. It is the basis of the common-depth-point method.

There are two major types of corrections made to seismic data. The normal move-out correction such as just described depends upon arrival time and is called a dynamic correction. Each trace has to be shifted by a different amount at different travel times to line up the primary reflections. Static corrections are independent of time, the amount of shift being the same for all points on any trace. Static corrections are needed because of irregularities in the near surface. They are made for three things:

(1) Elevation variations,

(2) Low-velocity or weathering layer variations,

(3) Datum.

Figure 7.8 shows the basic idea of static corrections. The actual geophones are on the surface of the earth at various elevations, and the energy source may be at another elevation. The material near the surface of the earth is highly variable both in velocity and thickness and reflection arrivals may vary more because of near-surface variations than because of the subsurface relief in which we're interested. We

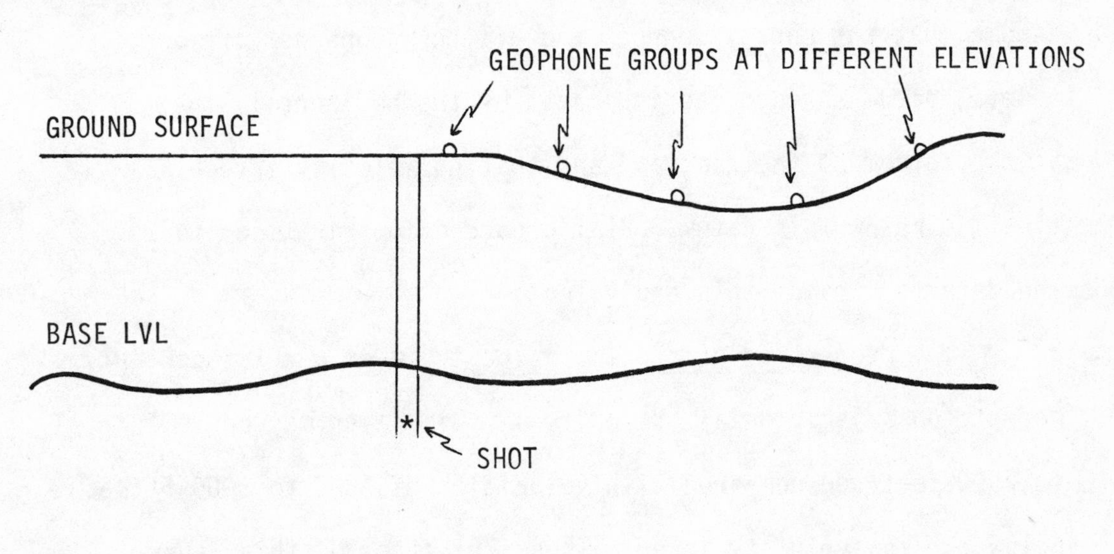

ASSUMED SITUATION:

SHOT AND GEOPHONES ON DATUM VERTICALLY BENEATH SURFACE LOCATIONS. SAME VELOCITY DISTRIBUTION BELOW DATUM.

Figure 7.8

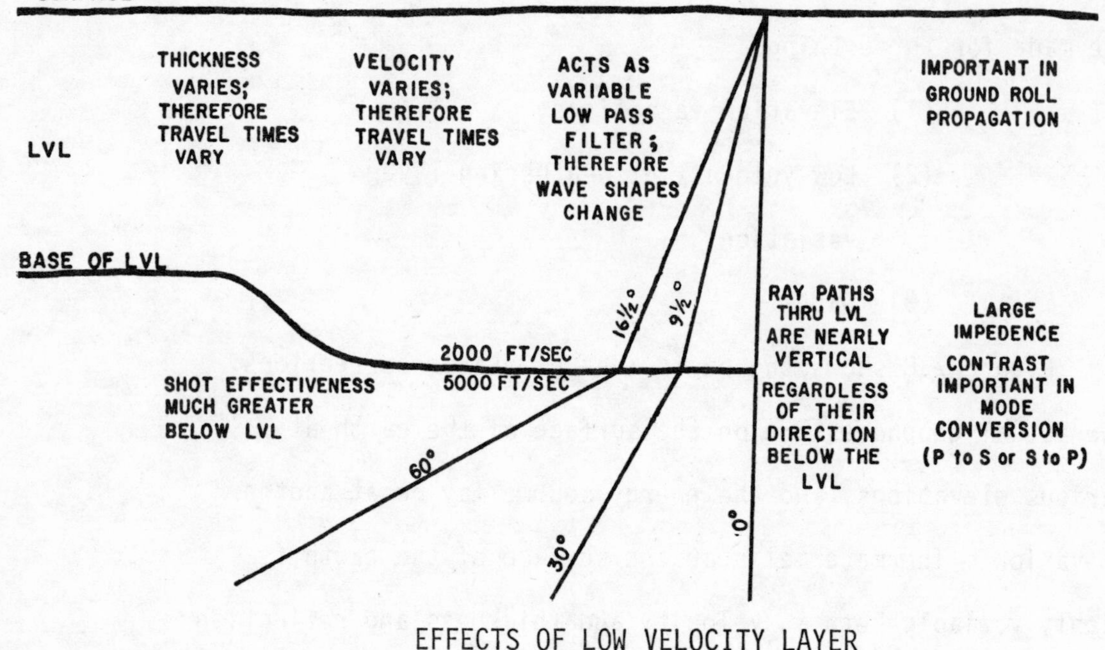

EFFECTS OF LOW VELOCITY LAYER

Figure 7.9

assume that the energy source and all the geophones are on a level datum surface, as indicated by the horizontal line at the bottom of the diagram, and that no velocity irregularities lie below this datum. Static corrections are made to the data to provide this equivalent.

The <u>properties of the low velocity layer</u> exert important effects on seismic data. Velocity in this layer may be 1000 to 2500 feet/sec compared with velocities of 5000 to 8000 ft/sec below the low velocity layer. Thus the bottom of this layer is an important contact where a large change in velocity occurs. The base of the low velocity layer is often, but not always, the water table. The low velocity layer sediments are usually aerated, with gas bubbles in the pore spaces. The thickness of this layer can vary from 10 to 200 feet, occasionally more. The thickness of the layer and the velocity within the layer may change in both horizontal and vertical directions rapidly and erratically.

Some of the <u>effects of this low velocity layer</u> are shown in Fig. 7.9. The large velocity contrast at the base of the layer bends raypaths considerably and makes the energy travel almost vertically through the low velocity layer, as shown for three raypaths. This large contrast is also a strong producer of multiples. We usually assume that raypaths travel vertically in the low velocity layer but the error in travel time which this assumption involves is only a fraction of a millisecond, even for energy traveling at extreme angles below the low velocity layer.

The low velocity layer is also called the weathering layer, but weathering here has a different meaning than its usual meaning in geology. <u>Types of near surface conditions</u> are:

(1) More or less uniform;

(2) Thicker on hills because water table is deeper there;

(3) Thicker in valleys because alluvial fill is greater there;

(4) Deserts where layer is abnormally thick with gradational velocity;

(5) Double-layer weathering, types of which are shown in Fig. 7.10.

(6) Glacial drift, which produces highly variable conditions, sometimes independent of water table;

(7) Frozen near-surface (permafrost); characterized by high velocity, discussed below.

(8) Alluvial fans, stream channels, etc.;

(9) Marine problems because of variations in water depth;

(10) A low-velocity aerated layer below water bottom.

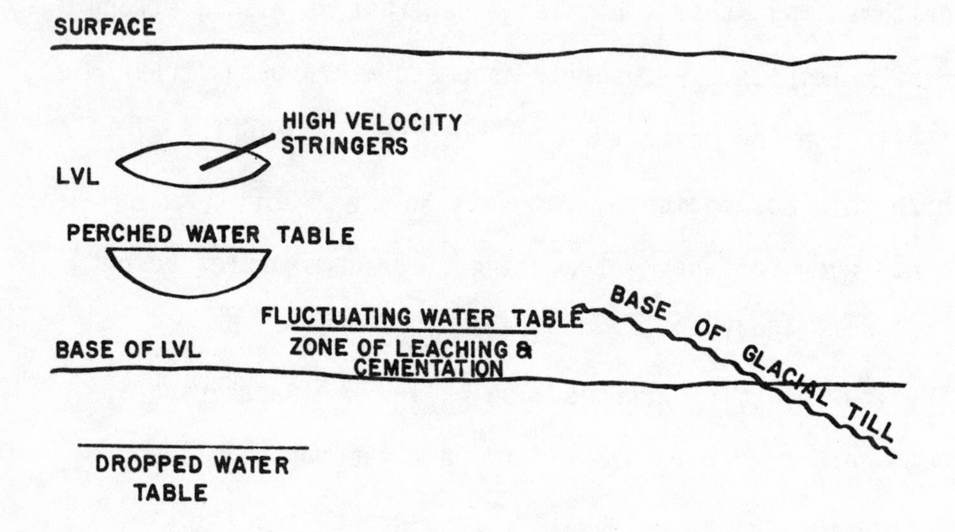

DOUBLE-LAYER WEATHERING

Figure 7.10

Permafrost presents one of the most difficult of statics correction problems. Velocity in the permafrost layer is appreciably larger (12-14,000 ft/sec) than the velocity beneath the permafrost (7-9,000 ft/sec), and energy travels at a greater angle in the upper layer, accentuating the effects of the layer. Permafrost is variable in thickness, often in an erratic way, producing distortions in travel paths. Where surface lakes are present, the permafrost may be absent. Local permafrost thins can be of the same dimensions as the seismic spread length. Permafrost may distort reflections in different ways so that the effects do not lineuup vertically on the seismic section.

There are several methods of determining static corrections:

(1) Use of uphole data;

(2) Use of refraction breaks;

(3) Smoothing of reflections

 a. By trial-and-error optimization or

 b. Statistically;

(4) Electrical measurements, surface wave dispersion, etc.

The uphole methods measure near-surface effects directly, but require holes drilled below the low velocity layer and these are often not available and almost never available in sufficient numbers to check lateral variations adequately. The first break or refraction break methods often suffer where geophone groups are large. The computer statics programs are based mainly on smoothing reflections. Examples of computer-derived statics corrections are shown in Fig. 16.3 to 16.6. The electrical and surface wave methods of near-surface examination are used very rarely.

If static corrections are not done properly, several adverse effects may occur:

(1) Poor event continuity;

(2) Poor stacking velocity determination (see Fig. 16.5 and 16.6);

(3) Waveform instability;

(4) Unwanted amplitude variations.

It is important that both dynamic and statics corrections are the very best possible so that the resulting record sections will tell us what we need to know about the subsurface.

CHAPTER 8

LAND OPERATIONS

The diagram on the Soviet Union postage stamp shown in Fig. 8.1 indicates ray paths leaving the vicinity of an explosive shot, reflecting from bedding interfaces in the earth, and reaching geophone stations along the surface. Land geophysical exploration is carried on this way worldwide by organizations of various nationalities.

The first step in on-site operations, after scouting the area to determine how best to carry out the program, is locating positions accurately on the surface. Standard surveying methods are employed, the precise method depending on the specific situation. Surveying is often not given the attention it deserves because the attention

Figure 8.1

is usually concentrated on optimizing record quality. Errors in surveying affect subsequent operations and can create very serious processing and interpretation errors.

About half of present land seismic operations still use detonation of dynamite in shotholes as energy source, about a third use vibrators (Vibroseis*), and the remaining sixth use other sources - weight dropping, confined explosive gases such as Dinoseis**, air guns, explosive detonating cord plowed into the earth, etc. The choice of source is mainly a matter of economics and different source types produce essentially identical results providing the amount of energy is adequate. Thus difficulties in drilling shotholes, restrictions on the use of explosives, the need for source arrays, and similar factors usually determine the best source to use.

Reflected seismic waves are detected by geophones laid on the surface. Most geophones are of the moving-coil type in which a coil of wire suspended by a spring tends to remain stationary when the surrounding case is moved. The surrounding unit is a magnet and motion of the case and magnet in response to the passage of a seismic wave generates an electrical voltage across the coil. Hydrophones are also used as seismic detectors. Pressure deforms a piezoelectric ceramic which produces a voltage. Moving-coil geophones usually respond only to vertical motion whereas hydrophones respond to pressure changes from any direction.

Geophones must be well coupled to the earth to detect the motion involved in a seismic wave faithfully. Geophones often have a spike on their base which is pushed into the earth so that the phone moves

* Tradename of Continental Oil Company
** Tradename of Sinclair Oil Company

with the earth. On hard surfaces base plates are used to give the geophone a larger area of contact with the surface. The coupling of phones to the earth should be as uniform as possible; where the surface varies it is better to have all the phones sitting on rock, or all planted in sand, or all in mud, than to have them differ in coupling. Hydrophones are usually placed under a few feet of water to achieve coupling with the surrounding fluid.

Geophones are usually grouped in arrays with several connected together electrically so that an entire group acts as a single phone (Fig. 8.2). The elements of a group are distributed over sufficient distance that waves with appreciable horizontal component of travel are attenuated. Such waves affect the various phones in a group differently at any time, so that the effects partially cancel, whereas vertically-travelling waves affect all components in the same way so that the effects add. An array is typically 25-50 meters across, sometimes 2 or 3 times this distance. Arrays are most commonly in-line, sometimes areal.

Many types of equipment are available for carrying out operations. Trucks are the preferred transport method wherever they can be used.

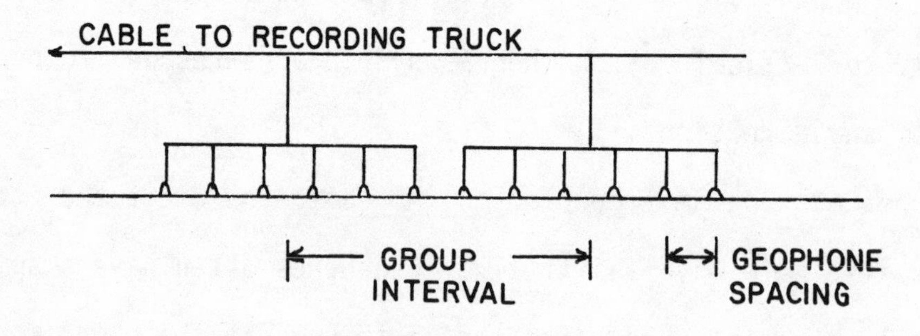

Figure 8.2

Tracked vehicles which exert relatively little pressure on the ground permit moving across soft surfaces such as marshes, mud flats, and muskeg. In many areas operations are partially or completely portable, men carrying the equipment along the line. Other types of equipment sometimes used include pull boats for moving through heavily timbered swamps, hovercraft for mud flats and coastal regions, helicopters and barges and boats of various kinds.

Selection of Field Parameters

Field parameters should be determined in a logical way, although existing equipment (number of channels and geophones and how cables and geophones are wired) may somewhat prejudice decisions.

(1) The maximum offset, the distance from source to the farthest group, should be comparable to the depth of the deepest zone of interest. A range of offset distances is required to obtain sufficient normal moveout to distinguish primary reflections from multiples and other coherent noise, but offsets should not be so large that reflection coefficients change appreciably and that conversion to shear waves becomes serious. If data quality in the deepest zone of interest does not require maximizing reflection quality, then the maximum offset should be increased up to the value of the basement depth.

(2) The minimum offset distance should be comparable to the depth of the shallowest section of interest. Getting sufficiently far from source-generated noise sometimes dictates a greater distance.

(3) The array length is determined by the minimum apparent velocity of reflections. An array should not be so long that reflected energy at opposite ends of the array differ by more than one cycle for the highest frequency and maximum dip expected. The minimum apparent velocity usually occurs at the maximum offset.

(4) The minimum useful in-line geophone spacing within arrays is determined by the ambient noise. Ambient noise characteristics can be determined experimentally by recording individual geophones spaced 0.5 to 1 m apart. The geohpone spacing for which the noise

appears to be incoherent is the minimum spacing required. This spacing is often 2 to 5 m, the smaller value being where noise is fundamentally of local generation (such as noise caused by wind blowing grass or shrubs) and the larger value where noise is mainly caused by distant sources (such as microseisms, surf noise, traffic noise, etc.). If more geophones are available, an areal distribution of phones at this minimum spacing will be more effective than crowding the phones closer together in-line. Areal arrays are rarely required to attenuate noise coming from the side of the line. Airwave attenuation may require a maximum geophone separation of 5 to 3 m (for 4 ms and 2 ms sampling, respectively).

(5) Group interval should be no more than double the desired horizontal resolution, thus providing sub-surface spacing equal to the desired resolution. However the group interval should not exceed the maximum permissable array length indicated by (3) above.

(6) The minimum number of channels required is determined by the combination of spread length and group interval decisions already reached.

(7) The minimum charge size is determined by the ambient noise late on the record. If two shots are repeated, the records should be nearly identical to a time which corresponds to a depth below the deepest section of interest, often basement. If this is not the case the charge size (or source effort) should be increased.

(8) Special circumstances may require variations from these guidelines. For example, mapping a zone of maximum interest may require recording in an offset window between wavetrains following the first arrivals and those caused by surface waves.

CHAPTER 9

MARINE OPERATIONS

Marine operations differ from land operations primarily in the pace of the operations. Costs are so much more at sea than on land that efficiency assumes primary importance. Marine operations are carried out with a routine efficiency which makes special adaptations, variations or experimentation difficult. The emphasis on efficiency has been so successful and productivity is so high that cost per mile may be only one-tenth the cost per mile of land data, despite very large per-day costs.

A typical marine ship, such as shown in Fig. 9.1, is of the order of 50 meters in length. It may tow three kilometers of streamer which

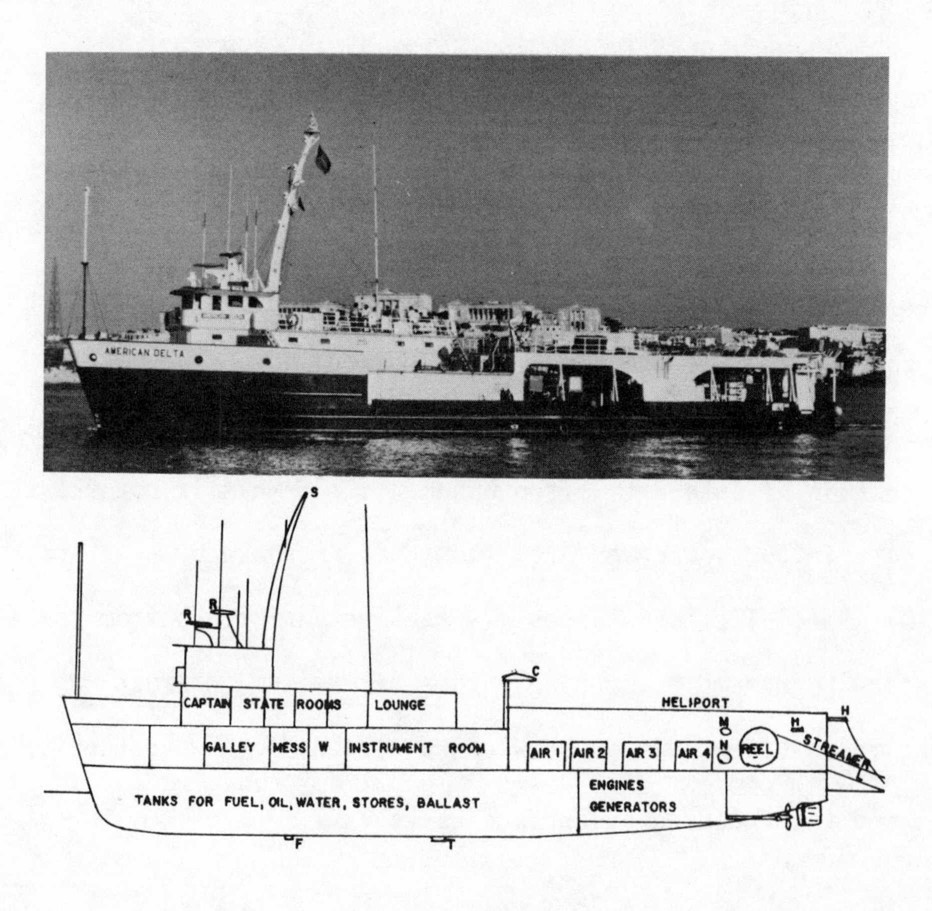

Figure 9.1 SD

contains the detectors of seismic waves. It has to generate seismic waves at a rapid rate and so must include large generating capacity. Airguns are the most common marine energy source and much of the space on a ship may be occupied by the compressors (Air 1, 2, 3, 4 in Fig. 9.1) which provide the volume of high pressure air required. The ship also has to be equipped with a number of navigational devices so that its precise location can be determined at all times. This is apt to include antennae for receiving one or more types of radio-navigation signals and for observation of navigation satellites (S), transducers for sending and receiving sonar signals (T) plus precision fathometer (F), gyrocompass and other navitation gear. The ship also will be equipped with radar (R) and several radiocommunication devices. In Fig. 9.1, H indicates airgun handlers, M magnetometer reel, N ministreamer reel.

Arrays of marine sources are often used. Figure 9.2 shows the layout of an array of 14 airguns. The component airguns

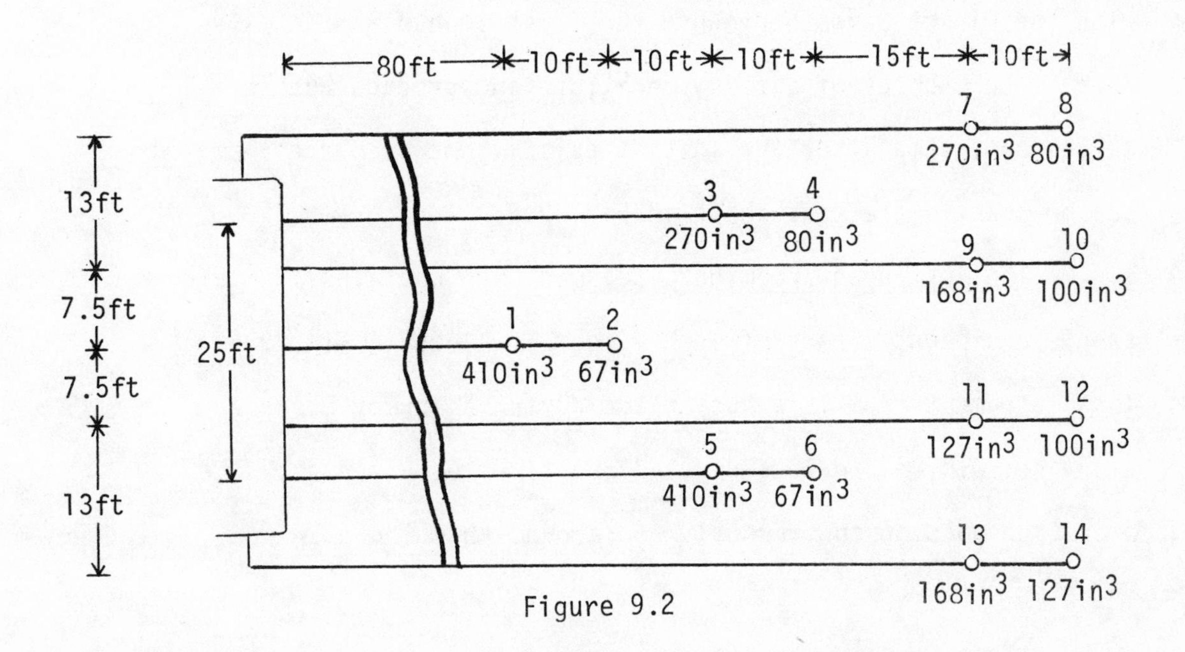

Figure 9.2

inject different volumes of air into the water in order to generate a seismic wave with a broad frequency spectrum. The frequency spectrum generated by an array of airguns is shown at the top of Fig. 9.3 and wavelet shape is shown at the bottom. The initial shock produced by the injection of air into the water produces the initial peak. Some of this energy travels upward, reflects off the surface of the water, and then joins

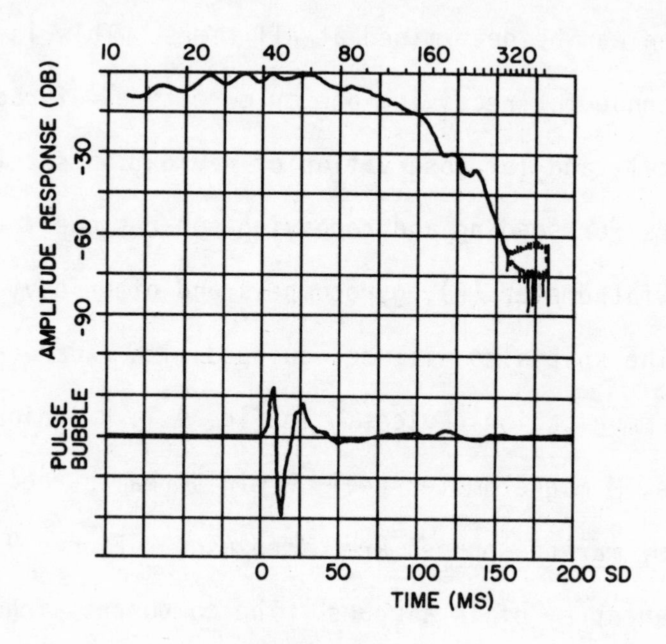

Figure 9.3

with the direct wave to produce the first trough in the waveform. The bubbles of air may oscillate and produce bubble pulses some time after the initial pulse; some bubble effect can be seen in the wave shape in Fig. 9.3.

The instrument room on a seismic ship contains much electronic equipment. Part of such a room is shown in Fig. 9.4. The instrument room is a busy place during a survey because a new seismic record is generated 2-3 times each minute. The operation goes on continuously and around the clock.

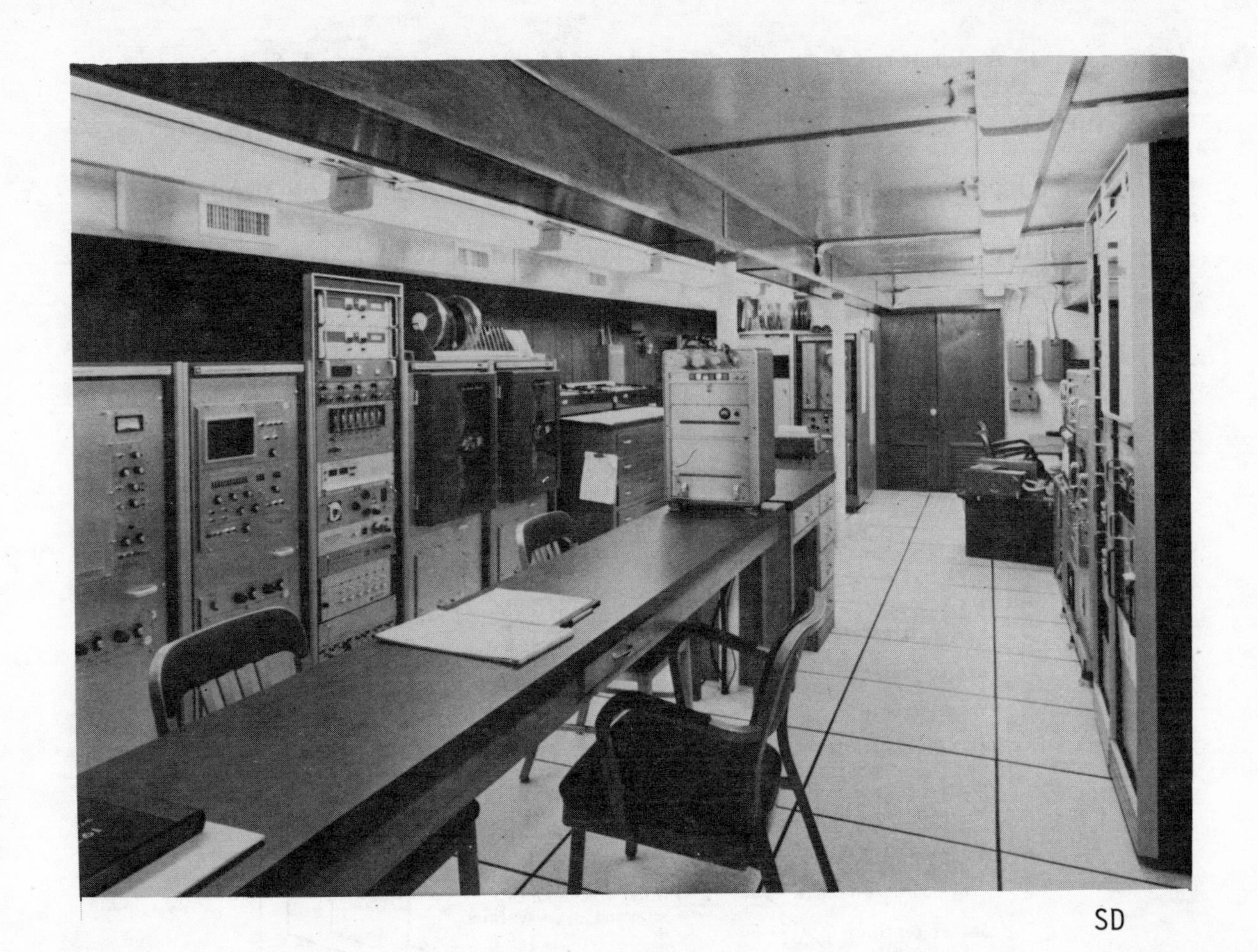

Figure 9.4a

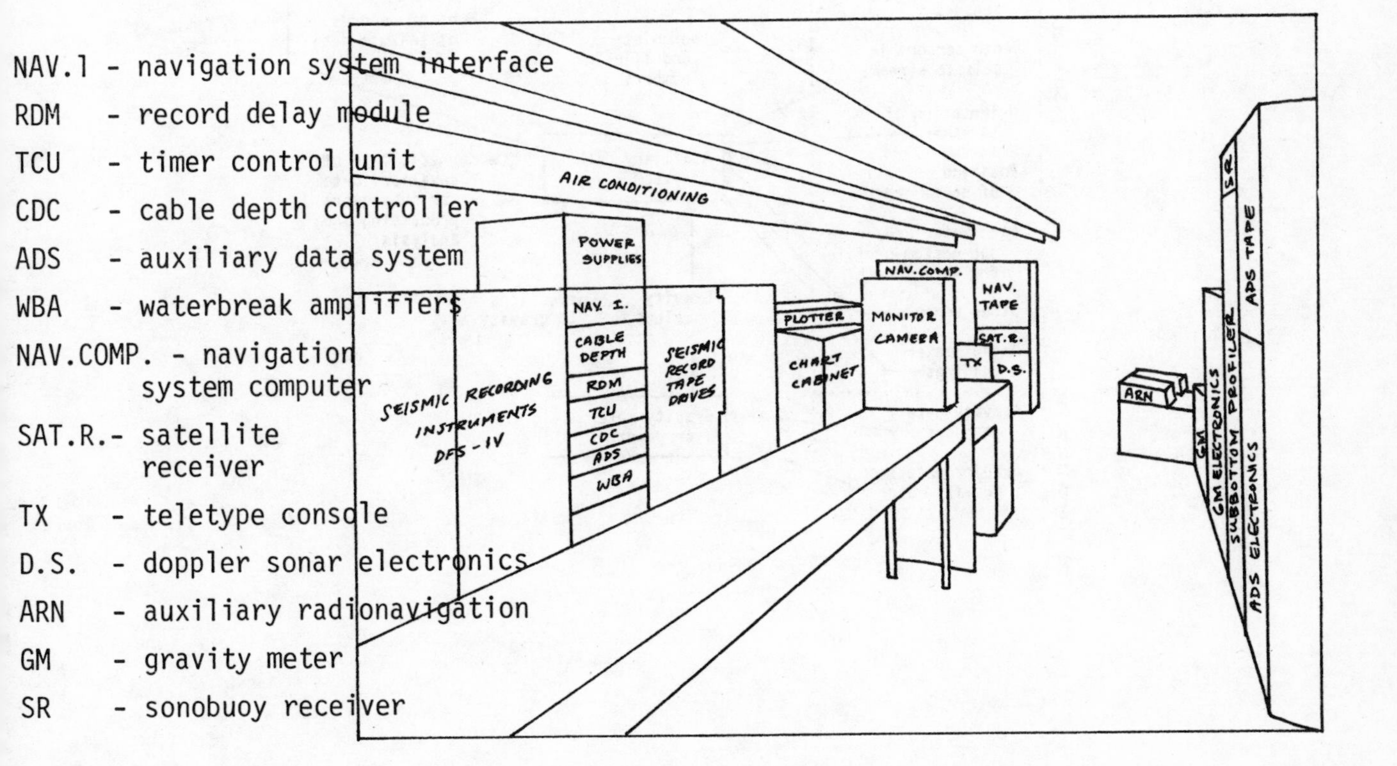

Figure 9.4b

A number of types of information are recorded from a survey. Some of the information flow is diagramed in Fig. 9.5. The output of the survey is contained mainly on three types of magnetic tapes: (1) a tape of navigation data, (2) a tape of the seismic results, and (3) a tape of auxiliary information from the survey.

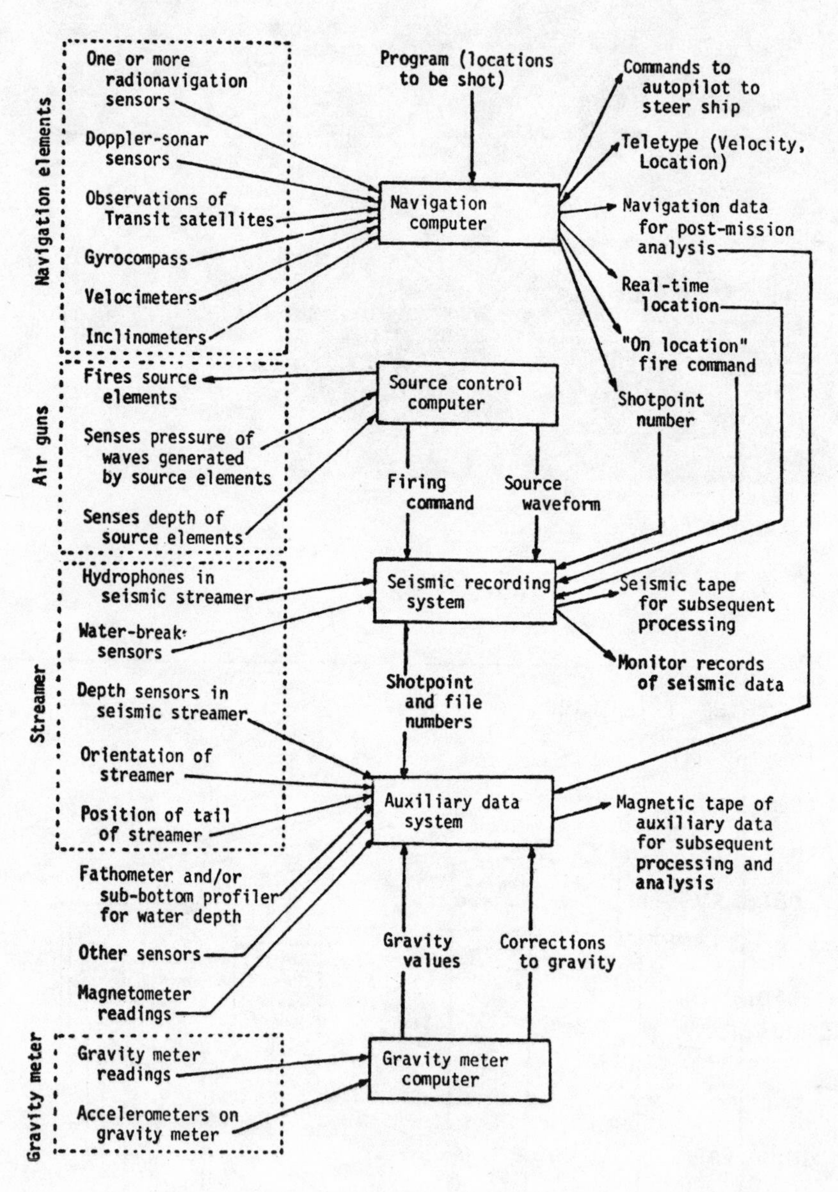

Figure 9.5

CHAPTER 10

SEISMIC REFLECTIONS

Let's now examine the composition of a reflection. Consider an interface between two rock types with different velocities and densities. Part of the seismic energy will be reflected at the interface and will return to the surface as a reflected wave, but most of the energy will be transmitted right through the interface. Particle displacement and stress must be continuous at the interface. These physical conditions determine the partition of energy.

For a wave which strikes the interface at normal incidence, the equation for the ratio of the amplitude of the reflected wave compared to the amplitude of the incident wave is simple. This ratio is called the reflection coefficient, represented by the symbol R. The product of the velocity V and density ρ is called acoustic impedance; the subscript 1 means the incident medium, 2 the other medium.

$$R = \frac{\text{amplitude of reflected wave}}{\text{amplitude of incident wave}}$$

$$= \frac{V_2\rho_2 - V_1\rho_1}{V_2\rho_2 + V_1\rho_1}$$

$$= \frac{\text{change in acoustic impedance}}{2(\text{average impedance})}$$

$$= \text{change in log (acoustic impedance)}.$$

Most normal reflection work is fairly close to normal incidence and if the wave strikes the interface within about 20° of normal, this equation gives a reasonable approximation of the correct answer.

If the raypath strikes the interface at non-normal incidence, four waves are generated: reflected compressional wave, reflected shear wave, transmitted compressional wave and transmitted shear wave. It is difficult to generalize on the partitioning of energy under these conditions because so many quantities are involved - both shear and compressional velocities in both media and density in both media. The graph in Fig. 10.1 shows the energy partitions as the angle of incidence varies for one set of variables. The left side represents normal or perpendicular incidence, the right side represents grazing incidence. Beyond about 20^0 however, the partitioning can change considerably and near the critical angle $\theta_c = \sin^{-1}(V_2/V_1)$ the partitioning curves can change rapidly with incident angle.

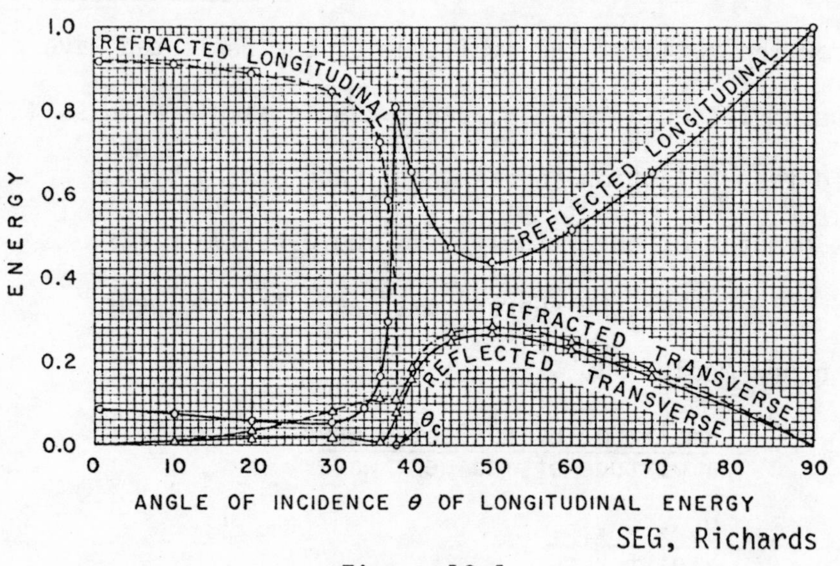

SEG, Richards

Figure 10.1

A synthetic seismogram is a calculation of what a seismic record would look like for given conditions in the earth. The procedure is shown schematically in Fig. 10.2. The earth is assumed to be made up of a number of layers, each of constant properties, but the velocity or density or both may differ from layer to layer. The acoustic impedance is calculated for each layer and then the reflection coefficients from the changes in acoustic impedance. A seismic wave traveling down through the earth will be reflected back at each of these interfaces, the magnitude of the reflected wave being proportional to the magnitude of the reflection coefficient. The seismic record will be the superposition or sum of the many component reflections.

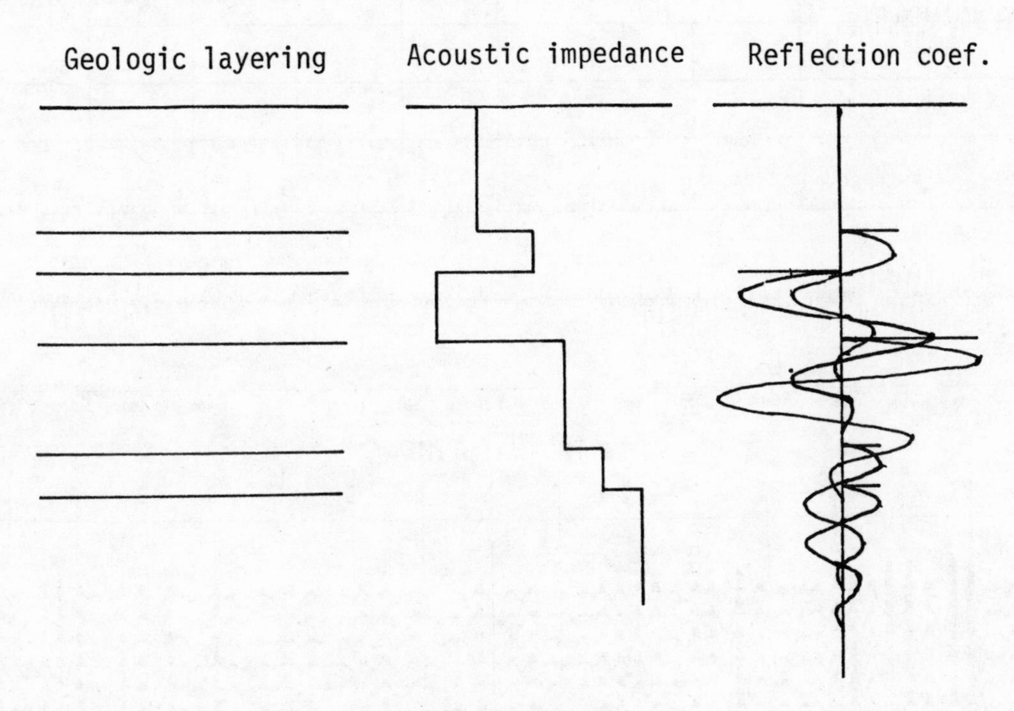

Figure 10.2

An example synthetic seismogram is shown in Fig. 10.3. The input information, the graph at the top, is a sonic log in a borehole. Constant density is assumed; this is an over-simplification, but a common one because density information often is not available and the adverse effects of neglecting it are usually small. The log is plotted to a time rather than depth scale. From the variations in the sonic log, a reflection coefficient log (middle of the grapn) is calculated and this is then combined

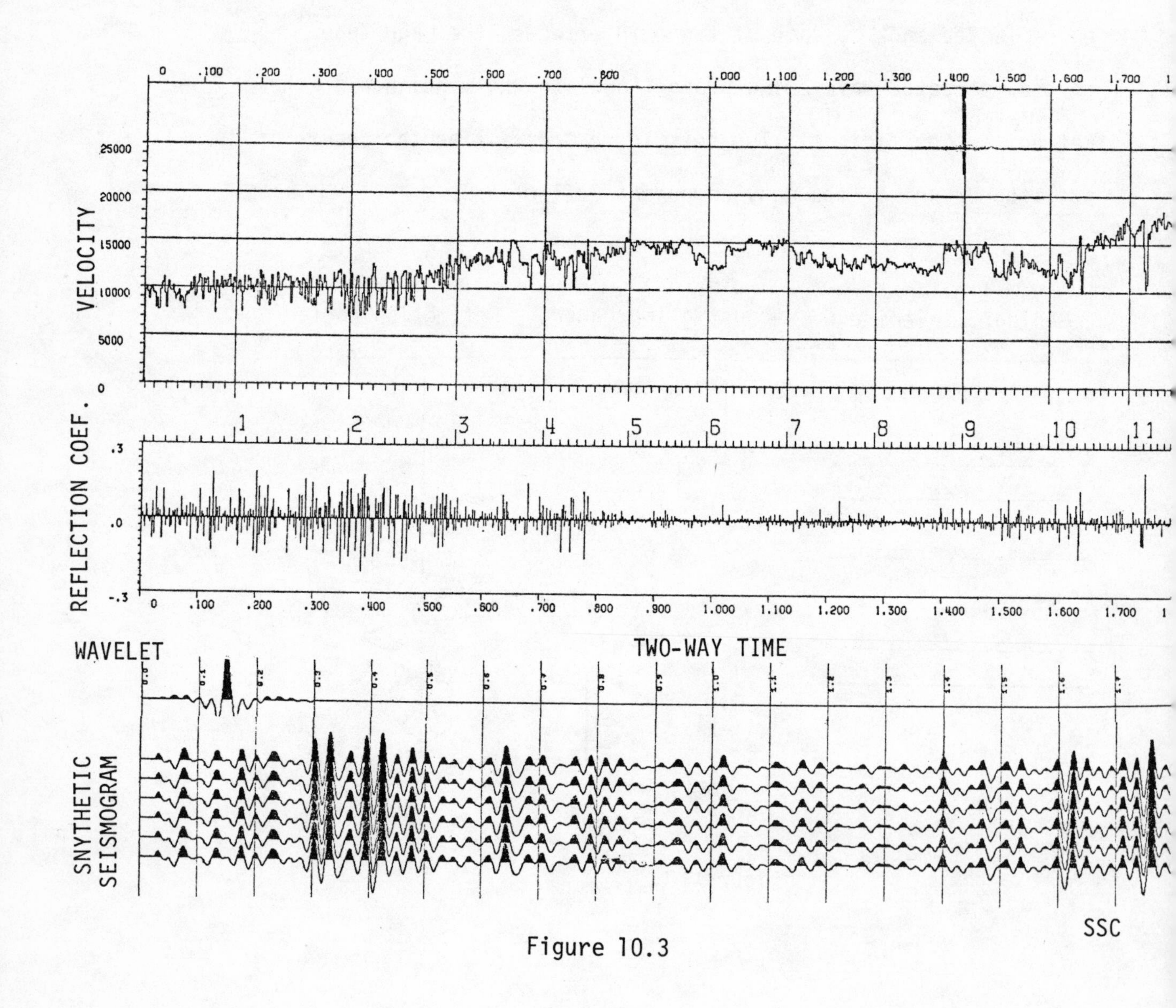

Figure 10.3

with an assumed wavelet shape to give the seismic record we would expect for this particular velocity distribution, the bottom of the graph. Note that reflections from the interfaces usually overlap each other considerably. The process of combining the reflection coefficients with the wavelet shape is called convolution; it involves replacing each reflection coefficient element with the wavelet scaled in magnitude according to the magnitude of the reflection coefficient and summing the results. This process happens naturally in the earth.

Synthetic seismogram manufacture is a tool to help us relate features on a seismic record to interfaces in the earth. We can take the data from an actual borehole, for example, change the thickness or lithology of specific beds and see how such changes affect a seismic record.

Exercises:

1. Reflections from the shallow part of the earth where the velocity is relatively low often are of reasonably high frequency, whereas those from the deeper part where the velocity is high are often of low frequency. Calculate the wavelengths for the following situations:

   Shallow: $V = 2000$ m/s, $f = 50$ Hz, $\lambda =$ ______

   Deep: $V = 6000$ m/s, $f = 20$ Hz, $\lambda =$ ______

   (wavelength = velocity/frequency.)

2. Given the following properties, calculate acoustic impedance values.

At depth of 2000 m	velocity km/s	density g/cc	acoustic impedance
Shale, ϕ (porosity) = 0.25	2.8	2.25	______
Shale, ϕ = 0.25, 10% lime	3.2	2.25	______
Shale, ϕ = 0.20	3.1	2.3	______
Sand, ϕ = 0.30	3.2	2.15	______
Sand, ϕ = 0.30, 10% lime	3.4	2.15	______
Sand, ϕ = 0.30, gas	2.7	2.0	______
Limestone, ϕ = 0.30	4.1	2.2	______
Limestone, ϕ = 0.30, gas	3.5	2.0	______

B. Calculate the reflection coefficients for:

Shale (ϕ = 0.25) over sand (ϕ = 0.30) ______ (______)

Shale (ϕ = 0.25) over sand with 10% lime ______ (______)

Shale (ϕ = 0.25) over gas sand ______ (______)

Shale (ϕ = 0.25) over shale (ϕ = 0.20) ______ (______)

Shale (ϕ = 0.25) over limestone ______ (______)

Shale (ϕ = 0.25) over gas in limestone ______ (______)

Gas sand over water sand ______ (______)

Limey shale over sand ______

Limey shale over limestone ______

Limey shale over gas in limestone ______

Shale (ϕ = 0.20) over sand ______

C. Recalculate (where indicated by the empty parentheses above) neglecting changes in density.

Work these problems before looking at the answers on the following page.

Answers to exercises 1 and 2:

1. Shallow: V = 2000 m/s, f = 50 Hz, λ = 40m.
   Deep: V = 6000 m/s, f = 20 Hz, λ = 300m.

2. At depth of 2000 m

	velocity km/s	density g/cc	acoustic impedance kmg/s cc
Shale, ϕ (porosity) = 0.25	2.8	2.25	6.30
Shale, ϕ = 0.25, 10% lime	3.2	2.25	7.20
Shale, ϕ = 0.20	3.1	2.3	7.13
Sand, ϕ =0.30	3.2	2.15	6.88
Sand, ϕ = 0.30, 10% lime	3.4	2.15	7.31
Sand, ϕ = 0.30 gas	2.7	2.0	5.40
Limestone, ϕ = 0.30	4.1	2.2	9.02
Limestone, ϕ = 0.30, gas	3.5	2.0	7.00

Reflection coefficients:

Shale (ϕ = 0.25)	over sand (ϕ = 0.30)	0.044	(0.067)
Shale	over sand with 10% lime	0.074	
Shale	over gas sand	-0.077	(-0.018)
Shale	over shale (ϕ = 0.20)	0.062	
Shale	over limestone	0.178	(0.188)
Shale	over gas in limestone	0.053	(0.111)
Gas sand	over water sand	0.121	(0.085)
Limey shale	over sand	-0.023	
Limey shale	over limestone	0.112	
Limey shale	over gas in limestone	-0.014	
Shale (ϕ = 0.20)	over sand	-0.018	

Comments on problem 2:

Of the above contrasts, the shale-limestone contrasts produce the strongest reflections, followed by contrasts involving the gas-sand. Shale-sand contrasts can vary over quite a range and involve either positive or negative contrasts depending on the porosities involved or lime content. The contrasts between shale and water-sand can be greater than between shale and gas-sand. The decrease in contrast at a shale-limestone contrast as gas fills the limestone porosity gives rise to the"dim spot" phenomenon. Neglecting the density in calculations causes the greatest error when gas is involved.

Continuation of exercises:

3. In the following, assume that the dip is so small as to have negligible effect. The reflection coefficients shown below are scaled up and rounded off to make calculation easier. Assume a minimum-phase wavelet (dominant frequency 50 Hz) digitized at 0.004 s intervals:

10, 9, -8, -9, 0, 5, 3, 0

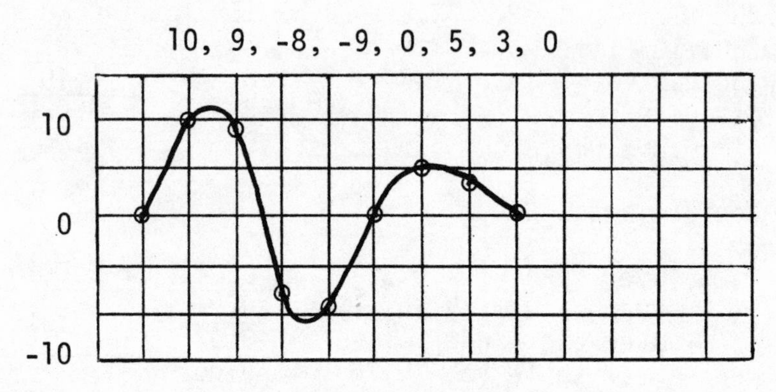

Figure 10.4

a. Assume a sand encased in shale; calculate and plot the reflection waveshape when the sand is 20 m thick. The two-way travel-time through the sand is 0.012 sec. Assume a reflection coefficient from the shale-to-sand of 0.1 (to make it easy to calculate).

b. Repeat for the sand 27 m thick (2-way time is 0.016 s).

c. Assume two sands each 20 m thick separated by 20m of shale and encased in shale. Plot the reflection waveshape.

d. Assume a dipping sand 41 m thick with gas in the upper 14 m. Two-way time through the gas sand is 0.008 s and 0.016 s through the water sand. Assume -0.2 shale to gas sand reflection coefficient, +0.3 gas sand to water sand, and -0.1 water sand to shale.

e. Repeat for gas in the upper 28 m.

f, g, h. Repeat situation a, c and d for the same waveshape but half the frequency. Such a wavelet is

5, 10, 12, 9, 0, -8, -11, -9, -4, 0, 4, 5, 4, 3, 1, 0.

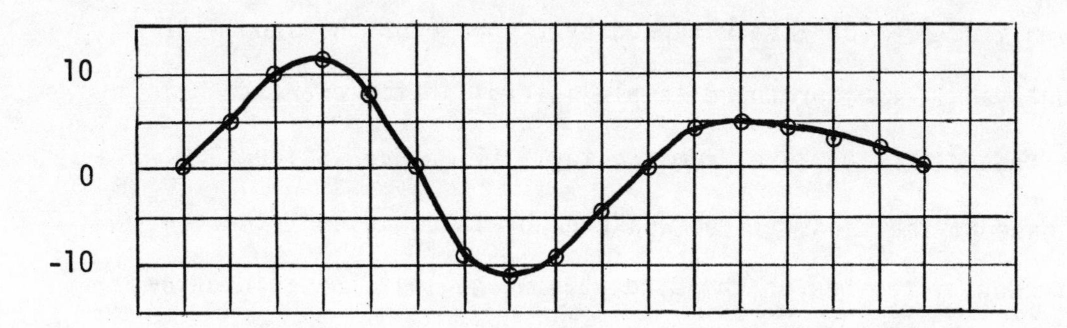

Figure 10.5

i,j. Repeat cases a and d for a zero-phase waveshape (Fig. 10.6) of same frequency as the wavelet of Fig. 10.4; this wavelet is: -4. -7, 3, 12. 3 -7, -4, 0.

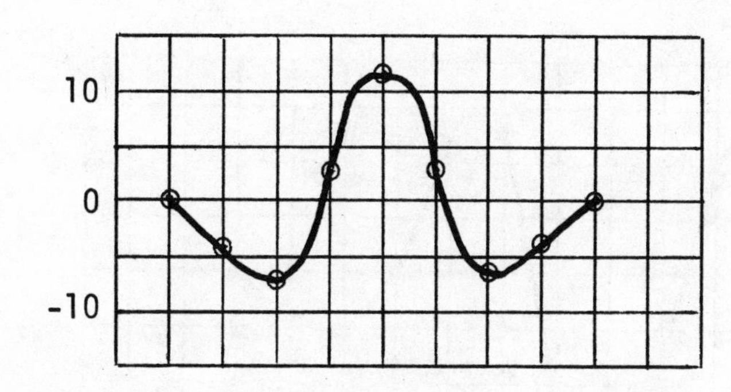

Figure 10.6

(A minimum-phase wavelet as shown in Fig. 10.4 (and 10.5) has its major energy toward the front of the waveform; it is sometimes referred to as "front-loaded." A zero-phase wavelet is symmetrical about a central point.)

Method of solving Exercise 3:

Each interface will return the same wavelet shape except its amplitude will be multiplied by the reflection coefficient (0.1 for the top of the sand and -0.1 for the base of the sand; these values are large for such an interface but are selected to be easy to multiply by and the end result will be of correct waveshape, which we're trying to illustrate, although too large in amplitude). The wave returned by the base of the sand will be delayed 12 msec or three sample intervals (for exercise 3a) with respect to the wave from the top. (The wave incident on the base of the sand will be weakened by 1% because of the energy reflected by the top of the sand, but we neglect this since the effect is so small.)

The geophone will see the sum of both waveforms:

```
1.0,  .9, -.8,  -.9,   0,  .5,  .3,  0
                -1.0, -.9, +.8, +.9,  0, -.5, -.3,  0
---------------------------------------------------
1.0,  .9, -.8, -1.9, -.9, 1.3, 1.2,  0, -.5, -.3,  0
```

We then plot these values to get the waveshape (Fig. 10.7):

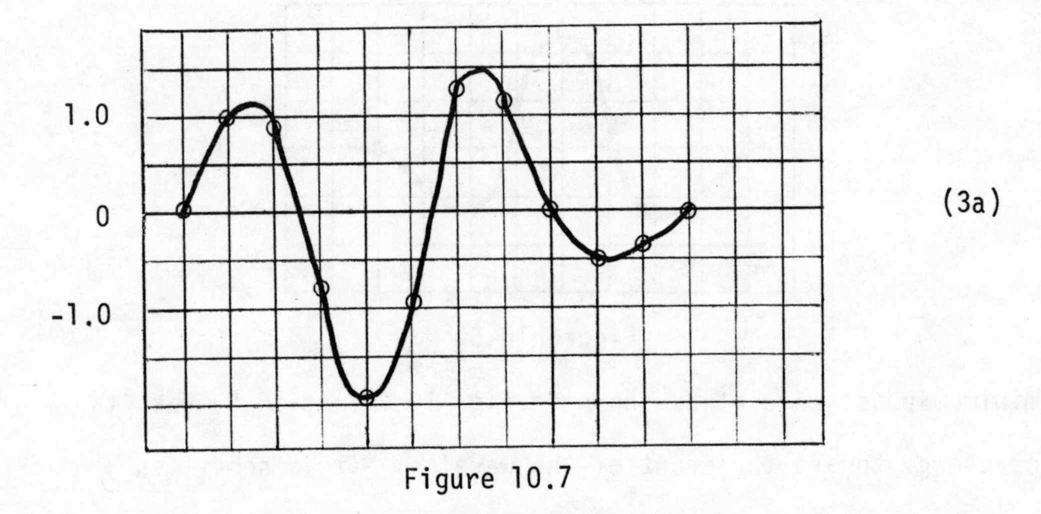

Figure 10.7

Work exercise 3 before going on to the following comments.

Comments on exercise 3:

Assume that exercises 3a and 3b refer to two points on a wedge whose taper is sufficiently gentle that migration effects can be neglected. The reflection from this wedge is shown in Fig. 10.9 The answers to parts a and b (Fig. 10.7 and 10.8) are two members of this set.

A quarter-wavelength is roughly 27 m. The appearance of this waveform makes it clear that more than one interface is involved, so we say that the top and base are resolved. If we knew the waveshape and background noise were small, we might be able to tell the sand thickness even if it were only 5 to 10% of the dominant wavelength from its amplitude.

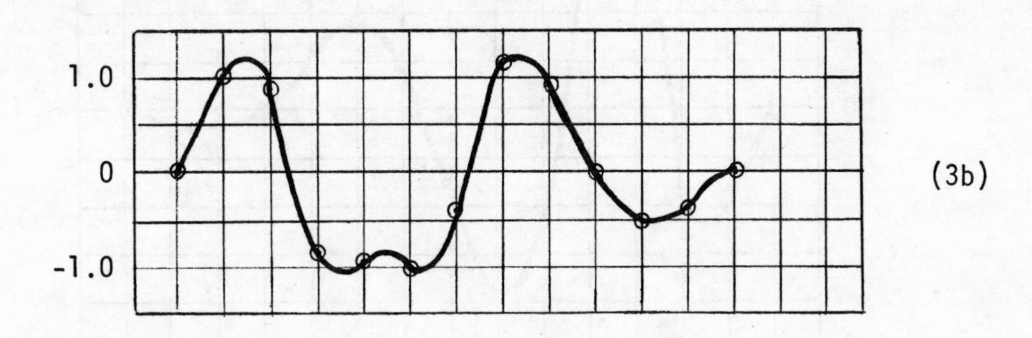

Figure 10.8

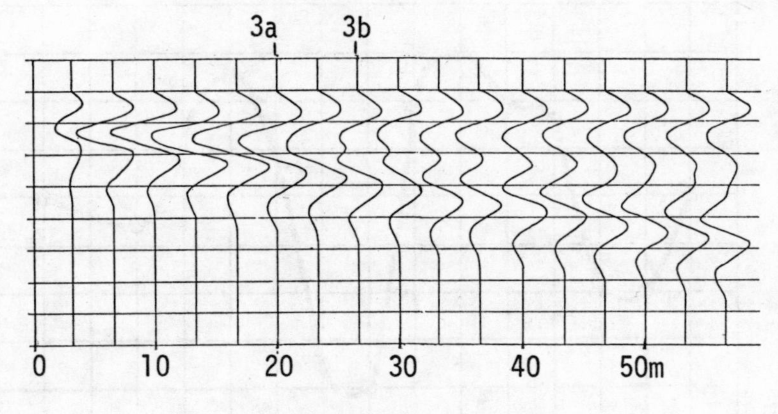

Figure 10.9

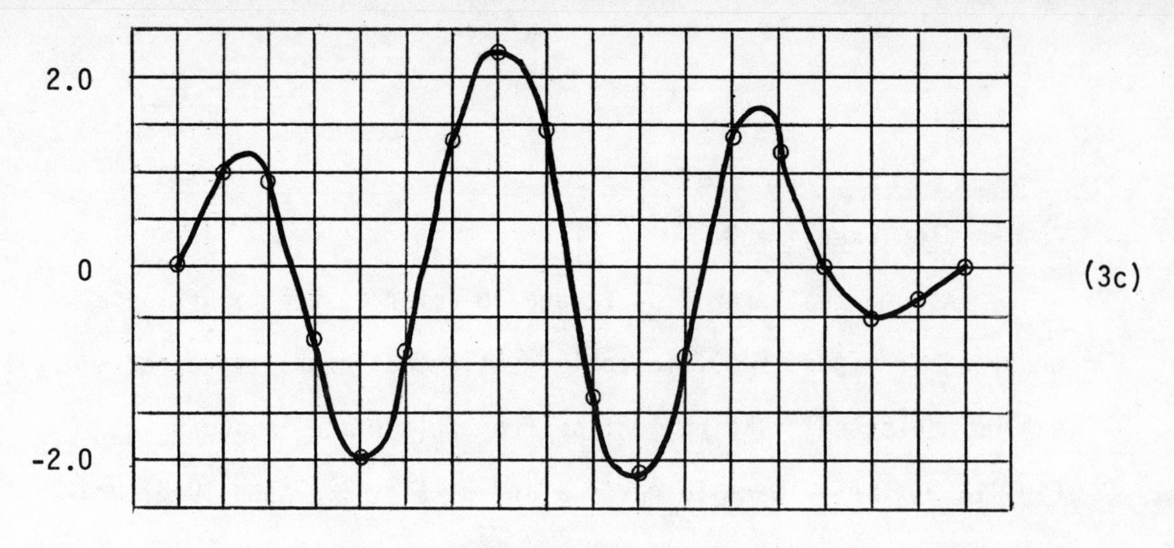

Figure 10.10

The sequence of four reflectors has a natural resonance at 42 Hz, which is close to the dominant frequency of this reflector (50 Hz; see Fig. 10.10). The result is some buildup of energy.

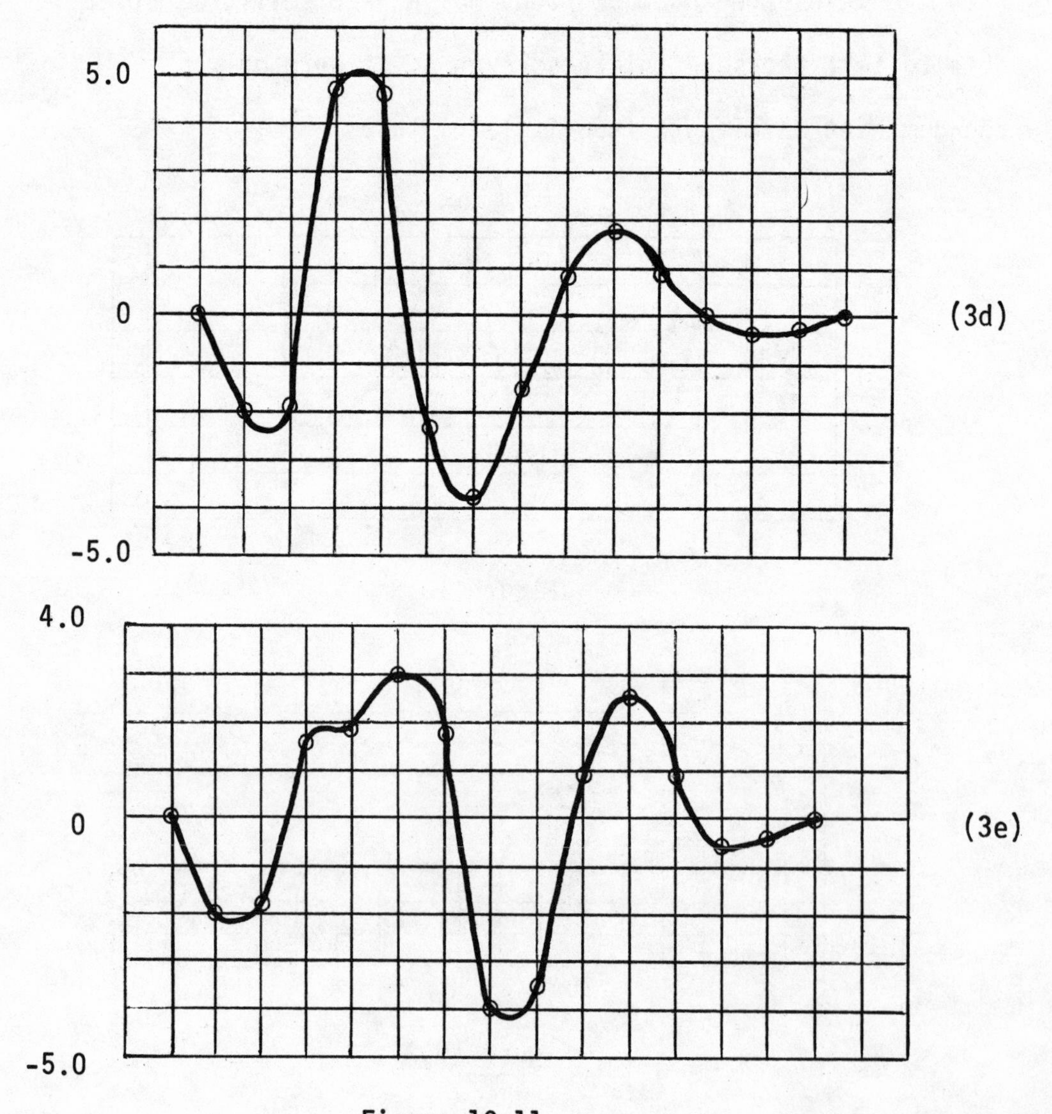

Figure 10.11

The two reflection waveforms from the dipping sand member as the gas-water level varies (calculated in exercises 3d and 3e) are shown in Fig. 10.11. These are two members of the set shown in Fig. 10.12 as the dipping sand crosses from water-filled to gas-filled at the right. The amplitude buildup illustrates the classical "bright spot" concept. The highest amplitude marks a fluid contact "flat spot."

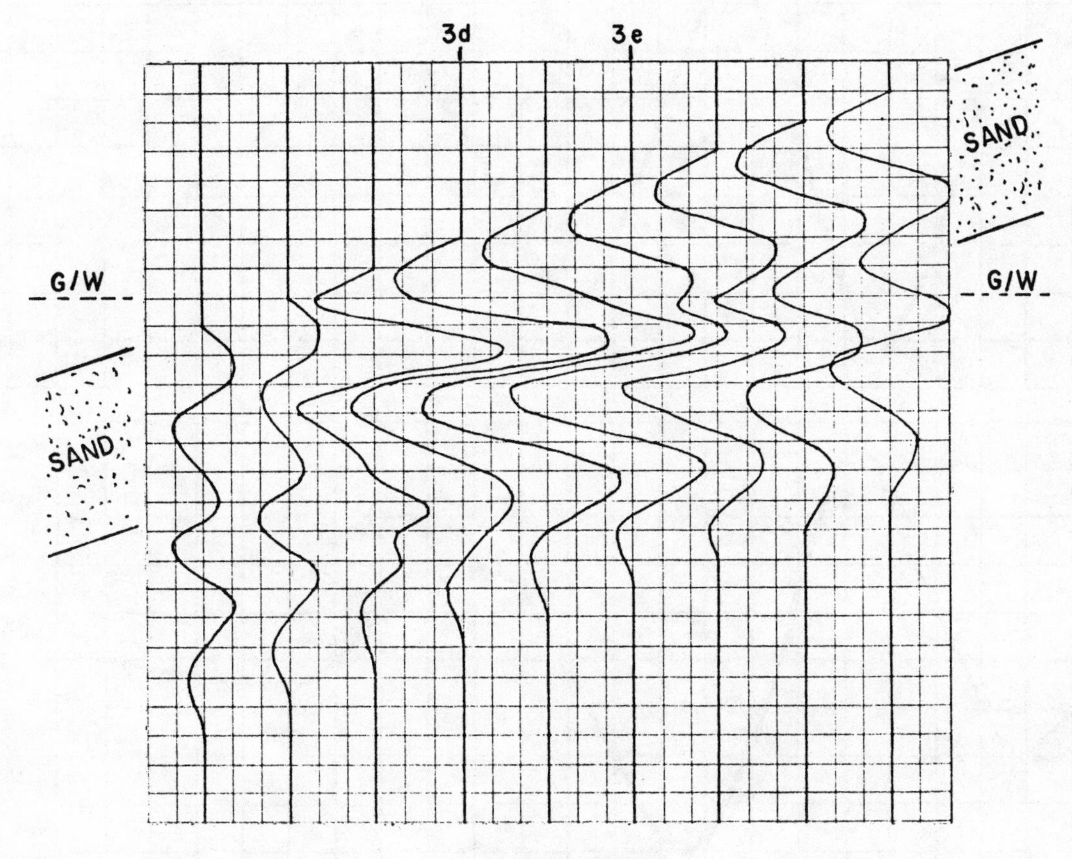

Figure 10.12

Waveforms for the same layering as calculated in exercises 3a, 3c and 3d were recalculated for the lower frequency (25Hz) wavelet. These are shown in Fig. 10.13. They can be compared respectively with those in Fig. 10.7, 10.10 and the upper wavelet in Fig. 10.11. The resolution is, of course, poorer with the lower frequencies.

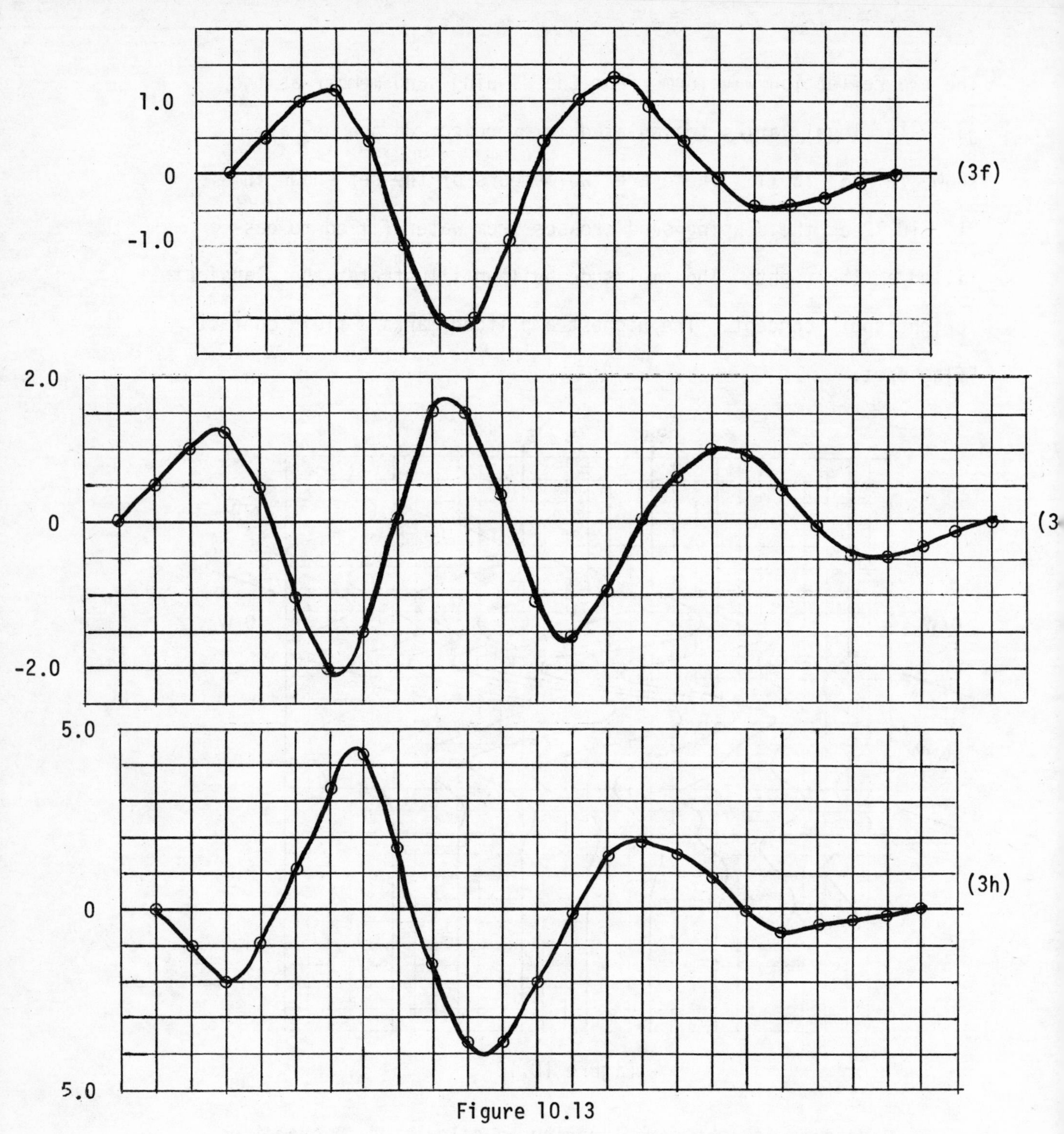

Figure 10.13

The response to the layering situations specified in exercises 3a and 3d to a zero-phase wavelet with the same dominant frequency is shown in Fig. 10.14. These should be compared with the minimum-phase response shown in Fig. 10.7 and the upper wavelet in Fig. 10.11. The resolution for the zero-phase wavelet is not markedly different than for a minimum-phase one. The dipping wedge case shown in Fig. 10.9 for minimum phase can also be compared with Fig. 10.15 for zero phase.

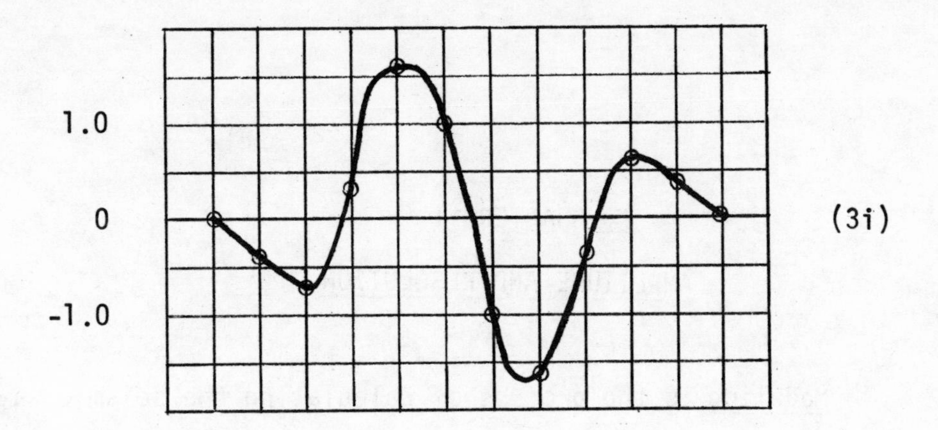

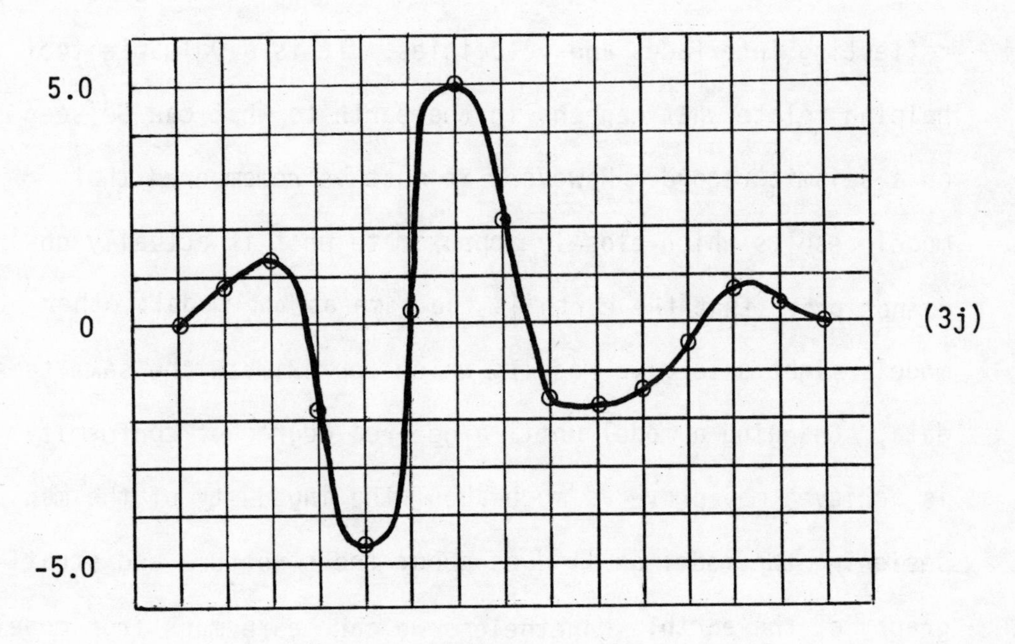

Figure 10.14

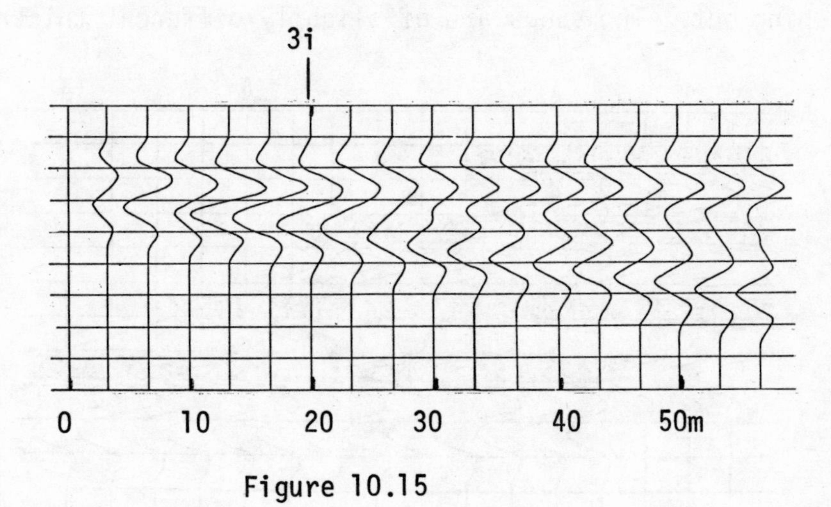

Figure 10.15

CHAPTER 11

AMPLITUDE AND RESOLUTION

Modeling is the process of calculating the seismic waveform or section that would result from a given configuration of reflecting interfaces and velocities. It is a valuable tool in helping relate what happens in the earth to what can be seen on a seismic record. However, it must be remembered that model results which closely approximate what is actually observed do not prove that the earth is the same as the model; other models might also give results which approximate the same seismic data. Changing a model until a desired degree of conformity is achieved may prove as much about the ingenuity of the man designing the model as it does about the structure and stratigraphy of the earth. Nonetheless we can learn much from models.

The model in Fig. 11.1 is of a series of sands that are pinching out. The sands are of slightly different thicknesses

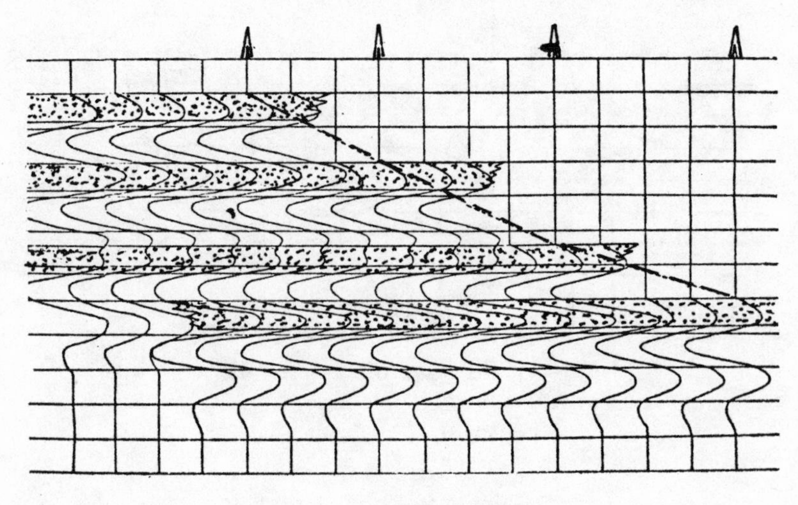

Figure 11.1

and separated by different amounts. The seismic events follow the depositional time lines because successive traces are so close together that little changes between them, until finally an event just disappears. Wells on the other hand sample spaces that are separated by much larger distances and well correlations tend to see facies units and hence facies lines like the "top of the sands".

A model from O'Doherty and Anstey shows the "peg-leg" phenomena (Fig. 11.2). Part of the energy in a wave reflects from the bottom of a layer and again from the top of the layer, the extra bounce in the layer delaying it slightly, and then it adds to the seismic waveform. This bouncing in thin layers gradually removes energy from the front of the seismic wave and adds tail to the wave.

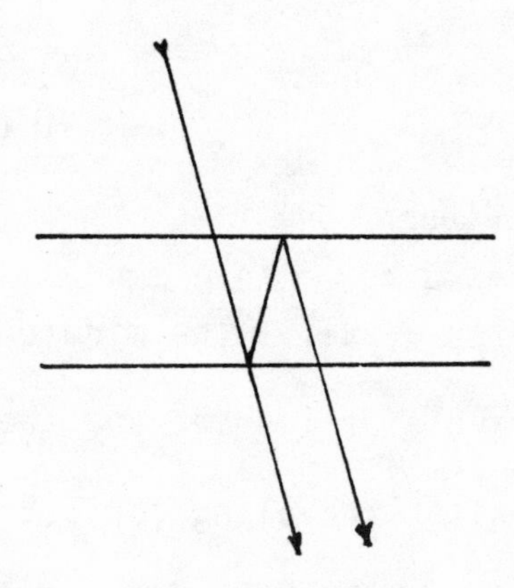

Figure 11.2

The top diagram of Fig. 11.3 shows a seismic wave which started as a sharp impulse after it has traveled through the earth for 0.7 seconds. Peg-leg multiples have taken some of the energy from the head end of the wavelet, delayed it and added

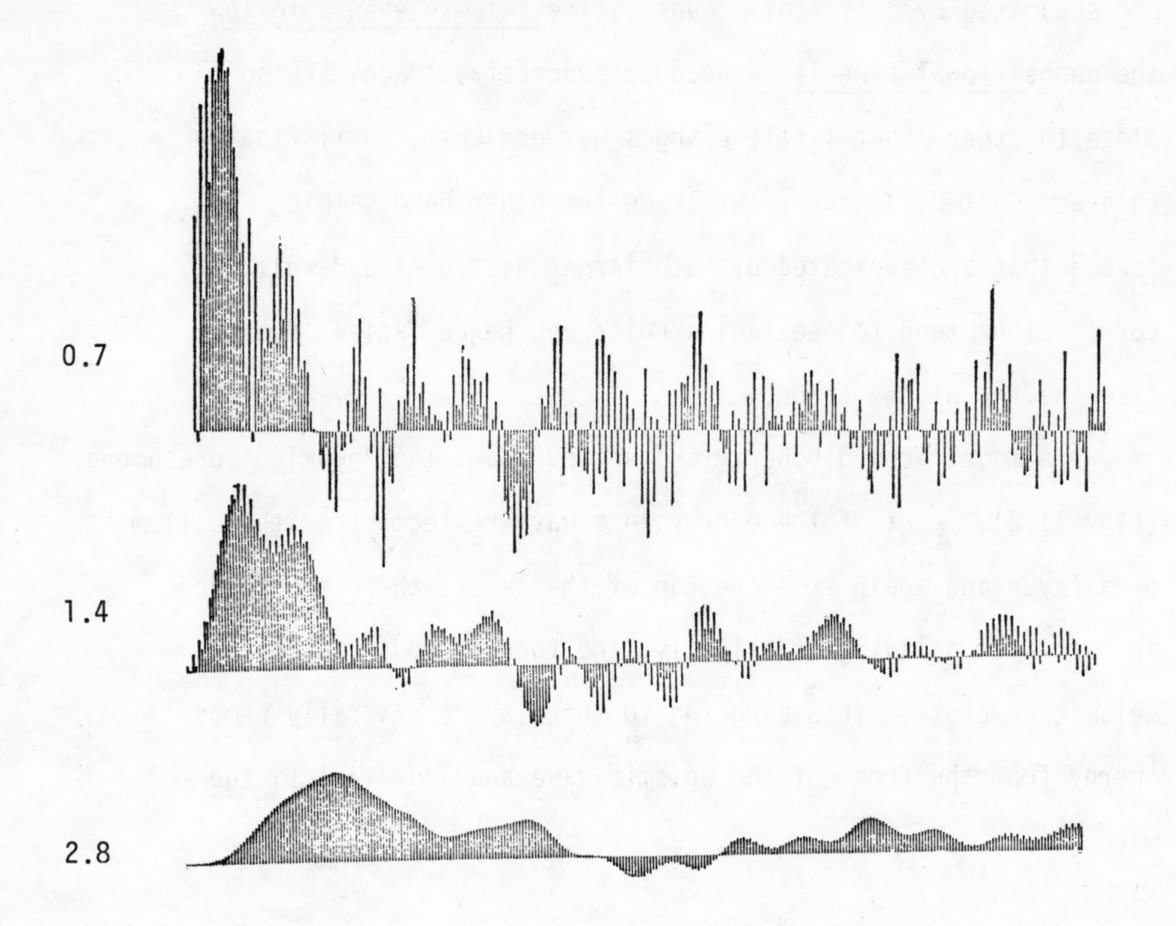

O'Doherty and Anstey

Figure 11.3

it back in to give the wavelet a tail. The middle diagram shows the waveform after travel for 1.4 sec and the bottom diagram shows the waveform after traveling for 2.8 sec. The "peg-leg" process has diminished the high frequencies and created low frequency waveform. Peg-leg multiples provide a vehicle for changing the spectrum of the waveform without invoking absorption. This effect is shown in Fig. 11.4 in the frequency domain, that is, amplitudes are shown as a function of frequency rather than of time.

Many <u>factors affect the amplitude</u> of seismic waves; some of these factors are shown in Fig. 11.5. Many of these do not

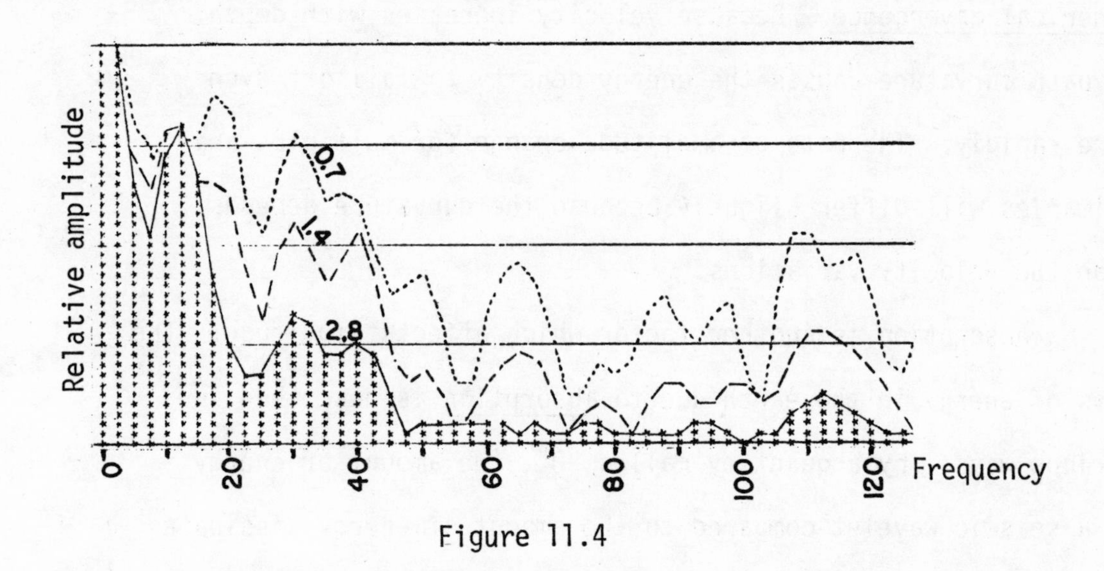

Figure 11.4

involve the subsurface of the earth. In order to do meaningful interpretation of amplitude variations, the effects of irrelevant factors need to be removed.

The energy density of a seismic wave decreases as it travels down in the earth. This decrease of energy density is inversely proportional to the square of the distance over which the wave

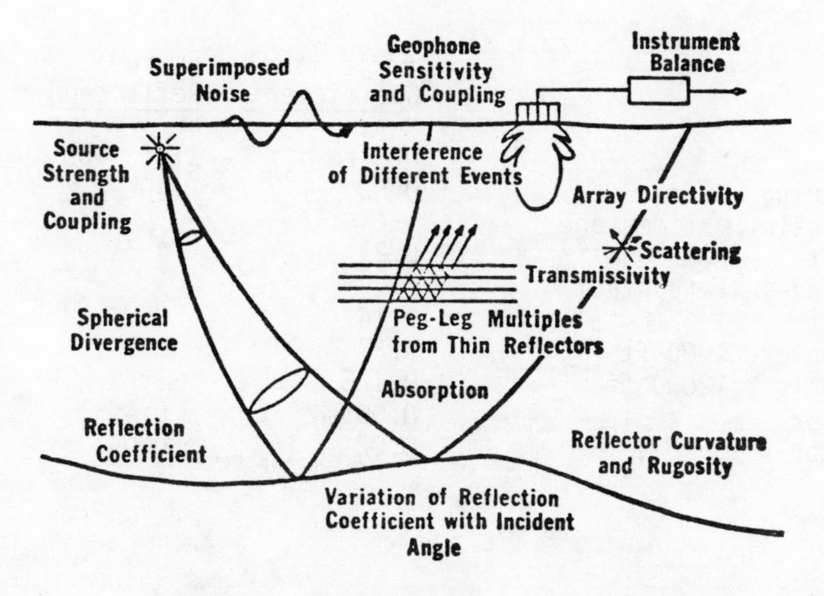

FACTORS WHICH AFFECT AMPLITUDE

Figure 11.5

has traveled, assuming a constant velocity; this is called spherical divergence. Because velocity increases with depth, raypath curvature causes the energy density to fall off even more rapidly. The rate of amplitude change for multiples and primaries will differ slightly because the curvature depends upon the velocity variations.

Absorption is another factor which affects amplitude. The loss of energy in the earth due to absorption is described in various ways: by a quantity called "Q", the amount of energy in a seismic wavelet compared to the amount of energy dissipated in one cycle; by an absorption coefficient, which is an exponential decay factor; by logrithmic decrement, a measure of the change of amplitude between two successive cycles. The loss of energy by absorption increases with frequency.

The table below gives reflection coefficients for several situations. The large contrast between weathering and sub-

	Reflection Coefficient	Energy Reflected
Ocean bottom	0.3 to 0.7	11 to 45%
Base of weathering	0.63	40%
Sandstone or shale vs. limestone at 4000 ft	0.21	4.5%
Maximum for sand-shale contact at 4000 ft	0.14	2%
Gas sand vs. shale @ 4000 ft	0.27	5.5%
Gas sand vs. shale @ 12000 ft	0.125	1.5%
"fair" reflector	0.06	0.4%
"poor" reflector	0.02	0.04%

weathering will give rise to a very strong reflection. The base of the weathering is one of the strongest reflectors encountered and is the principle generator of multiples, seismic energy which has bounced more than once in the earth. The sea floor also has a large change in acoustic impedance and is important in multiple generation, as is the surface of the sea. Most of the other reflecting interfaces that are encountered in the earth reflect only a very small percent of the energy.

The shape of a reflector affects the energy density that is seen, as shown in Fig. 11.6 for a reflector of constant contrast. The curvature of the reflector acts to focus or defocus the energy. Velocity situations such as a gas cap can have similar effects, acting as a lens to distort reflections from underneath them.

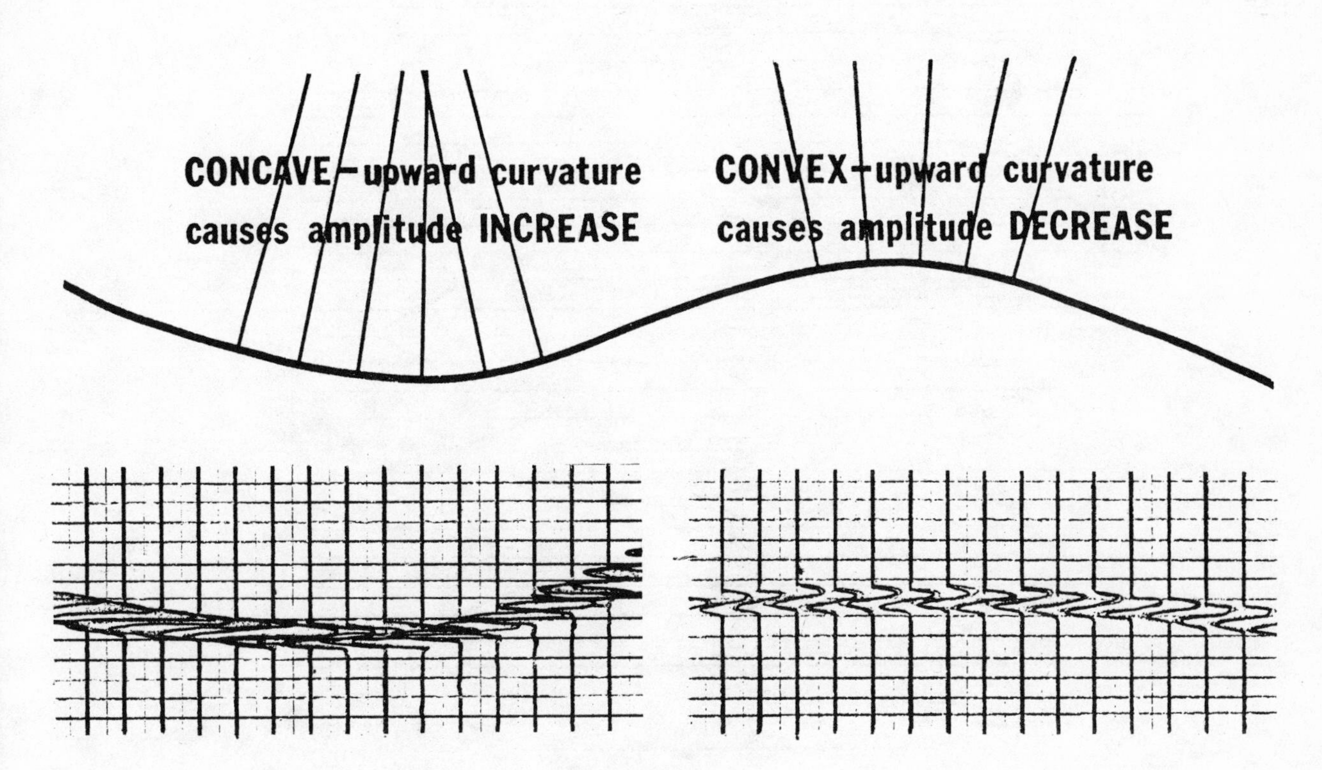

Figure 11.6

Resolution concerns the ability to tell that separate features exist. It depends upon the distance between features when compared to the wavelength. Shorter wavelengths have higher resolving power. The only practical way to get smaller wavelengths is to increase the frequency.

A major reason why we do not record higher frequencies is that ground mix, the combining of the outputs of several geophones on the ground, attenuates high frequencies. The use of just one geophone in a group makes it easier to record higher frequency

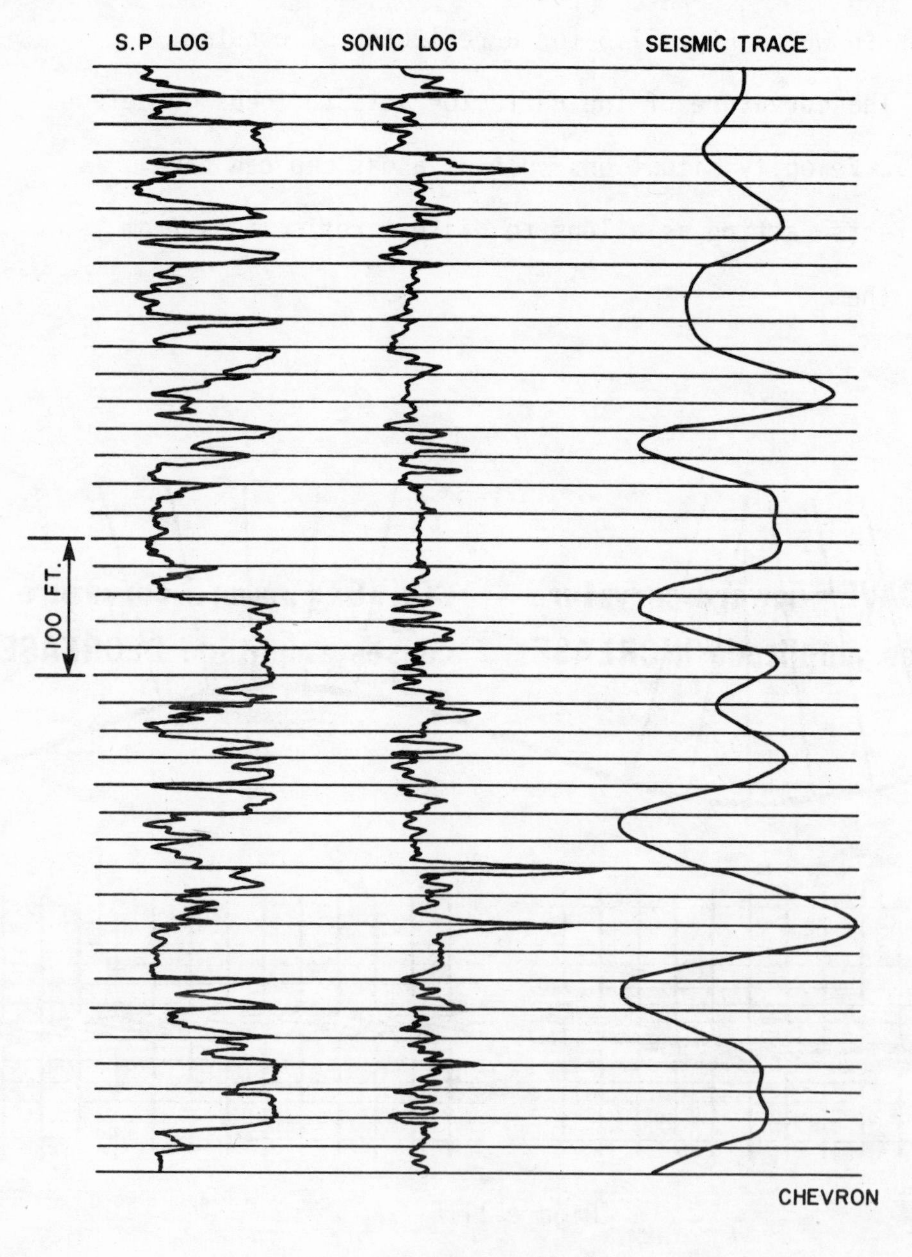

Figure 11.7

energy, which involves shorter wavelengths and improved resolving power; unfortunately only one phone per group also results in lower sensitivity and decreased ability to discriminate against horizontally travelling energy such as ground roll.

The graph in Fig. 11.7 compares a seismic wave with a sonic log and a self-potential log. The seismic wave is of very low frequency compared to the well logs and much thin bedding detail disappears on the seismic record.

Figure 11.8 shows a series of faults with the throw on each fault measured in terms of wavelength. The <u>resolving power</u> is somewhere between 1/8 and 1/4 wavelength. Faults with throw larger than this can be seen fairly clearly whereas the effects of smaller features are so small that they are apt to not be seen.

Resolving power limitations concerning the <u>ability to detect pinchouts</u> was illustrated in Fig. 10.7 and 10.9; where the material above and below the wedge was the same. Figure 11.9 shows a wedge which has a velocity intermediate between that above and below the wedge. The wedge becomes obvious at about 1/4 wavelength.

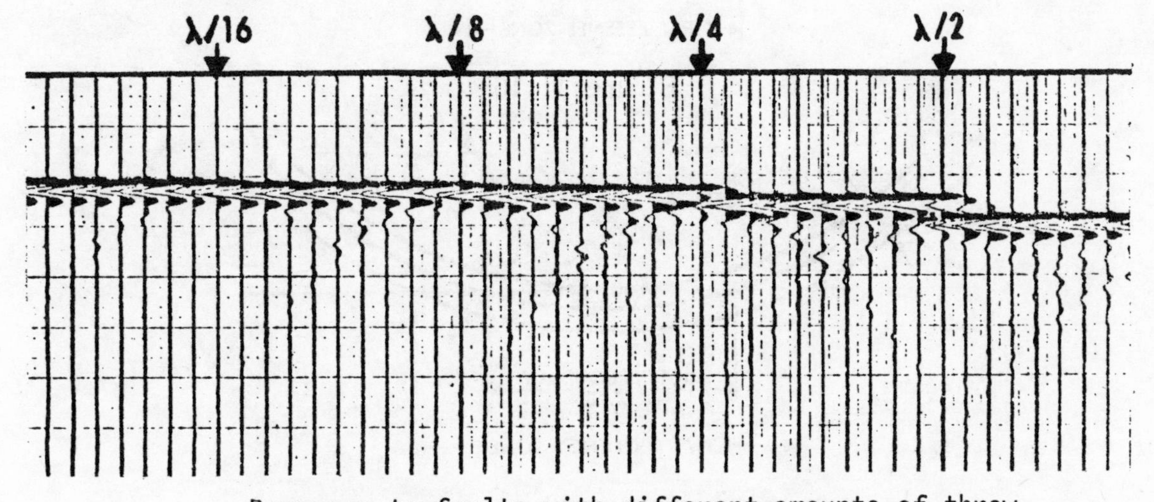

Response to faults with different amounts of throw

Figure 11.8

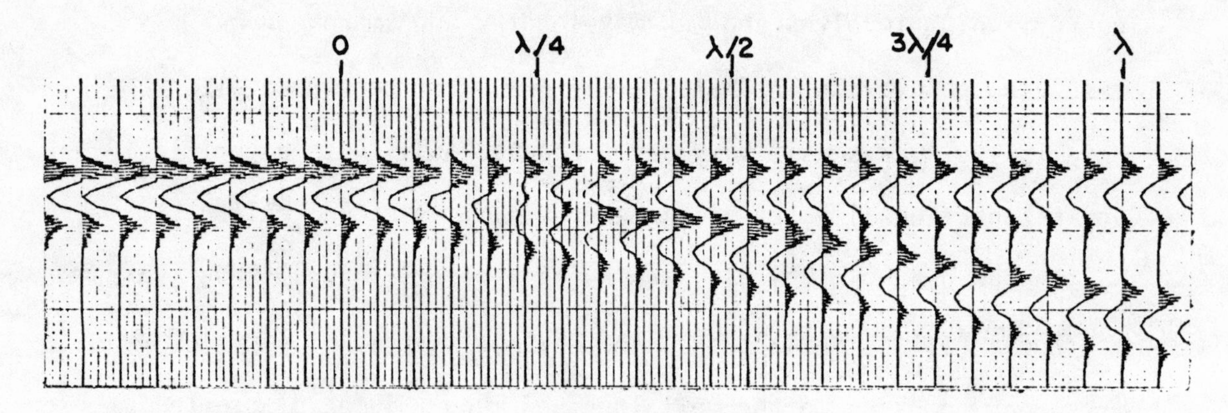

Figure 11.9

We are also concerned with <u>spatial resolving power</u>, that is, how large a structure must be in a horizontal sense in order to be seen on seismic data. This involves the concept of Fresnel zone. A <u>Fresnel zone</u> is that portion of a reflecting interface, such that the energy from that portion will arrive back at a detecting station within a half-cycle, so that it will produce constructive interference. Figure 11.10 shows two

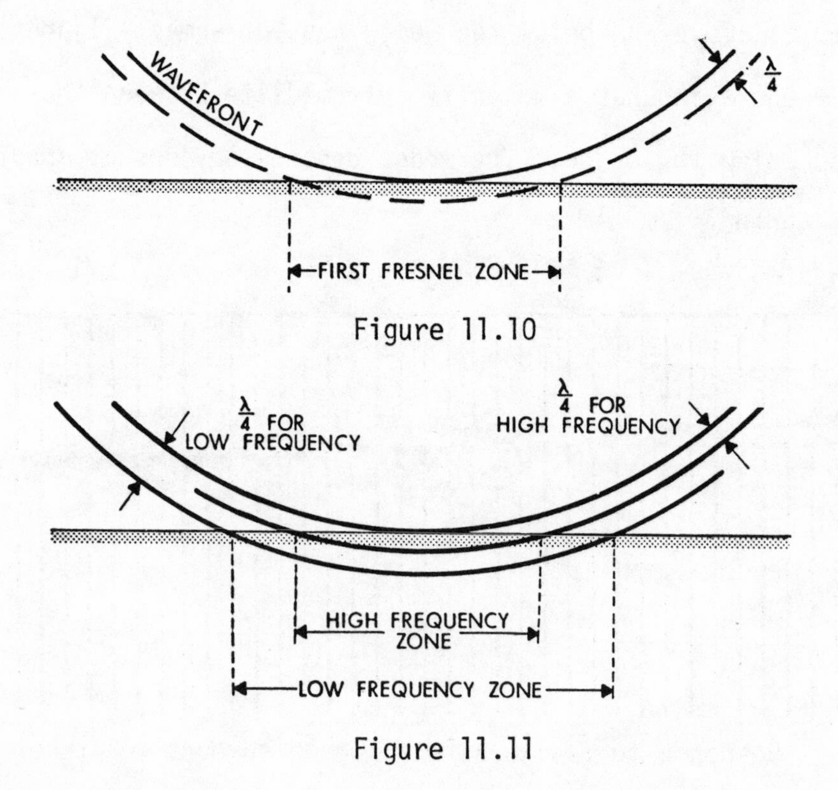

Figure 11.10

Figure 11.11

wavefronts separated by a quarter wavelength, one of which is tangent to an interface. The reflection from the outer edge of the first Fresnel zone will have gained 1/4 wavelength in coming from the source down to the reflector and another 1/4 wavelength in getting back to a detector on the surface and so will arrive a half wavelength after the reflection from the center of the zone. This first Fresnel zone is surrounded by a second Fresnel zone, a ring, from which the reflected energy is delayed by a half to one cycle, and that by a third Fresnel zone, and so on. The effect of the second zone will nearly cancel that of the third, the fourth will nearly cancel the fifth, etc., so that the net effect is that of the first Fresnel zone. The dimensions of this first Fresnel zone are, therefore, the important factor in determining what portion of the reflecting interface gives rise to a reflection. The table below gives the magnitude of this zone for a couple of specific instances.

Reflector depth	Velocity	Frequency	Zone radius
1000 m	2000 m/s	60 Hz	130 m
1000 m	2000 m/s	30 Hz	183 m
4000 m	3500 m/s	50 Hz	375 m
		20 Hz	600 m

The size of this zone depends on frequency, as illustrated by Fig. 11.11. The wavelength is smaller for a high frequency wave, and hence the zone is smaller. Note that for a reflection which contains both high frequency and low frequency energy, a different portion of the reflecting interface is responsible for

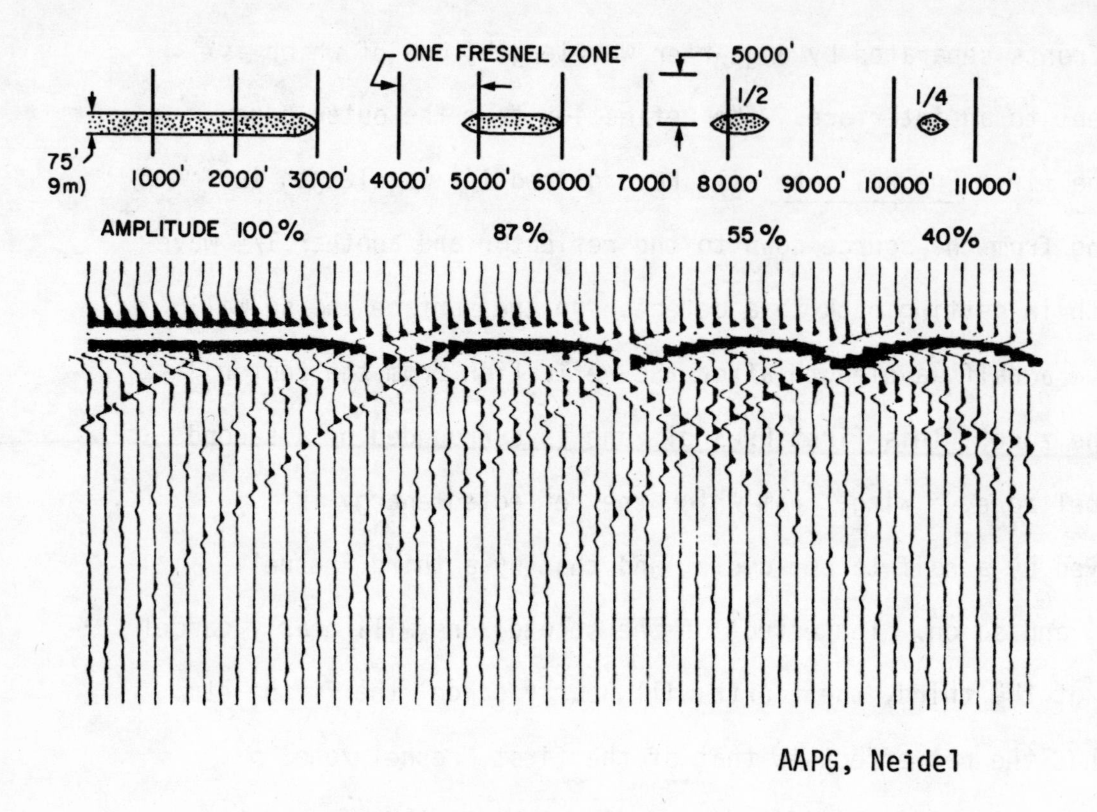

AAPG, Neidel

Figure 11.12

the different frequency components. If a change occurs to the reflecting interface, it could affect one frequency component more than another and result in a change in wave shape. Because of the large area of the first Fresnel zone, a change to a reflector will affect a number of geophones, so that instead of seeing a sharp feature the affect of a sharp change will be distributed over a number of traces.

Figure 11.12 shows the record resulting from reflecting surfaces whose lateral dimensions are measured in Fresnel zone dimensions. As the body becomes smaller than a Fresnel zone, it becomes in effect a point reflector and nearly indistinguishable from a diffraction.

Figure 11.13 shows a fault edge. In the left diagram the reflecting point is remote from the Fresnel zone for the high frequency component is the small circle, and that for a

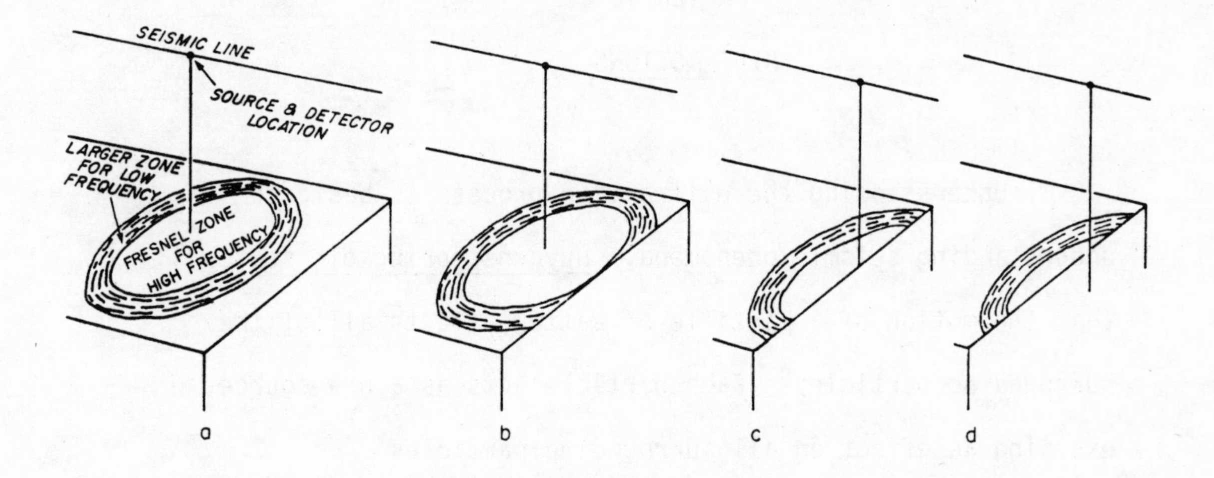

Figure 11.13

lower frequency component is the larger circle (including the smaller circle also). As the observing point approaches the edge (second diagram), the Fresnel zone for the low frequency component will be affected more than the Fresnel zone for the high frequency component, which will change the proportion of high frequency to low frequency energy and result in a change in wave shape. Thus the wave shape will anticipate approach to the edge of the reflector. At the edge half the high frequency Fresnel zone and half the low frequency zone will reflect, so the reflection will be of half strength but the same wave shape as when remote from the edge, because the ratio of high and low frequency components is the same. As we observe from beyond the edge, indicated by the right diagram, a portion of the Fresnel zones will still see the reflecting interface so we will still see some effects of the reflecting surface. In the next section this same effect will be seen from the diffraction viewpoint.

CHAPTER 12

DIFFRACTIONS

Understanding the diffraction process is basic to understanding seismic phenomena. Huygens' principle states that the motion of a particle of matter affects all of the surrounding particles. Each particle acts as a new source, exerting an effect on all surrounding particles.

Huygens' principle helps explain how seismic waves are propagated in the earth. Particles are linked to surrounding particles by elastic forces and motion of a particle changes the distances to, and hence the elastic forces on surrounding particles. The neighboring particles then begin to move, feeling the changes in the forces. If a number of particles are moving, some of the forces may cancel. If a row of particles are moving together as a wavefront, their influences will be in phase on the common tangent to the new wave fronts which each generates but their effects in other directions will cancel out. The combined result is that the next row of particles begins to move and thus the wavefront moves forward. Huygens' principle also suggests that it does not matter what caused the wave or what type of wave it is; once a wave is set in motion, it will propagate itself through the medium.

The sketch in Fig. 12.1 shows reflections from three half-planes. These half-planes lie at different depths and extend from the center of the diagram to the right. The seismic section shows the reflections from these half planes (the flat events) and diffractions from the edges of the planes (the curves). The reason for the diffractions can be understood in terms of Huygens' principle, or in terms of the Fresnel zone. Huygens' principle has the particles near the edge of the reflecting planes radiating energy in all directions and the portions of this energy with lateral components will not cancel as would be the case if the reflecting planes were continuous.

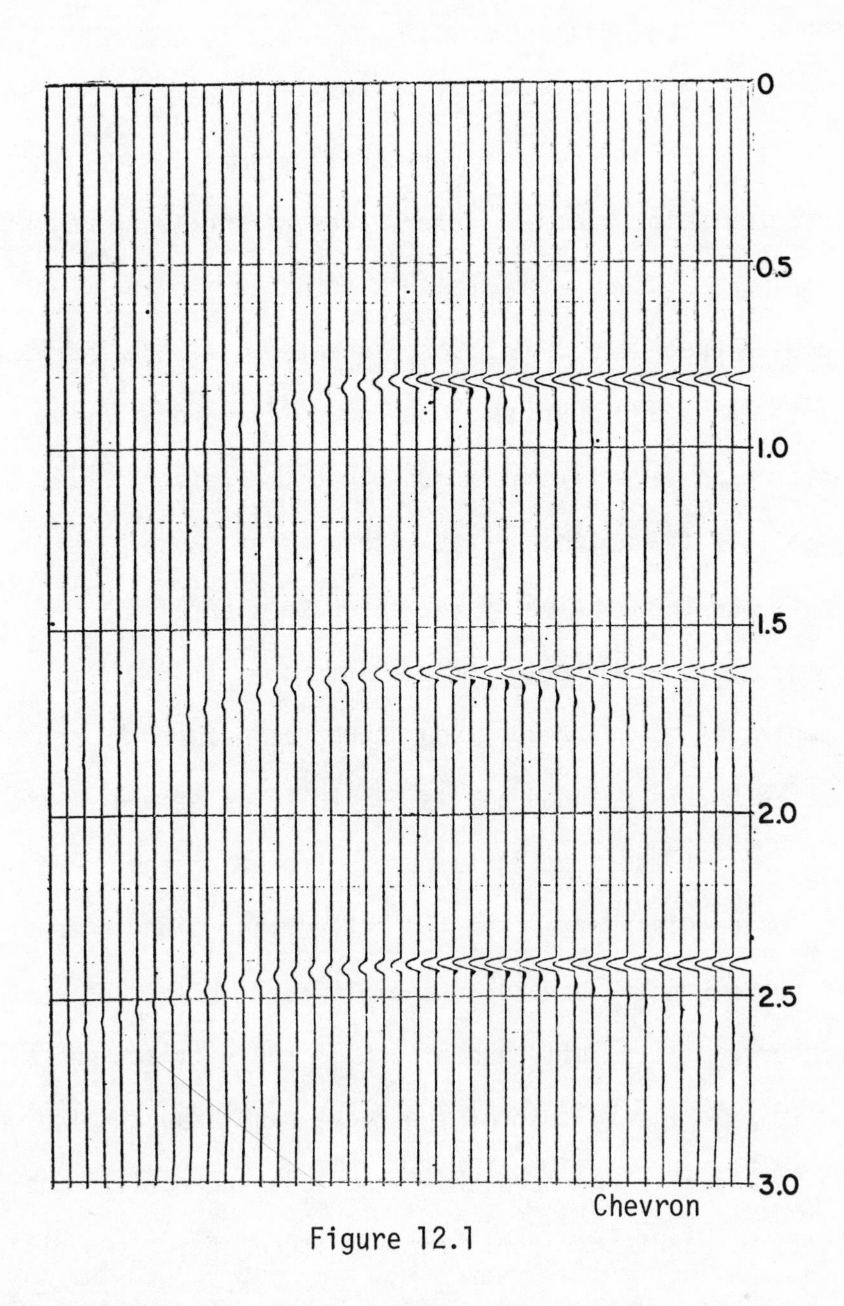

Figure 12.1

The Fresnel zone concept states that a large region is responsible for reflected energy rather than just a point on the reflector. As an edge is approached, the Fresnel zone region sees the edge before the reflecting point reaches the edge and thus affects the reflection (see Fig. 11.11). Likewise, after the edge is passed, a portion of the Fresnel zone may still see the reflector and therefore still return energy to the geophone.

Some of the features of diffractions can be seen in Fig. 12.1. The diffraction curvature becomes less as the diffracting point becomes deeper. The Fresnel zone may be thought of as a cone with the shot point at its apex, and as one goes deeper the cone spans a greater distance and consequently a larger portion of the subsurface contributes effects. The wave is continuous where the reflection is tangent to the diffraction curve. There is no phase break to indicate the termination of the reflector; amplitude and wave shape are smooth and continuous at this point. The amplitude of the reflection decreases before the end of the reflector is reached and is only half as strong at the end of the reflector as it is remote from the edge. The energy which disappears from the reflection is the energy which subsequently appears in the diffraction so that energy is conserved. The left-hand branch of the diffraction has opposite polarity to that of the right-hand branch. The wave shape in Fig.12.1 has been shortened to an unrealistic half-cycle to make this clearer. The magnitude of the two diffraction branches are the same, however, when one considers equal distances from the point at which the reflection is tangent.

We can imagine two half-planes, one extending to the left and the other to the right so that they join to form a continuous reflector. The continuous reflection must be the sum of the effects of the two half-planes. No diffraction would be expected at the junction of the component parts and the fact that the two diffraction branches have the same amplitude but opposite polarity thus accounts for their cancelling each other completely. At the junction, half of the energy is contributed by the right half-plane and the other half by the left half-plane.

Figure 12.2 shows a <u>dipping half-plane</u>, its location indicated by the shaded portion, with the seismic record superimposed upon it. The termination of the half-plane is labeled P; the termination of the reflection from this point is seen at P', on the seismic trace found by following the raypath from this point back to the surface. The diffraction crest is located at the terminus of the reflector; the reflection is tangent to the diffraction at the amplitude symmetry point of the diffraction. The back limb of the diffraction tends to get lost in the tail of the reflection event and so is not very clear; this is often the case. Figure 12.3 is a similar diagram but for a half-plane dipping more steeply in the opposite direction. If these two half-planes are superimposed on each other, as in Fig. 12.4, the sum is a <u>bent reflector</u>. The diffractions between the two reflections interfere constructively and so fill in the region between them so that the seismic reflection appears to have a gentle bend whereas the reflector has a sharp bend.

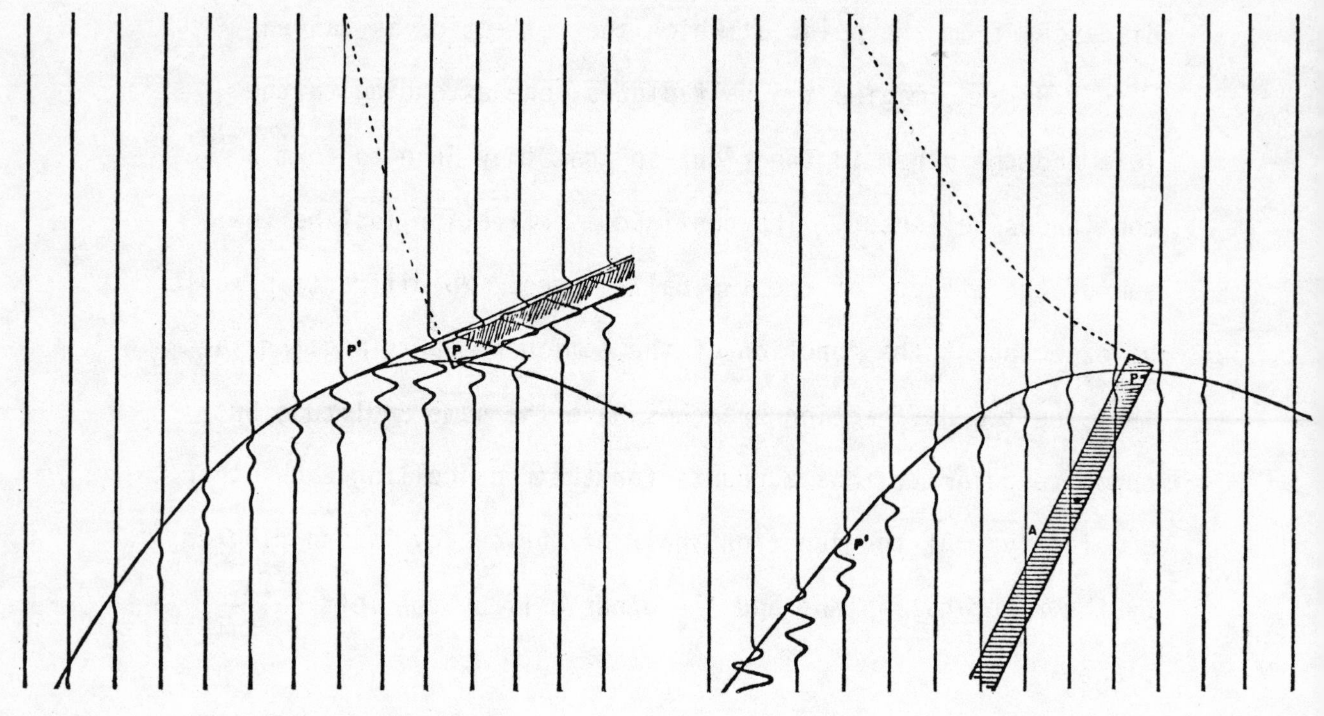

Figure 12.2

Figure 12.3

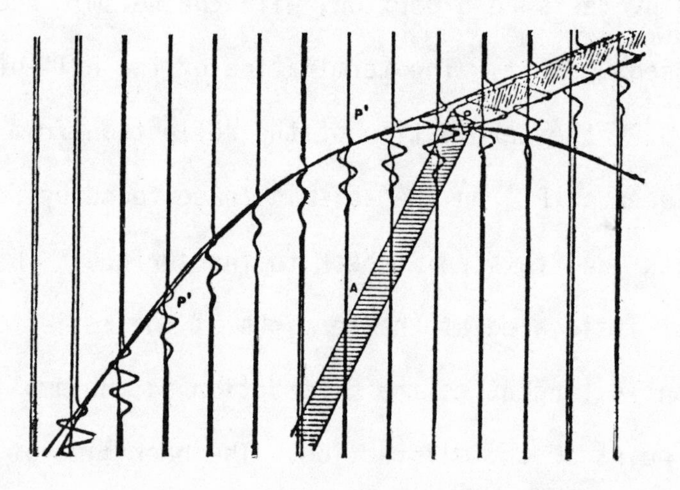

Figure 12.4

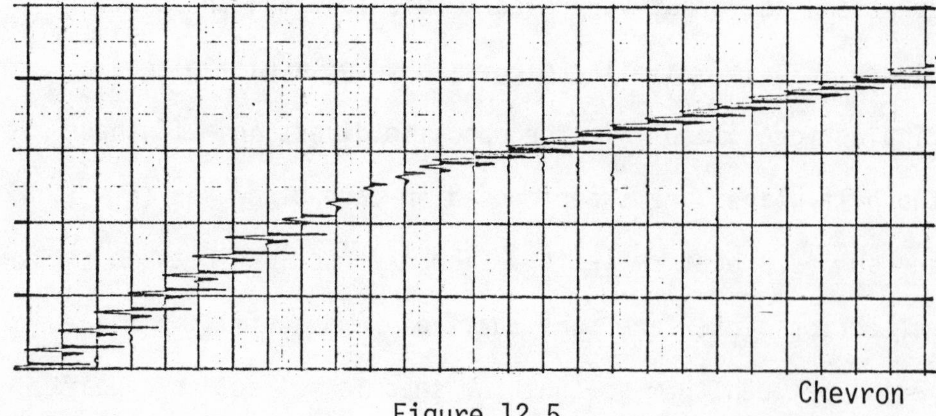

Figure 12.5

Diffractions limit the sharpness which can be seen. Fig. 12.5 shows the seismic reflections and diffractions from a sharp bend in a reflector in true amplitude. The information that there is a sharp bend in the reflector is contained in the data even though it is not obvious and usual picking procedures, such as following a peak, are apt to pick smoothly and continuously across without any recognition of the sharp feature.

Reflections from three half-planes dipping in different directions is shown in Fig. 12.6. The terminus of the reflectors is indicated by the arrow. In each case, the end

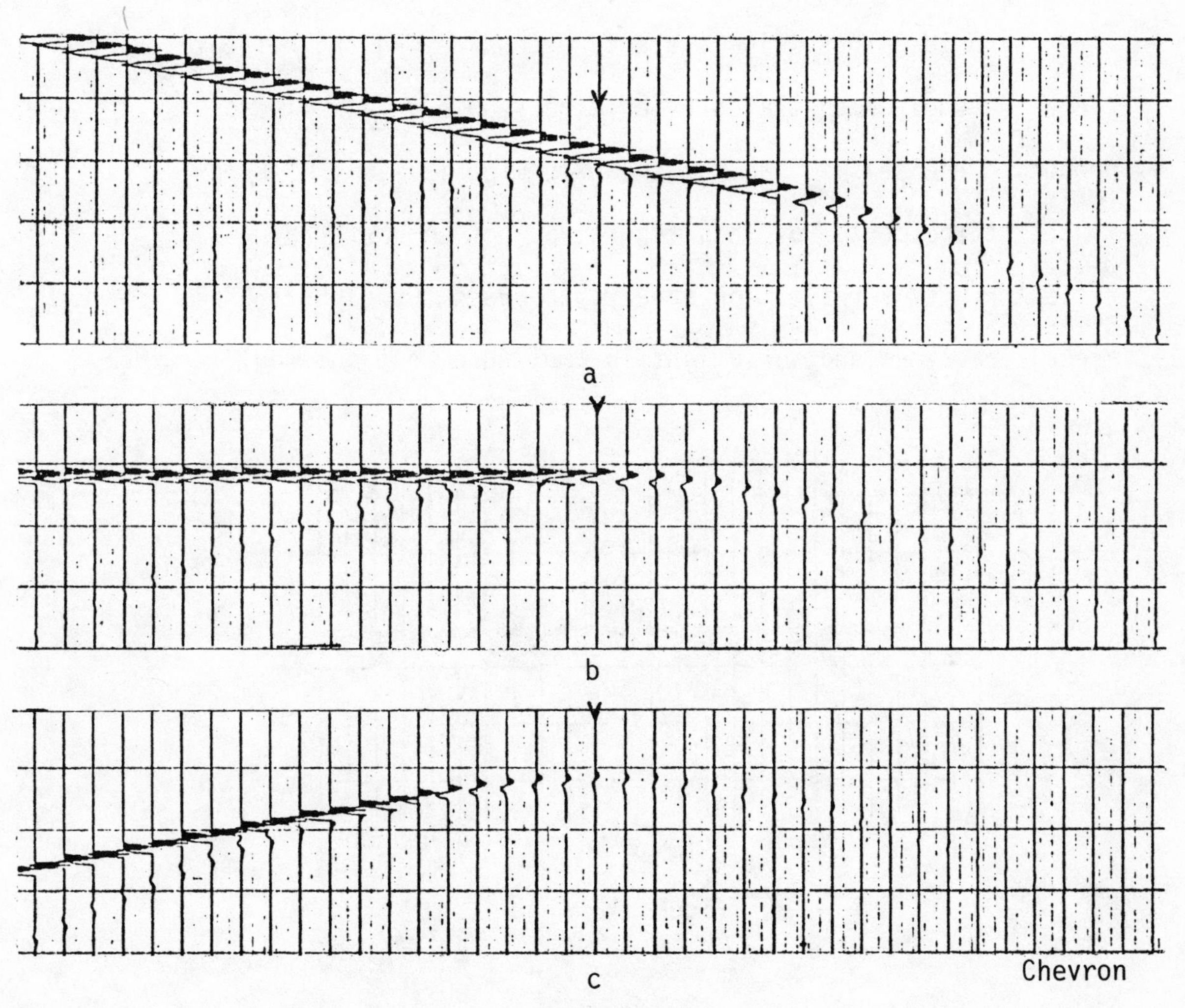

Figure 12.6

of the reflection is seen down-dip from the arrow. If these three reflectors terminated at the same point, each would give rise to a diffraction curve at exactly the same location with the crest at the termination point, but the distribution of energy would be different for each of the diffraction curves, in each instance the diffraction amplitude being symmetrical about the points of tangency of the reflections. The fact that the diffraction curve crest marks the reflector termination provides a tool that is useful in locating faults and the flanks of salt domes where reflectors terminate. This statement should be qualified as implying a simple velocity distribution; complicated velocity distributions can distort diffractions.

The idea that seismic events extend beyond the reflectors can be further demonstrated by cutting a hole in a reflector, producing the seismic section shown in Fig. 12.7. The three traces at the center of this figure, do not encounter a reflector, yet a reflection appears to cross the gap. There are diffractions and variations in amplitude, but one would

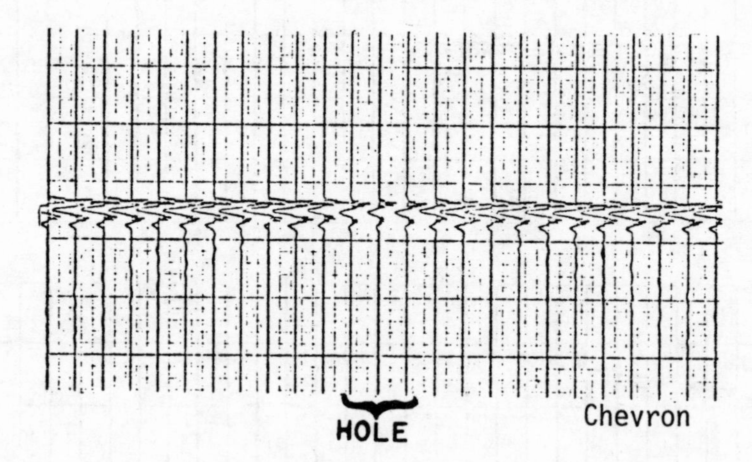

Figure 12.7

be apt to ignore these and map the event as continuous. This might be the model of a pinnacle reef poking through an otherwise continuous bed. If the reef were small, the reflection might appear to be continuous right through the reef.

Diffraction phenomena can be thought of in terms of the "pass-it-on" concept of transmitting seismic waves in the earth. Figure 12.8 shows where energy is located a short time after a plane wavefront reached the point of a perfect reflecting wedge. It shows a reflected wavefront accompanied

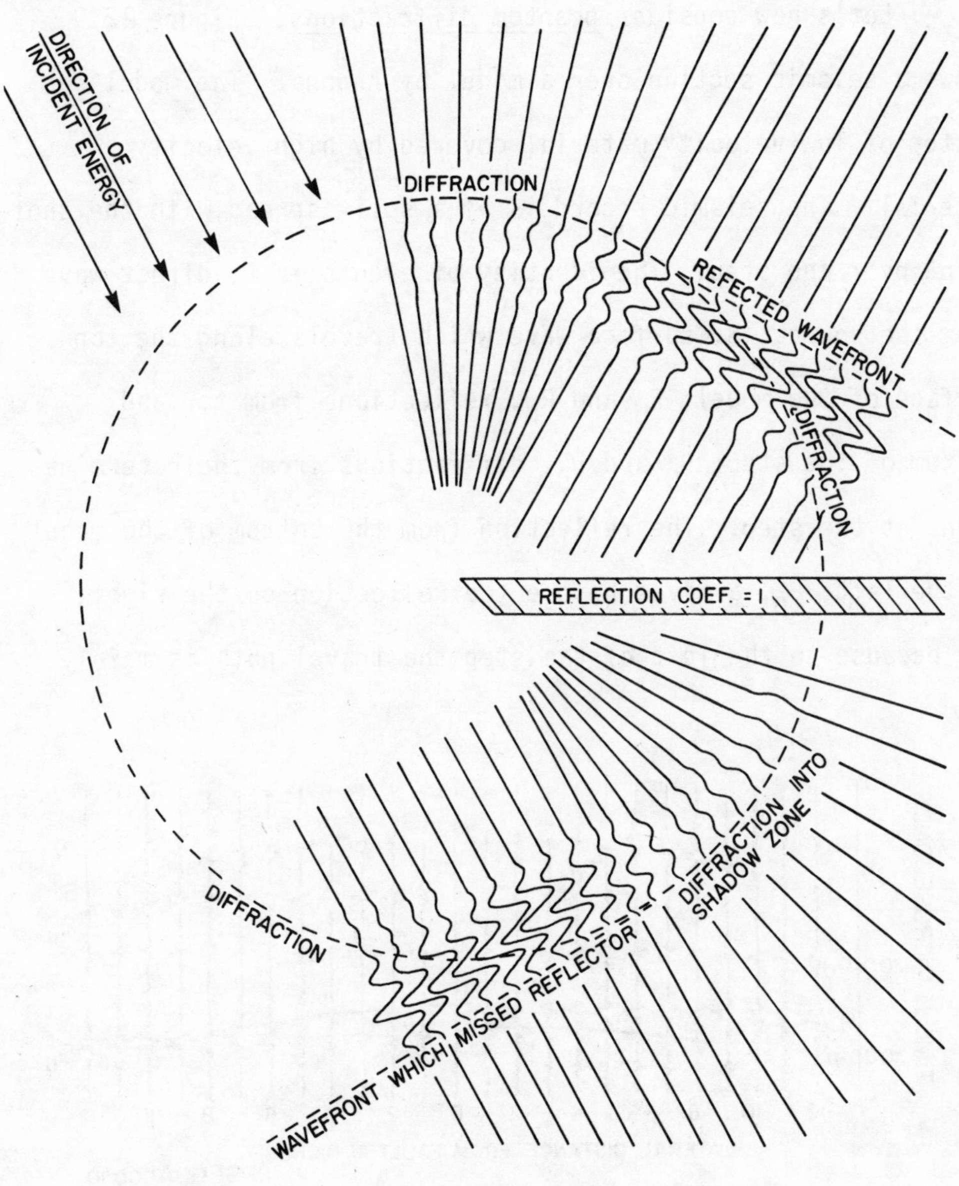

Figure 12.8

by a diffraction. Part of the wave missed the wedge entirely and is still traveling in the downward direction; this wave also has an associated diffraction with both forward and backward limbs. The forward limb involves energy moving into the shadow zone underneath the reflector. Diffraction provides a mechanism for getting energy into regions which cannot be reached on raypath theory. This effect is the same as for a wave on the surface of water traveling around an obstruction such as a breakwater.

Let's now consider phantom diffractions. Figure 12.9 shows a seismic section over a model by Angona. The model was a step of low velocity material covered by high velocity material. The seismic record is of a split spread with the shot-point over the step. The identity of events is P, direct wave from the source; S, surface wave which travels along the top surface of the model; R_U and R_D, reflections from top and bottom of the step; D_U and D_D, diffractions from their terminations at the step. The reflection from the bottom of the model on the left, B_D, arrives before the reflection on the right, B_U, because to the left of the step the travel path is mainly

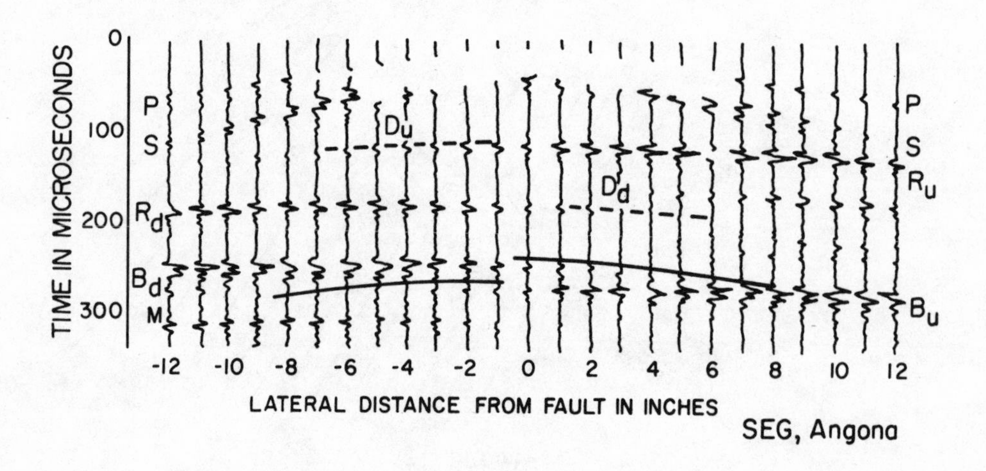

Figure 12.9

through the upper high velocity medium whereas to the right it is mainly through the lower low velocity medium. These reflection segments arrive at different travel times and so are not continuous despite the fact that the reflecting interface, the base of the model, is continuous. These reflection segments terminate in diffractions (shown by solid lines) despite the fact that the reflector is continuous. These are "phantom" diffractions. Diffractions from portions of continuous reflectors usually interfere destructively but in this instance, they travel through materials of different velocities, hence arrive at different times, do not interfere destructively, and therefore show as phantom diffractions. Fig. 12.10 shows an example of phantom diffractions on actual seismic data. This

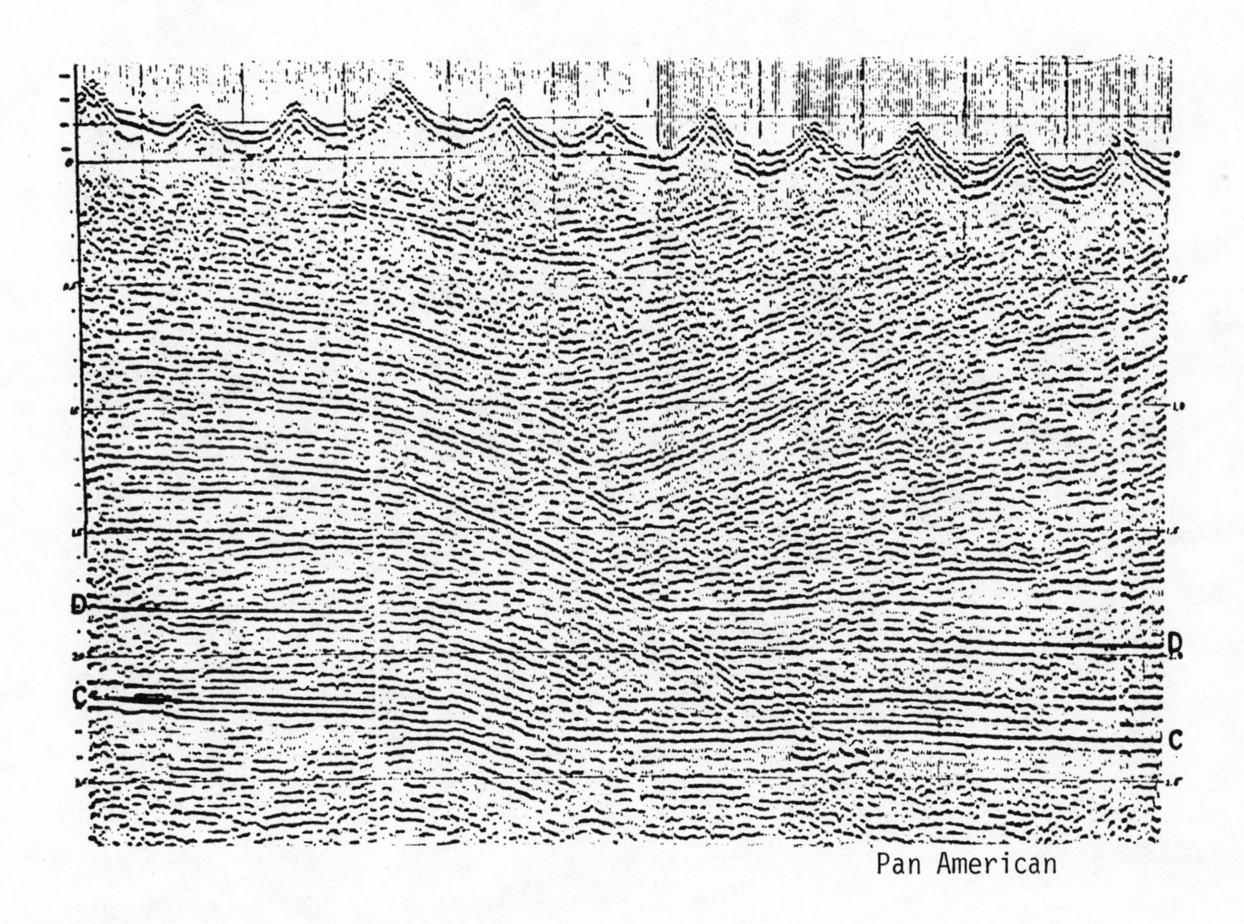

Pan American

Figure 12.10

section is across a thrust fault which has put extra high velocity rock on the left side. The reflections marked D and C are from continuous unfaulted reflectors. These reflections arrive early on the left compared to on the right, and the reflection segments terminate in phantom diffractions.

CHAPTER 13

SHARP SYNCLINES: BURIED FOCI

Synclinal reflectors may produce focusing of seismic energy and associated phenomena on seismic records. Let us consider a curved reflecting interface as given by the dashed line in Fig. 13.1a; the dot locates the center of curvature, taken for convenience as constant for the curved portion. We assume source and observing points on the horizontal dashed line which indicates the surface of the earth, and to make the geometrical considerations more obvious, we assume constant velocity so raypaths are straight lines. The solid lines show the arrival time of seismic events which would be seen on a line shot across this syncline. The syncline in the seismic data is appreciably sharper than the syncline in the earth, because of the focusing effect. Reflecting points are always updip from where reflections appear on a seismic section.

If the center of curvature is at the surface of the earth, as in Fig. 13.1b, all of the energy from the curved portion of the syncline focuses at the center of curvature and so would appear on one seismic trace, resulting in a burst of energy. The burst of energy would probably be interpreted as noise and the syncline not recognized. The left side of the seismic section shows by the solid line the reflection from the left flank of the syncline and the right side the reflection from the right flank.

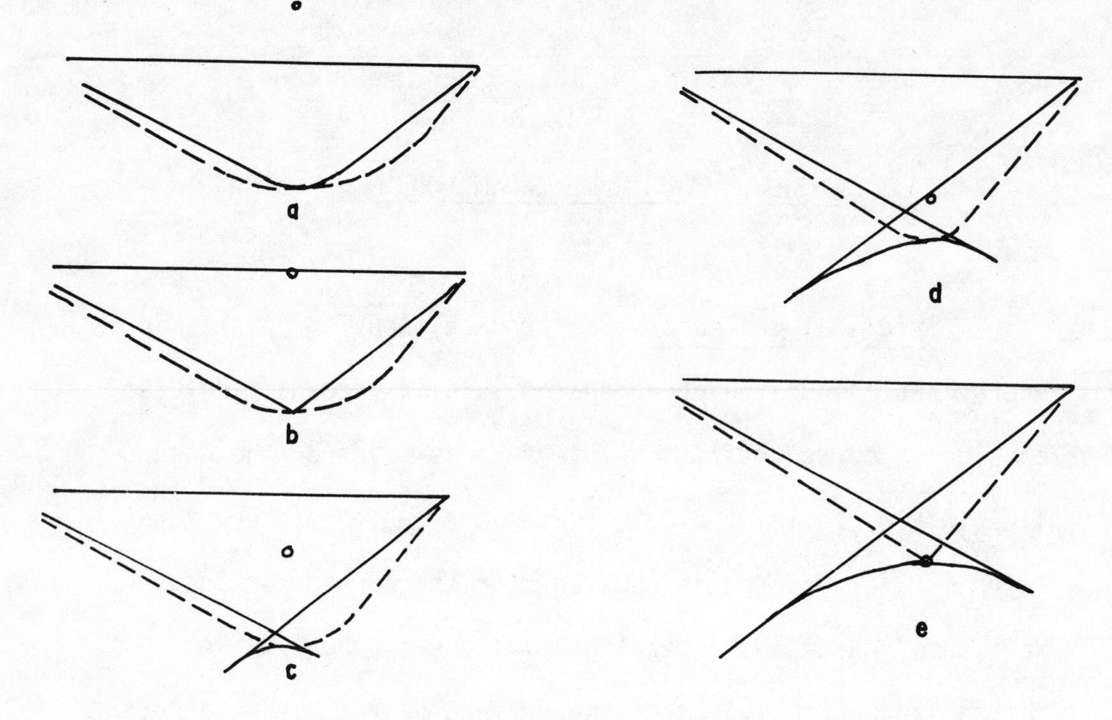

Figure 13.1

The center of curvature is below the surface of the earth in Fig. 13.1c. The reflections from the right and left flanks now overlap and over the center of the line each can be seen from the same surface location. We now have two "reflection branches" from the same reflector. The reflection from the bottom of the syncline is inverted in shape and appears on the seismic record as a convex-upward reflection although it is caused by the concave-upward reflector in the earth. This "reverse branch" is a third reflection branch. An observing point near the center of the syncline will see reflections from both flanks and also from the curved bottom of the syncline.

As the center of curvature is farther below the surface, as in Fig. 13.1d, the energy focuses more deeply and the extent of the buried focus phenomena increases. A tight syncline thus appears as a broad anticline on a seismic record.

The extreme case occurs when the center of curvature is on the reflecting surface itself, that is, where the reflecting surfaces bend sharply, as in Fig. 13.1e. In effect, the syncline becomes a diffracting point. The concepts of diffraction and buried foci give the same results.

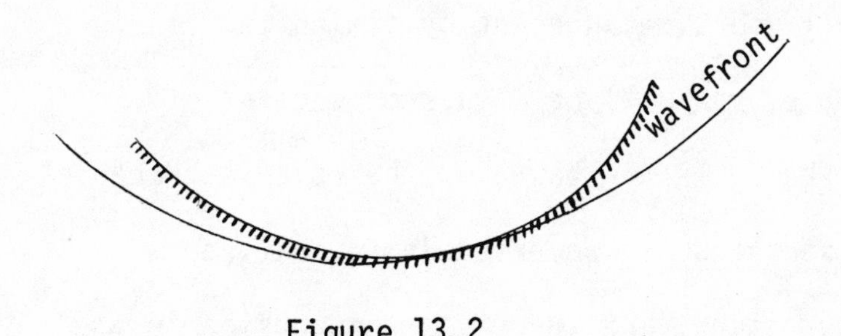

Figure 13.2

The curvature of a syncline which produces buried focus effects is sharper than the curvature of the wave front being reflected, as indicated in Fig. 13.2. The result is that scattered energy from the surrounding region reaches the observing point on the surface before the energy from the reflecting point where the law of reflection is obeyed. Fermat's <u>Principle of Least Time</u>, as it is usually called, specifies that the ray path to a reflecting point takes an extreme value (usually a minimum) of travel time compared to raypaths to adjacent points. This example involves a <u>maximum time path</u>, that is, the reflecting point is the last point reached by the wave front before reflection and the energy reflected at perpendicular incidence takes longer to return to the recording station than energy from other points on the reflecting surface.

In the more usual situation, such as reflection from a flat surface, the energy which travels along a raypath perpendicular to the reflector returns to the source vicinity before the energy from the peripheral area of the reflector. Since the buried focus situation is different, we might expect a change in wave shape and indeed the reverse branch involves a phase shift. If you time an event by timing the first trough, for example, you will be a quarter wavelet off in measurements to the reflector based on the reverse branch.

The reverse branch is reversed in several senses. It is reversed in curvature, and there is an inversion in the sense of right-leftness, that is, the sequence of reflection points in the subsurface is backwards with respect to the sequence of observing points on the surface.

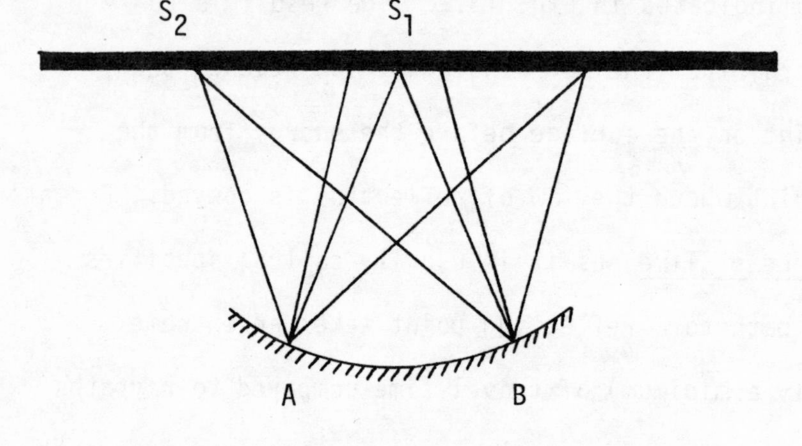

Figure 13.3

The examples so far have assumed that the observing station is at the source location. Consider, however, Fig. 13.3 for shot point S_1 with the geophone at the same location, there will be no buried focus effect. However, if the source is offset as at point S2, then the reflected raypaths cross before reaching the observing stations. Thus offset traces may involve buried focus effects whereas near traces do not. In CDP stacking, we mix long and short offset traces; if these differ in phasing because some involve buried focus effects, the result will be smeared data.

Usually we think of buried focus as involving just three branches, reflections from the right and left flanks and bottom of the curved syncline. More complicated structures are, however, possible, as shown in Fig. 13.4. This particular model has five reflection branches.

So far, we have assumed that the seismic line is perpendicular to the structure. Figure 13.5 shows a seismic line at an angle to the synclinal axis; the syncline is indicated by the contour lines. Reflecting points in the subsurface

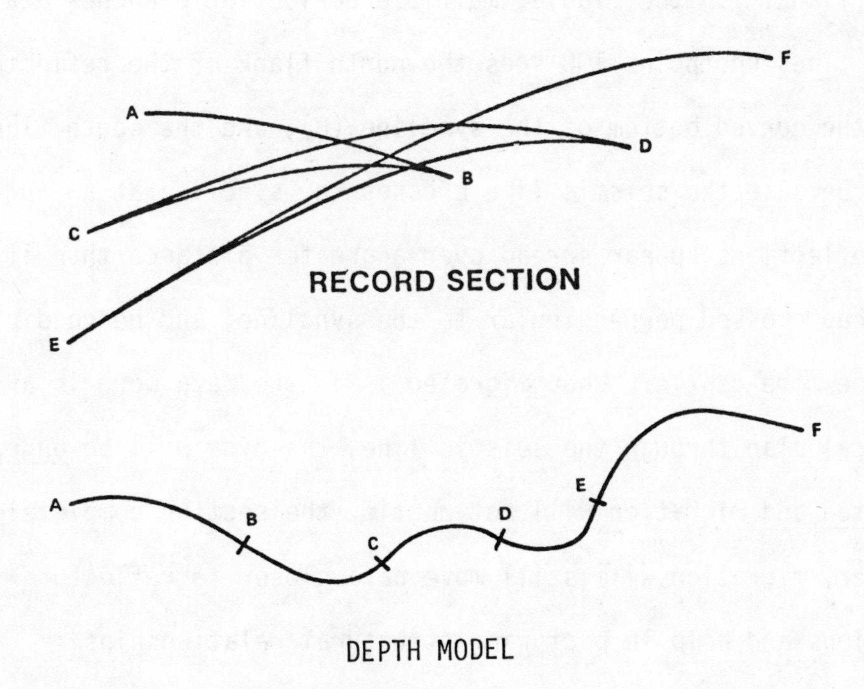

Figure 13.4

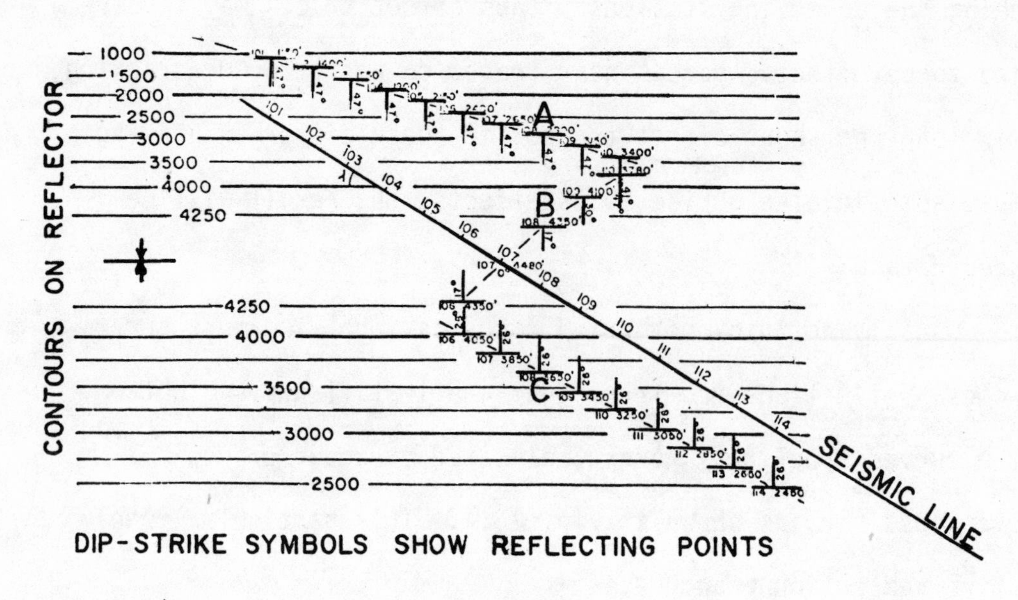

Figure 13.5

lie updip from observation points on a seismic line. A line connecting reflecting points is called the subsurface trace. The left end of the seismic line in Fig. 13.5 looks updip at the north flank of the syncline and the right end updip at the south flank. In the middle, multiple reflection branches are seen. Thus shotpoint 108 sees the north flank of the reflector (A), the curved bottom of the syncline (B), and the south flank (C). Because the seismic line crosses the syncline at an angle, the reflections appear spread over a greater distance than if the line had crossed perpendicular to the syncline, and hence dips will be more gentle. When migrated as if the data were in a vertical plan through the seismic line, the dips will be under-migrated and migration will not rectify the section completely. However, migration will still move data closer to reflector locations and help in picturing structural relationships.

Buried focus phenomena also affect <u>multiples</u>. As we have seen, buried focus phenomena occur when the reflector curvature is greater than the curvature of the wavefront. Multiples, having traveled further than primaries for the corresponding reflector, have less wavefront curvature than primaries and so curved reflectors may cause buried focus phenomena for multiples even where they do not cause such phenomena for primaries. Where dealing with dipping beds, multiples also involve different reflecting points from corresponding primaries on the same trace and hence the simple criteria of multiples arriving at reflecting times which are composites of the reflecting times of corresponding primaries is not correct.

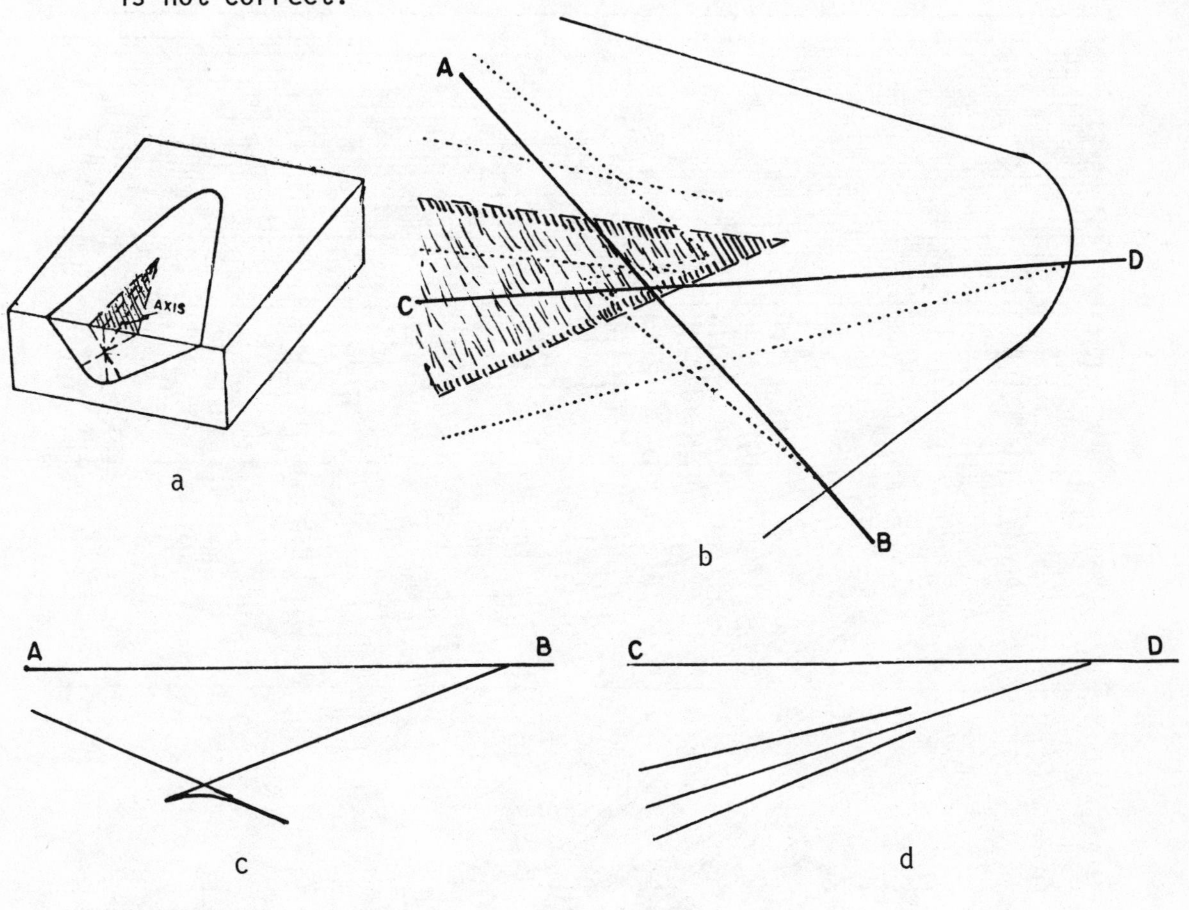

Figure 13.6

Figure 13.6 a is a model of a plunging syncline. The shaded zone indicates the portion which involves buried focus effects. The dotted lines indicate the subsurface trace for two seismic lines which cut across this region and below the plan view (b) are seismic sections of these lines. The several reflection branches may appear as separate arrivals with different dips rather than the usual bow-ties.

Figure 13.7 shows an actual seismic section containing bow-tie effects and Fig. 13.8 shows the data resolved by migration. Another example of buried focus effects in actual data is Fig. 19.12 and its migrated version, 19.13. Figures 19.2 to 19.6 show buried foci in model sections.

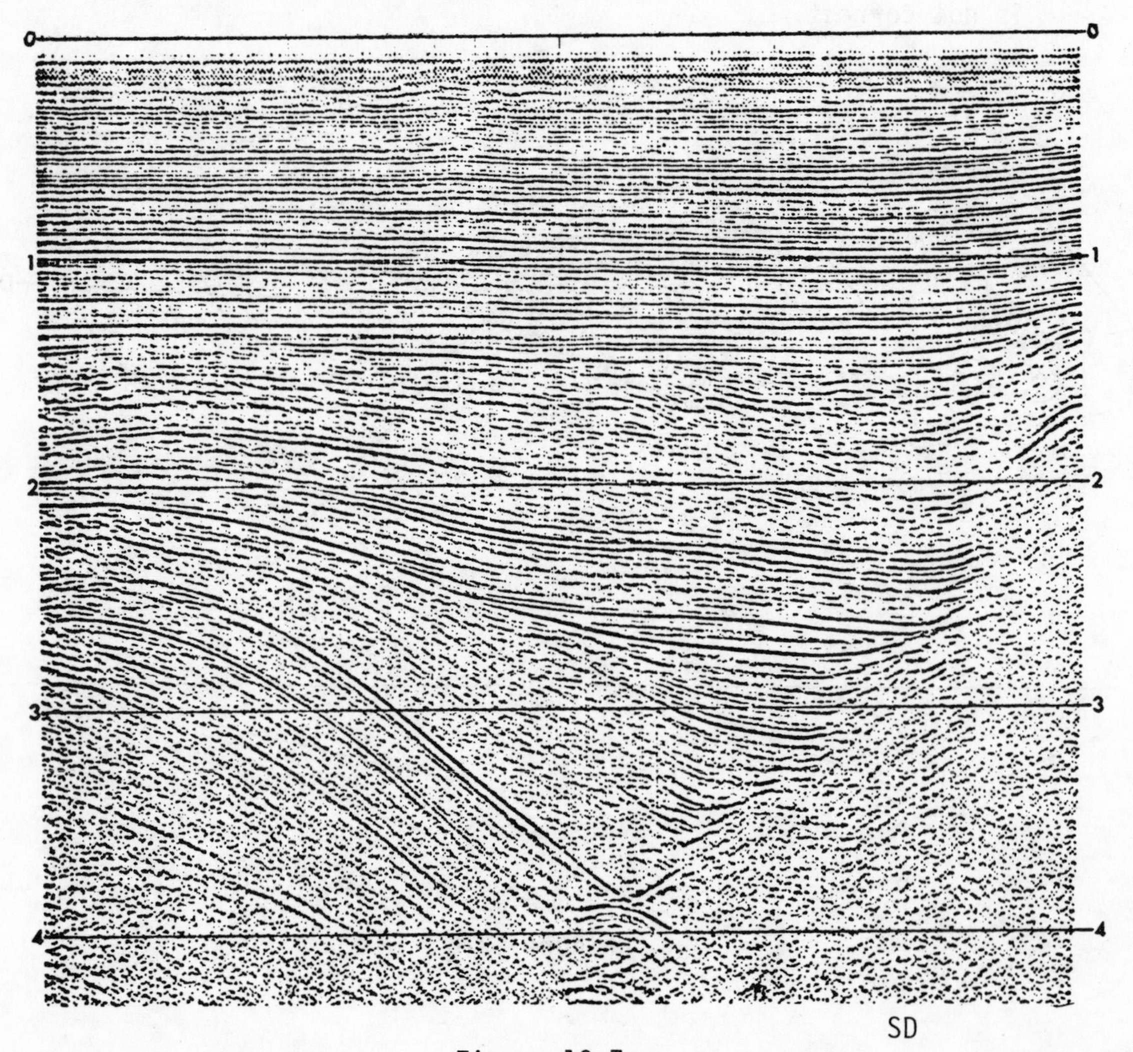

Figure 13.7

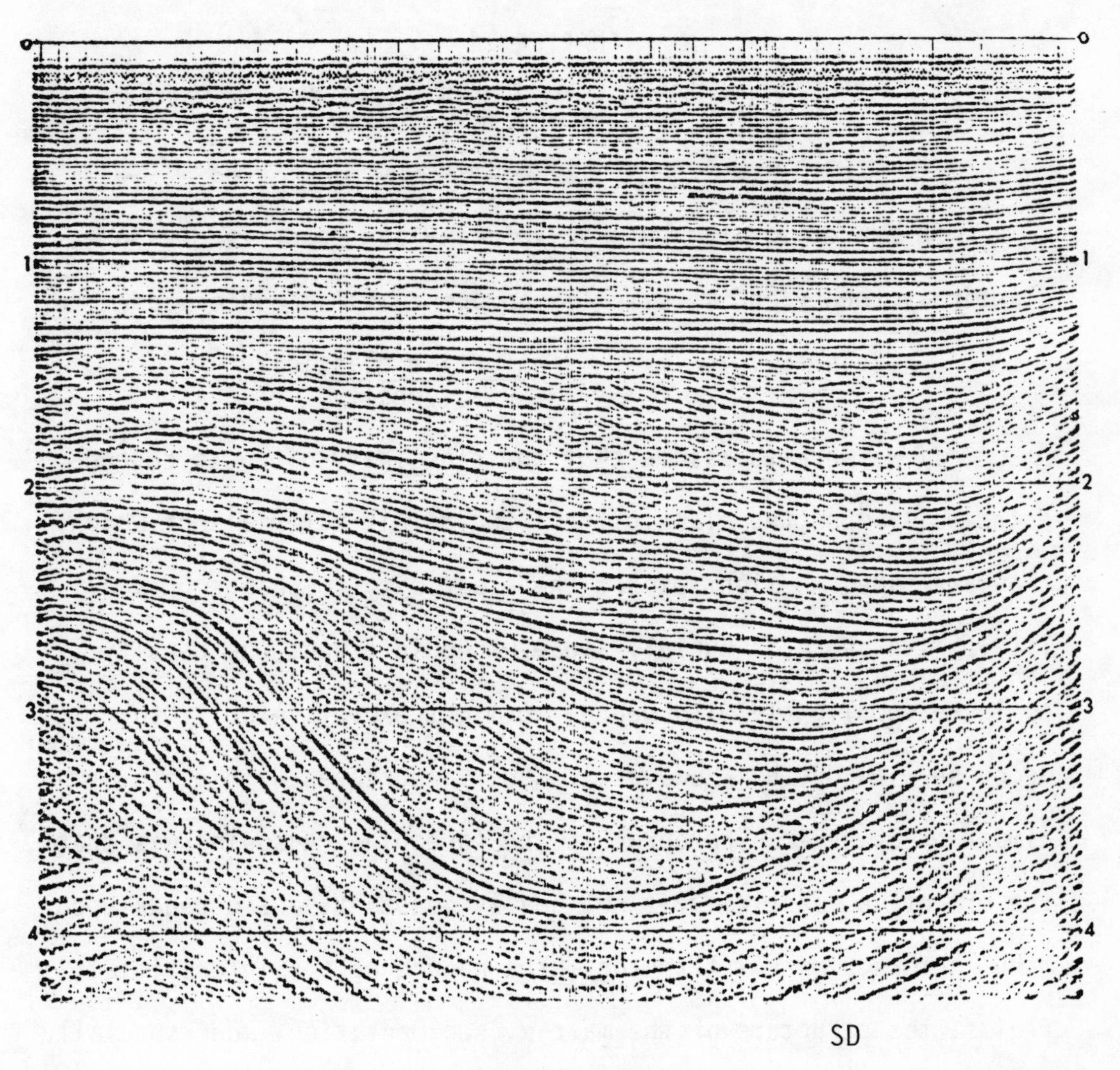

Figure 13.8

CHAPTER 14

VELOCITY I: THE FACTORS INVOLVED

An expression for seismic velocity can be derived using elasticity theory and assuming an homogeneous isotropic medium:

$$\text{Seismic velocity} = \sqrt{\frac{\text{effective elasticity}}{\text{density}}}$$

If we were to assume that the elasticity of rocks depends principally upon the elasticity of the matrix materials of which they are made, we might expect the elasticity to change only a little with depth of burial; since we would expect the density to increase slightly with depth, we might expect velocity to decrease with depth. However, velocity increases with depth almost everywhere. The error in the foregoing reasoning was that the effective elasticity is not nearly constant, but depends upon many factors: matrix lithology, the nature of the interstitial fluids, the structure of the matrix, sedimentation, and especially on porosity and differential pressure.

Velocity data from the GSA Handbook of Physical Constants is shown in Fig. 14.1. The data are 80% fiducial limits, meaning that 10% of the measurements fall below and 10% above the bars. The velocity for each class of rock encompasses a fairly broad range and velocity for the different classes of rock overlap.

The graph in Fig. 14.2, from Lindseth, shows velocity for various rock types with the addition of porosity information. The shaded curve at the bottom of each bar indicates porosity, The lower velocities for any rock type are the high-porosity

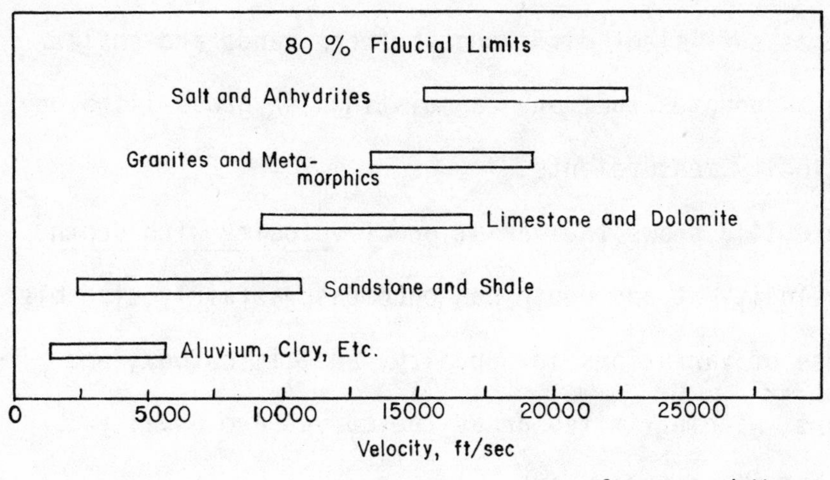

Grant and West

Figure 14.1

samples and the higher velocities are the low-porosity samples. This indicates that porosity is one of the most dominant factors involved in velocity. It is porosity which is primarily responsible for the large range in velocity for any given rock type.

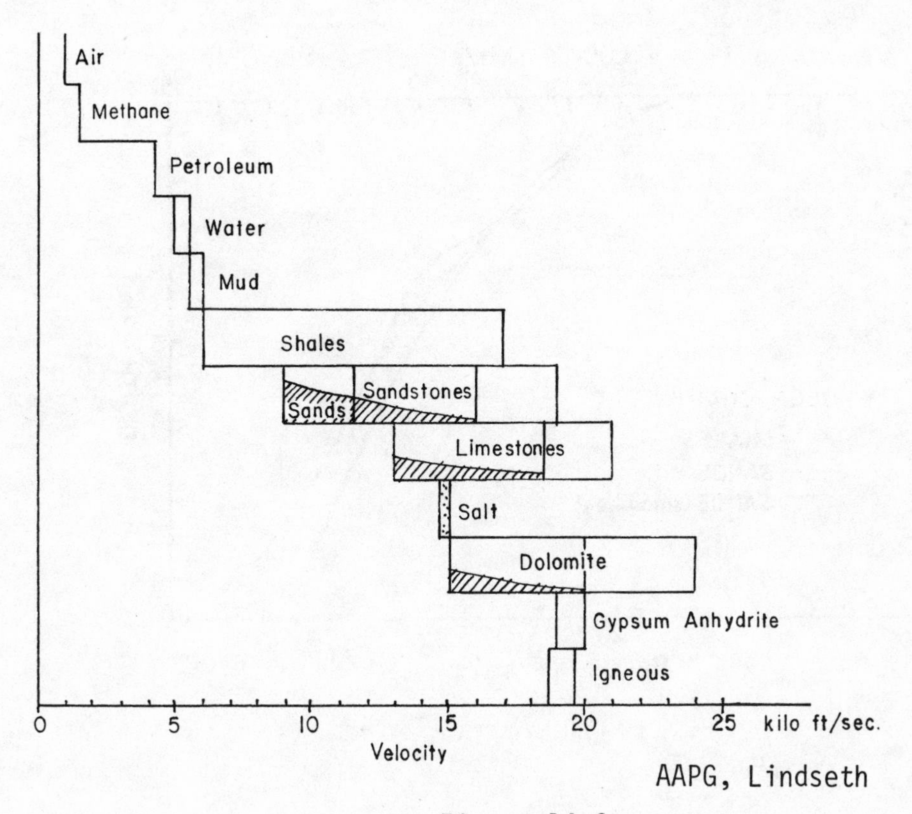

AAPG, Lindseth

Figure 14.2

The bars for different rock types overlap so much that it is difficult to determine lithology from velocity values. However, often there is sufficient difference between sands and shales compared to carbonates that one can distinguish gross lithology based on velocity measurements.

Figure 14.3 shows the variation of velocity with depth. While the velocity at any depth can encompass a fairly sizeable range because of variations in porosity, age, lithology, and other factors, within limited areas the curves are usually systematic. These data are taken from Gregory but they have been smoothed somewhat. While there is slight separation of the curves, in general the information, which is based on gross averages, is not sufficient for distinguishing between shales and sands, although on a statistical basis it might be useful within a limited area.

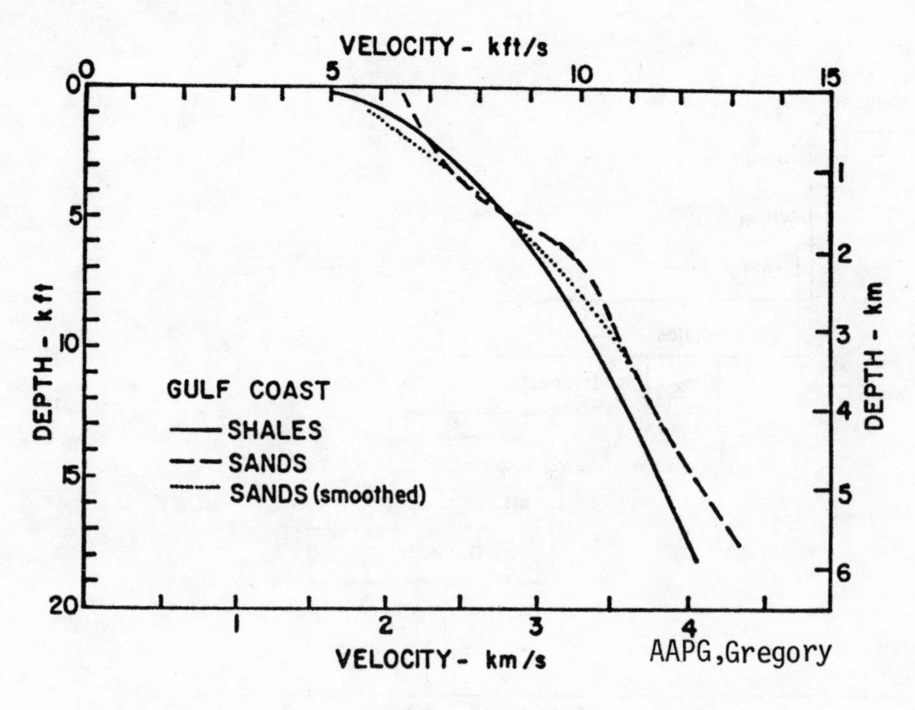

Figure 14.3

The graph in Fig. 14.4 is a modification by Neidell of data from Gardner, Gardner, and Gregory. Velocity is plotted against density on logarithmic scales. The curves for various rock types are almost linear when plotted in this logarithmic domain. Most rock types cluster about a single line, the equation for which is often used for relating density P to velocity V: $P = k\,V^{\frac{1}{4}}$.

Neidell added the lines which show acoustic impedence, because it is acoustic impedance that is responsible for reflectivity. The acoustic impedance lines are almost at right angles to the trend lines for various lithologies.

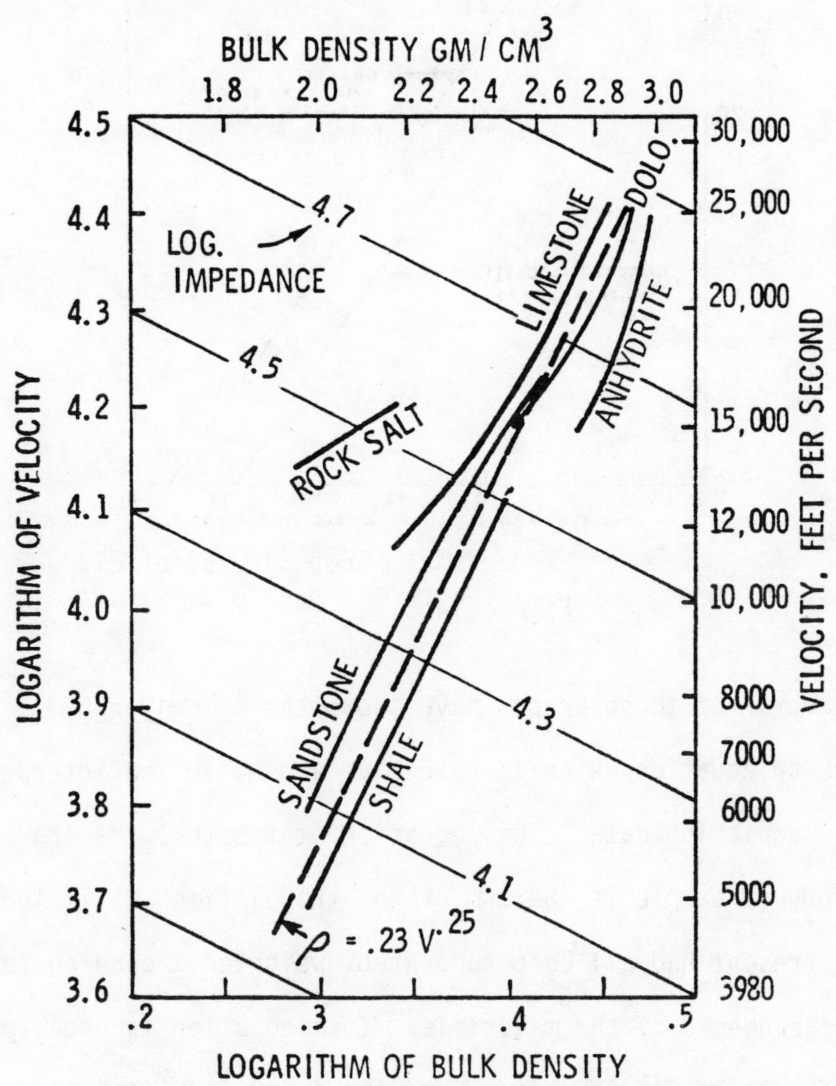

SEG, Gardner et al., Neidell

Figure 14.4

Figure 14.5 relates porosity and velocity directly. Two lines are shown because the samples were mixtures of calcite and quartz and sometimes the continuous matrix was formed by the quartz with the calcite predominantly in interstices in the quartz matrix, and sometimes the converse was true. While individual measurements depart somewhat from the lines, the overall relationship between velocity and density is quite clear.

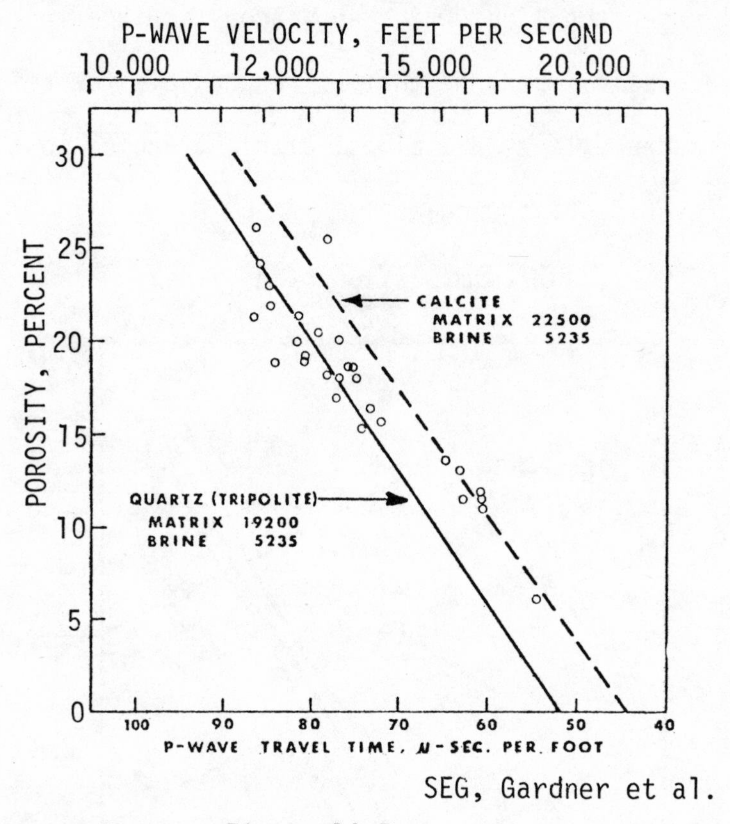

SEG, Gardner et al.

Figure 14.5

Several of these graphs have shown the time average equation, an equation which is extensively used in the interpretation of sonic log data. This equation states that the travel time through a sample is the sum of the travel times as if the material present had all been separated, weighted according to relative abundance of the materials. This equation is usually expressed in terms of porosity where the two "materials" are the matrix and the fluid. The amount of fluid present is indicated by the porosity, ϕ. Expressed in this way, the equation is

$$\frac{1}{V} = \frac{\phi}{V_f} + \frac{1-\phi}{V_m} ,$$

where V is the velocity in the rock, V_m is the velocity in the matrix material, and V_f is the velocity in the interstitial fluid. The time average equation is empirical rather than theoretical and provides a useful approximation within a limited range of porosity values. It can be made to fit various limited ranges by adjusting the values of the velocity constants empirically rather than using their true values.

Normally porosity decreases with depth of burial. As a rock is buried deeper, the overburden pressure increases and the porosity gets "squeezed out". Coupling this concept of decrease of porosity with depth of burial with the increase in velocity with loss of porosity, and the more familiar increase of velocity with depth follows.

The increase of velocity with depth can also be thought of as variation of velocity with pressure. The upper curve in Fig. 14.6 shows the variation of velocity with pressure as a rock sample is squeezed when the pore fluid is not constrained. This is, however, not a realistic picture of rock in the earth. While the weight of the overburden rock tries to compress a rock and eliminate its porosity, the fluid in the pore space exerts a back pressure which tries to hold the pore space open. The effective pressure on the rock is the difference between the weights of the overburden and the fluid pressure. The horizontal lines in Fig. 14.6 illustrate that the velocity remains constant

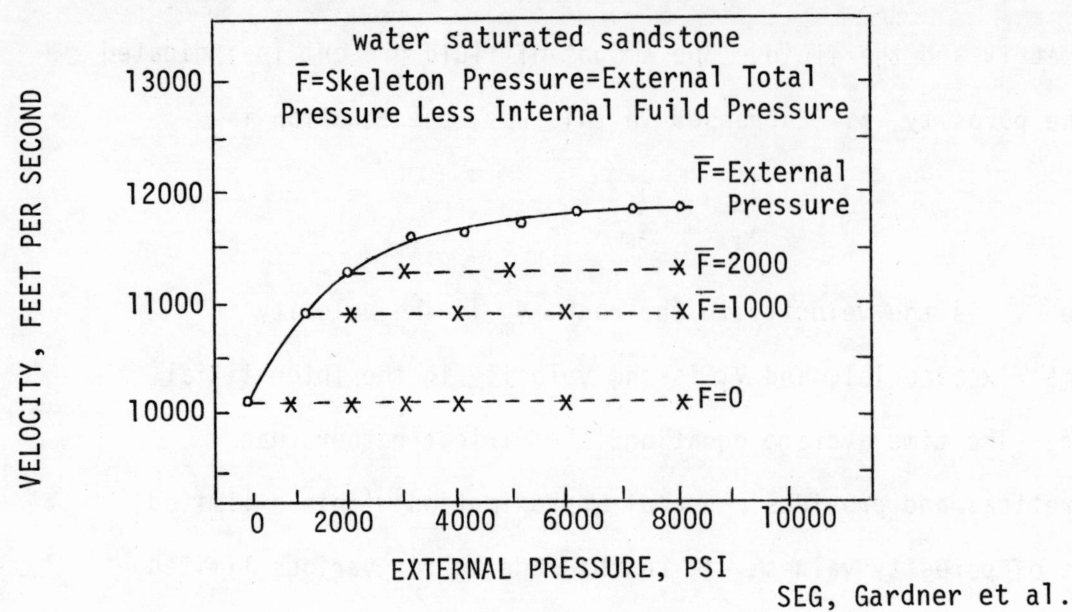

SEG, Gardner et al.

Figure 14.6

with overburden pressure when the fluid pressure is increased at the same rate as the overburden pressure, so that the differential pressure $\overline{F}$ is constant.

Non-porous rocks also increase their velocity with depth, although not to the same extent as porous rocks. Gardner, Gardner and Gregory postulated that the increase of velocity with pressure for such rocks was because of <u>micro-fractures</u>, very small frac-

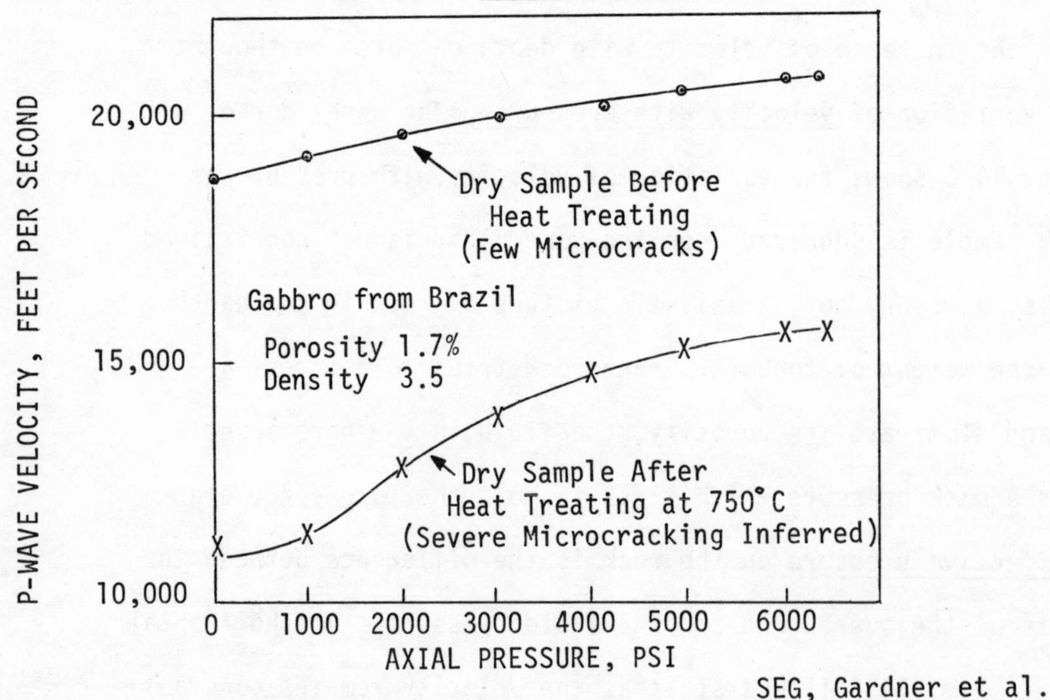

SEG, Gardner et al.

Figure 14.7

tures which delay the seismic wave from what it would have been in truly homogeneous material. Such micro-fractures are expected to be more prevalent in the shallow part of the earth and overburden pressure tends to heal the fractures. To demonstrate this hypothesis, Gardner, Gardner and Gregory subjected a sample of gabbro to a shock heat treatment to induce micro-fractures and then measured its velocity vs pressure relationship (Fig. 14.7). Micro-fractures significantly affected the velocity and the velocity-pressure slope. Presumably if the rock were put under sufficient pressure, eventually all of the micro-fractures would heal and the velocity would return to that for the rock before the shock treatment.

Faust made a statistical study of the effect of age of rock on seismic velocity. His conclusions are shown in Fig. 14.8. He found that velocity increased with the age of rocks according to the 1/6 power of the age. We might not expect a

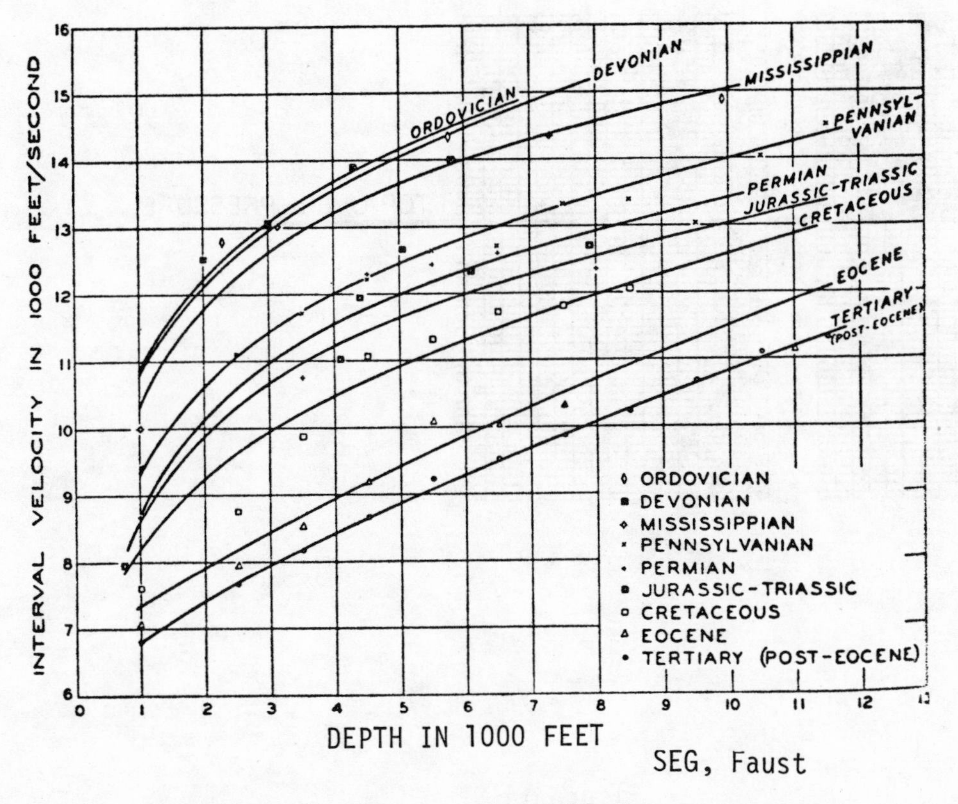

Figure 14.8

variation simply because of age. However, older rocks have been around longer and have had more opportunities for things to have happened to them, including cementation, being subjected to various stresses, etc. There also may be a time dependent deformation, recrystallization or long term compliance.

The graph of a Gulf Coast well in Fig. 14.9 shows an S.P. log and a sonic log through a portion of the well. A line indicates the top of a over-pressured shale section. The higher pressures are associated with significantly lower velocity. Such velocity reversal can sometimes be detected from surface measurements of velocity and used to predict when a high pressure section will be reached. The drillers can then be warned to expect high velocity and the well casing program designed better. The reason for this over-pressured phenomenon is that the interstitial fluid

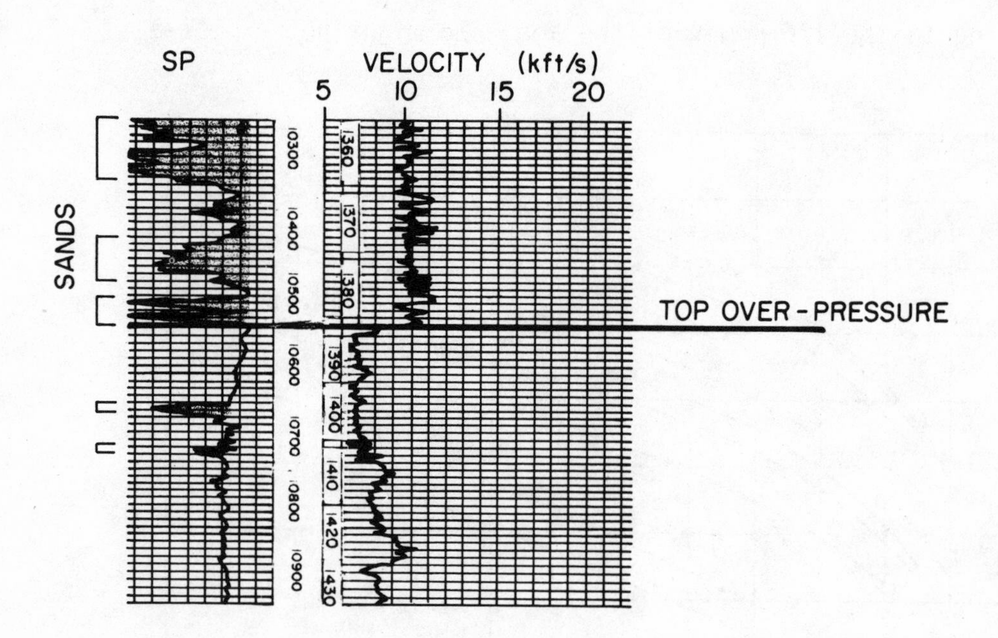

Figure 14.9

cannot escape from the impermeable shale and so tends to support the overburden. Ordinarily this fluid escapes until a balance is achieved with respect to the hydrostatic pressure appropriate for the depth of burial. Because the fluid pressure is abnormally high, the effective pressure on the rock is less than what would be expected for its normal depth of burial. Therefore, the rock feels a lower effective pressure and shows a lower velocity.

Figure 14.10 illustrates the problems in trying to determine sand-shale ratio from velocity data. Shown are sonic and SP curves for a portion of a well in the Gulf Coast area. Individual samples have a tendency to scatter. The SP curve shows sands where the curve is at its lowest and shales where the SP curve is high. A best-fit curve drawn through the sonic log values for the sand portions would show a higher velocity than a best-fit curve drawn through the values for the shale portions. However,

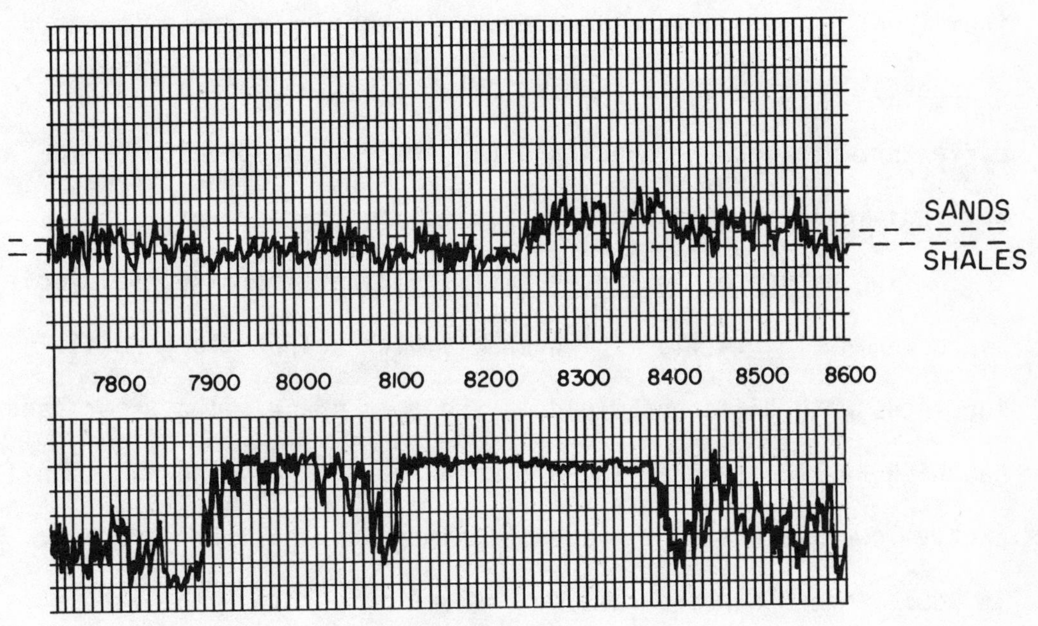

Figure 14.10

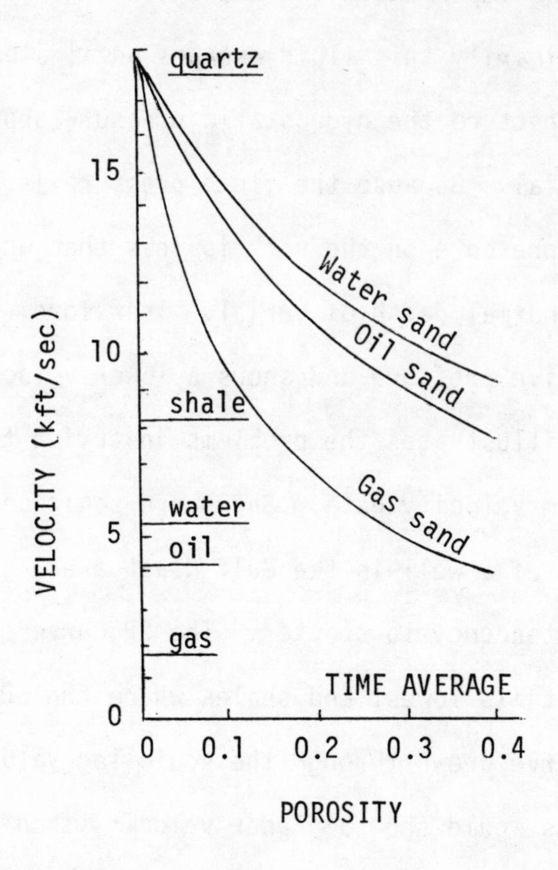

Figure 14.11

individual measurements have wide departures from these smoothed averages. On an overall statistical basis, there is a systematic difference between the sands and the shales, but prediction for any individual portion would be subject to large error.

The nature of the interstitial fluids also affects velocity as shown in Fig. 14.11. The graph is with respect to porosity for sands with different fluid in the pore space. The water sand has higher velocity than the oil sand and the gas sand has significantly lower velocity. If acoustic impedance has been graphed, it would have shown the same general relationship. The velocity of shale is also shown. The water sand or oil sand could have

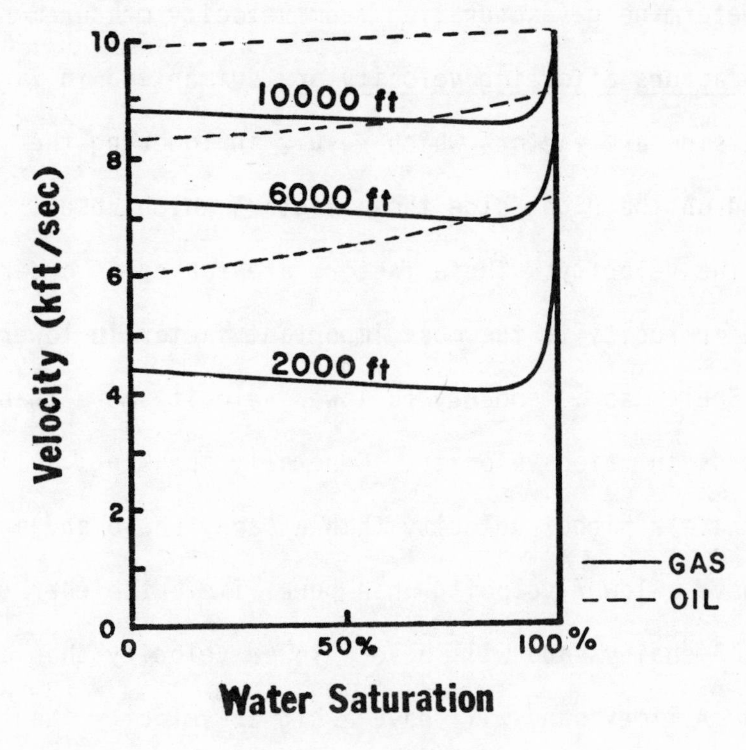

Figure 14.12

either larger or smaller velocity than the shale depending on its porosity. The curves are specific for a particular depth.

The effect of water saturation on velocity is shown in Fig. 14.12. The left side of the curves shows zero water saturation, which is equivalent in this case to 100% gas saturation or 100% oil saturation. Velocity changes gradually with water saturation except for very small amounts of gas. A few percent gas saturation produces a very large effect on velocity, whereas further increases in gas saturation have relatively little effect. These curves tell us that we will have difficulty trying to determine gas saturation from velocity measurements.

The factors affecting velocity are summarized in Fig. 14.13. On the left side are factors which result in lowering the velocity and on the right side those factors which result in increasing the velocity. These factors are ranked in order of importance. Porosity is the most important factor in lowering velocity. Shale has a tendency to lower velocity more than sand. Dolomite tends to raise velocity. Generally speaking, a limey shale will have a higher velocity than a sandy shale and a sandy lime will have a lower velocity than pure lime but higher than pure sand. A shaley sand will have a lower velocity than a clean sand. A limey sand will have a higher velocity than a clean sand. Porosity, the most important factor, depends especially upon the maximum differential pressure which the rock has undergone during its history. The effective stress on the rock is the differential pressure, the difference between overburden and interstitial pressures.

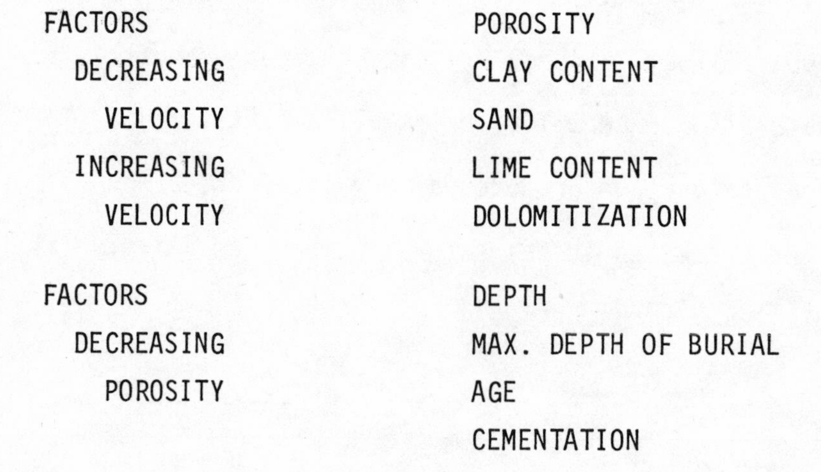

Figure 14.13

CHAPTER 15

VELOCITY II: THEORY AND MEASUREMENT (AND PROBLEM)

Let us consider a model of a clastic rock to help understand how various factors affect velocity. Clastic rocks, one of the most important rock types from the viewpoint of hydrocarbon exploration, are made of fragments of other rocks, occasionally of seashells.

Let us begin with well-sorted, well-rounded grains. Under gravitational packing force the first layer of grains resting on a flat surface would assume a regular arrangement, as indicated in Fig. 15.1a, and we might expect subsequent layers to also be regular. However, if we place some second-level grains in gravitationally stable positions, as shown in Fig. 15.1b, voids appear which are not big enough to hold another grain. The third level will not even be horizontal because some of the grains will sink partially into the second level voids. Thus even with uniform spherical particles, after a few layers we have a random pack with nearly 50% porosity. Packs of grains of other shapes but similar size would also lead to about the same porosity. We note that the porosity is a matter of geometry and whether the grains are of clay, sand or boulder size is not relevant.

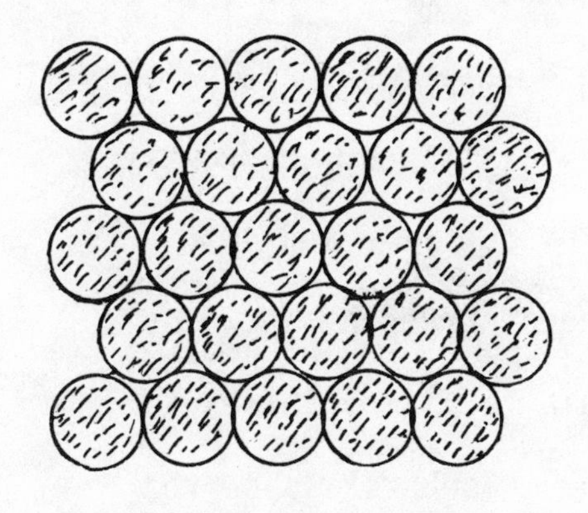

Figure 15.1a

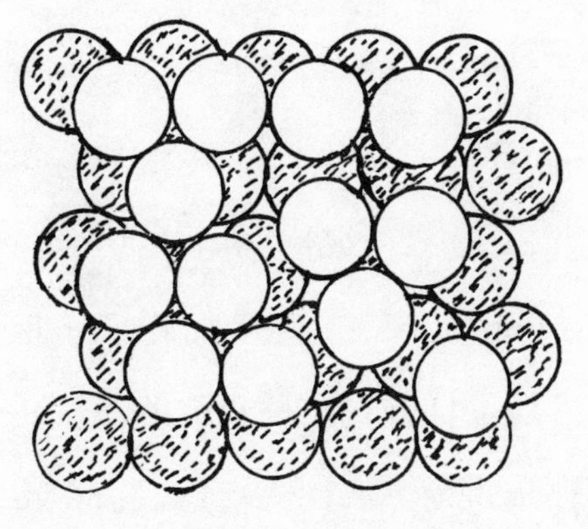

Figure 15.1b

We obtain decreased porosity by mixing together grains of different size, some of the little grains fitting in the spaces between the big grains. Porosity also decreases by changing the shape of the grains after they are in place.

Elasticity is a ratio of stress and strain. Stress is deforming force per unit area. The clastic rock model just explained has particles contacting each other only at certain places and much of the surface of a particle is in contact with the fluid which fills the interstices or pore space between particles. With no overburden pressure bearing on the matrix structure, the particles are in essentially point contact. The application of overburden pressure tends to change the point contact to a large area of contact. Because of this change in the area of contact with force, the stress will not be proportional to the force and we may expect the effective elasticity to change.

When the pressure is released, the rock particles may not regain their original shape so that they may retain a <u>memory of the maximum pressure</u> to which they have been subjected. This memory of the maximum pressure to which rock particles have been subjected was put to practical use by Jankowsky in studies in North Germany on the variation of velocity with depth of burial. Figure 15.2 shows some of his data. Curve A on the left, is a best fit velocity-depth relationship for shale samples and Curve B is the same relationship for limestones. Jankowsky reasoned that rocks which were composed of 60% lime and 40% shale would be 3/5 of the way between these curves as shown by Curve C. This velocity-depth relationship for limey shales is simply based on the time-average equation. Jankowsky observed

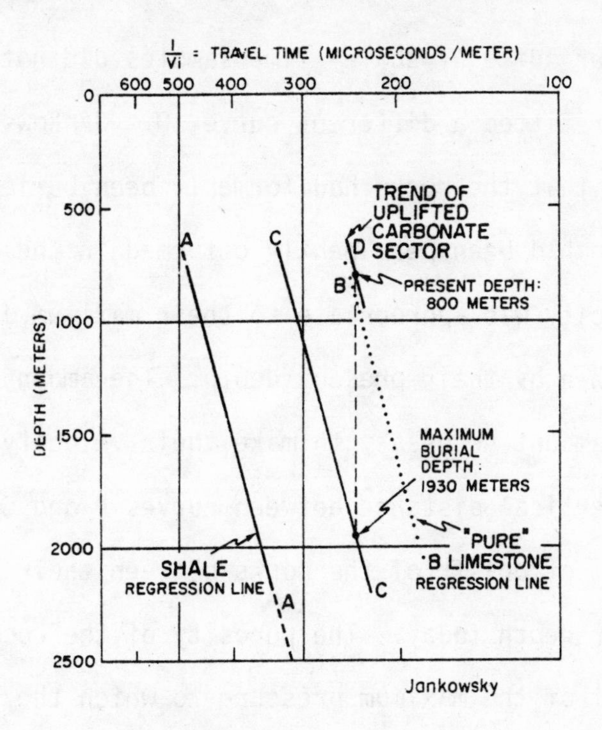

Figure 15.2

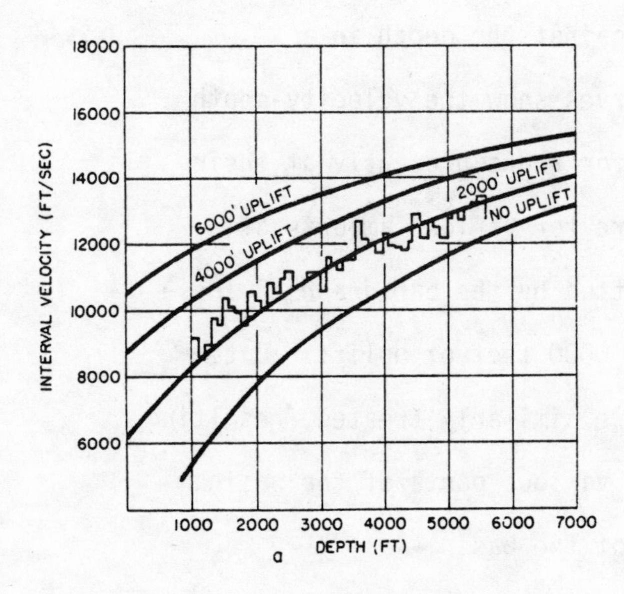

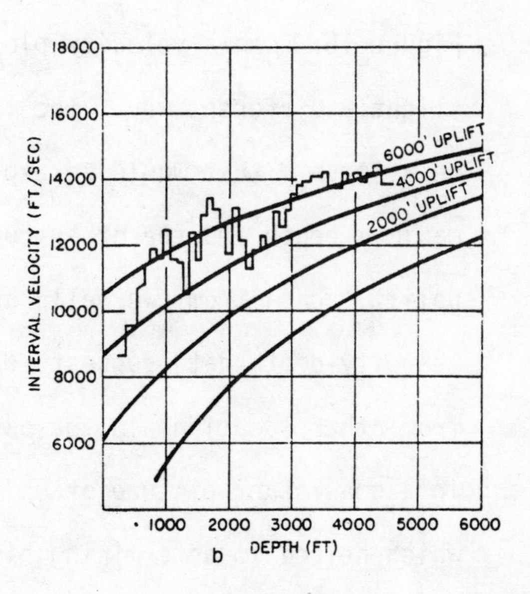

Western Australian
Petroleum Pty.
Limited, Perth,
West Australia

Western Australian
Petroleum Pty.
Limited, Perth,
West Australia

Figure 15.3

that a particular suite of shaley lime samples did not fit this curve but rather fitted a different curve, D. Jankowsky's explanation was that the rocks had formerly been buried deeper in the earth and had been permanently deformed in the process so that their velocity was appropriate to their maximum depth of burial rather than by their present depth. The amount of this uplift was the amount necessary to make their velocity fit curve C. The vertical distance between curves D and C thus gives the amount of uplift of the rocks between their maximum burial and their depth today. The porosity of the rocks is a remnant effect of the maximum pressure to which the rocks have been subjected, since velocity is dominantly governed by porosity, and porosity gives a record of the maximum depth of burial (in the absence of other effects on the porosity, such as cementation, secondary porosity development, etc.).

Similar studies have been carried on in other areas. Figure 15.3 shows velocity plotted against the depth in a slightly different way. The left curves show the velocity-depth relationship that would be expected for rocks presently at their maximum depth and the other curves are for various amounts of uplift. Data from two wells are plotted by the bar graphs. These velocity-depth data suggest 2000 and 6000 feet of uplift. Data from other locations in the basin were similarly treated, resulting in a consistant picture of uplift in various parts of the basin, which helped in determining history of the basin.

In an area which has been continually subsiding without any significant amounts of uplift, the rocks would now be at their maximum depth of burial. If the velocity depends mainly on the maximum depth of burial rather than on the nature of the rocks, as is the case in many areas like the U.S. Gulf Coast, almost the same velocity function should be appropriate over the entire region, with little regard for the age of the rocks. In such areas, surfaces of equal velocity (isovelocity surfaces) should run essentially horizontally without regard for local structure. This holds true in a general sense. Around salt uplifts, for example, the rocks may have appreciable amounts of dip but this does not necessarily indicate true uplift - perhaps the sediments surrounding the salt uplift have merely been subsiding more rapidly than the salt so that the uplift is only with respect to the surrounding sediments rather than with respect to absolute depth.

This is an important factor for seismic interpretation in such areas where the rocks are more or less at their maximum depth of burial. Relatively simple velocity asuumptions can be used, making it fairly easy to migrate and determine the depths associated with various features. Conversely, in areas that were formerly buried much deeper than today, we must expect velocity complications. In such cases isovelocity surfaces tend to follow the structure, though usually to a lesser amount than the amount of structure.

An area of considerable interest today, is the possible use of shear waves in exploration. This is also sometimes illustrated by Poisson's ratio, which is an elastic measure which depends on this ratio. Figure 15.4 is a graph of P-wave and of S-wave velocity for different rock types. Rock types differ in their behavior to compressional and shear waves and also as the fluid nature changes (S-wave velocity is relatively insensitive to the nature of the interstitial fluid). If

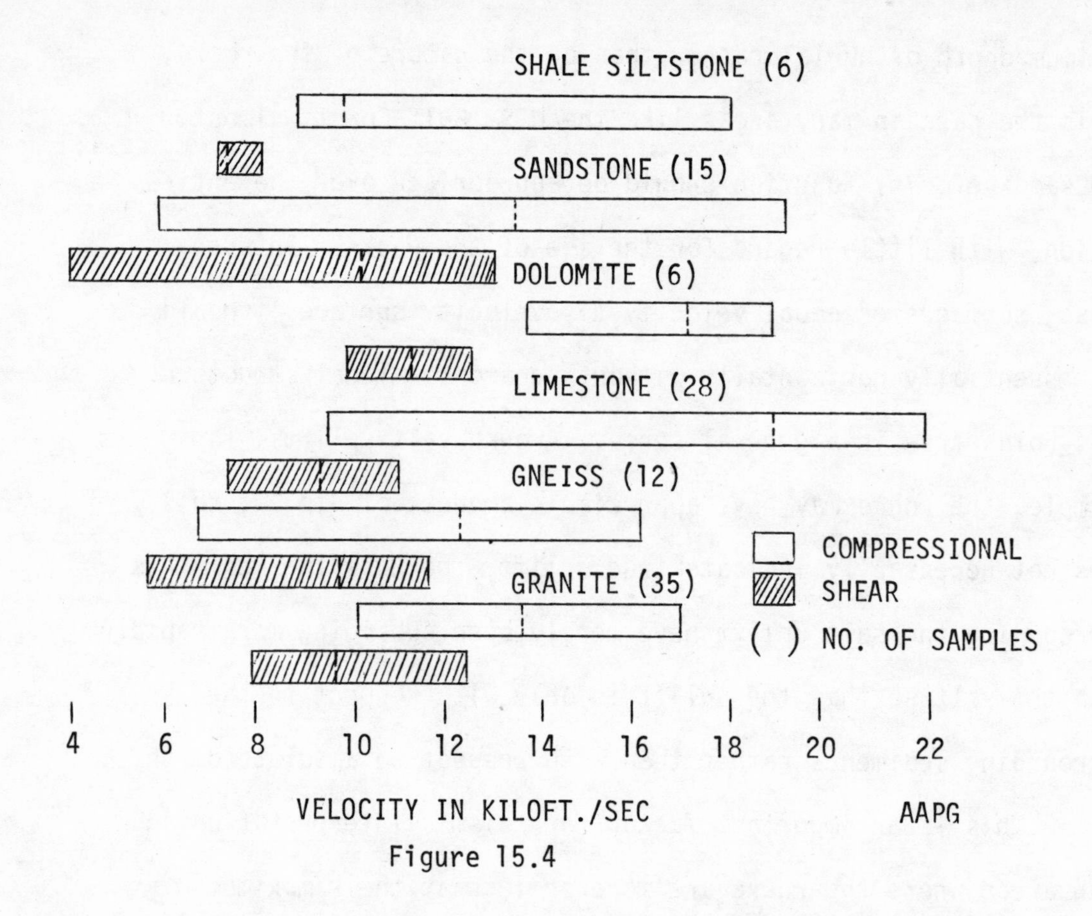

Figure 15.4

ways to measure S-wave velocity can be developed, we shall have another tool to use to tell us something about the nature of the rocks.

Velocity can be determined from seismic measurements of the variation of arrival time as the distance from shot to geophone changes. The measurement of velocity is based on the very simple sketch shown in Fig. 15.5. Let's assume a seismic source and a horizontal reflecting interface with rock of constant velocity V between the interface and the surface. For a geophone located near the shot point, the reflected energy will travel down to the reflecting interface and back to the geophone, so that t_o, the arrival time, will be simply the distance travelled, double the depth because the wave has to go both down and back, divided by the velocity. A geophone that is the distance X away from the

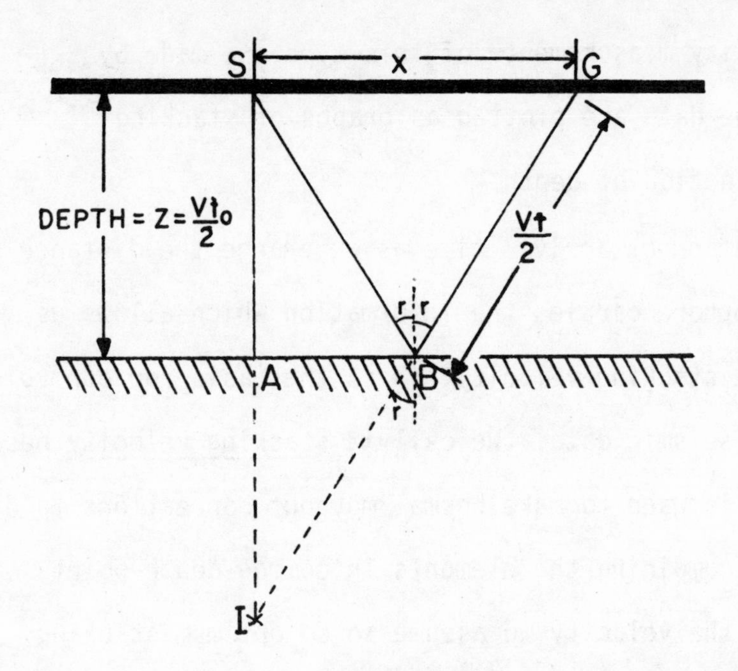

Figure 15.5

shotpoint, will detect reflected energy at a later arrival time, t, because the energy has traveled farther. This greater distance will be vt. The solution may be facilitated by flipping the right triangle SAB so that it lies as triangle IAB. Such a procedure in effect replaces the shotpoint with an image point. The Pythagorean theorem for the hypotenuse of a right triangle then gives the relationship

$$(2z)^2 = (t_0 v)^2 = (tv)^2 - x^2$$

$$v^2 = x^2/(t^2 - t_0^2)$$

$$= x^2/(t + t_0)(t - t_0)$$

$$= x^2/2\bar{t} \quad \Delta t$$

where $\Delta t = t - t_0$ = normal moveout,

$\bar{t} = (t + t_0)/2$ = average arrival time

Allowance can be made for the dip of the reflector or for velocity gradient by variations of the basic calculation.

Ordinarily velocity measurements of this type are made by computers and the data are plotted as graphs of stacking velocity as a function of depth.

The variation of arrival time as we change the distance from shot to geophone carries the information which allows us to determine the stacking velocity; it is the basis for our velocity measurements of seismic data. We call it stacking velocity because this same model is used to make normal moveout corrections to data elements before combining the elements in common-depth-point stacking; it is the velocity we assume to do optimum stacking. It is also sometimes called RMS velocity because it is a root-mean-square average of velocities in a model of parallel horizontal layers.

Velocity analyses are made by determining how much stacked energy (or some other criterion) would result from assuming a stacking velocity. This is done for a number of trials of stacking velocity over a series of narrow windows of data and the result is displayed in the form of the amount of stacked energy as a function of velocity and arrival time. Various amounts of smoothing are also involved. Figure 15.6 is a velocity analysis. presented as contours of semblance; semblance is simply the ratio of the energy resulting from a stacking velocity assumption compared to the amount of energy available to be stacked. Semblance is a coherence criterion which measures equally for weak and strong reflectors whereas merely displaying total energy favors strong reflectors. The contours in effect ring the values of stacking velocity as a function of arrival time which would result in bringing out events on a stacked section.

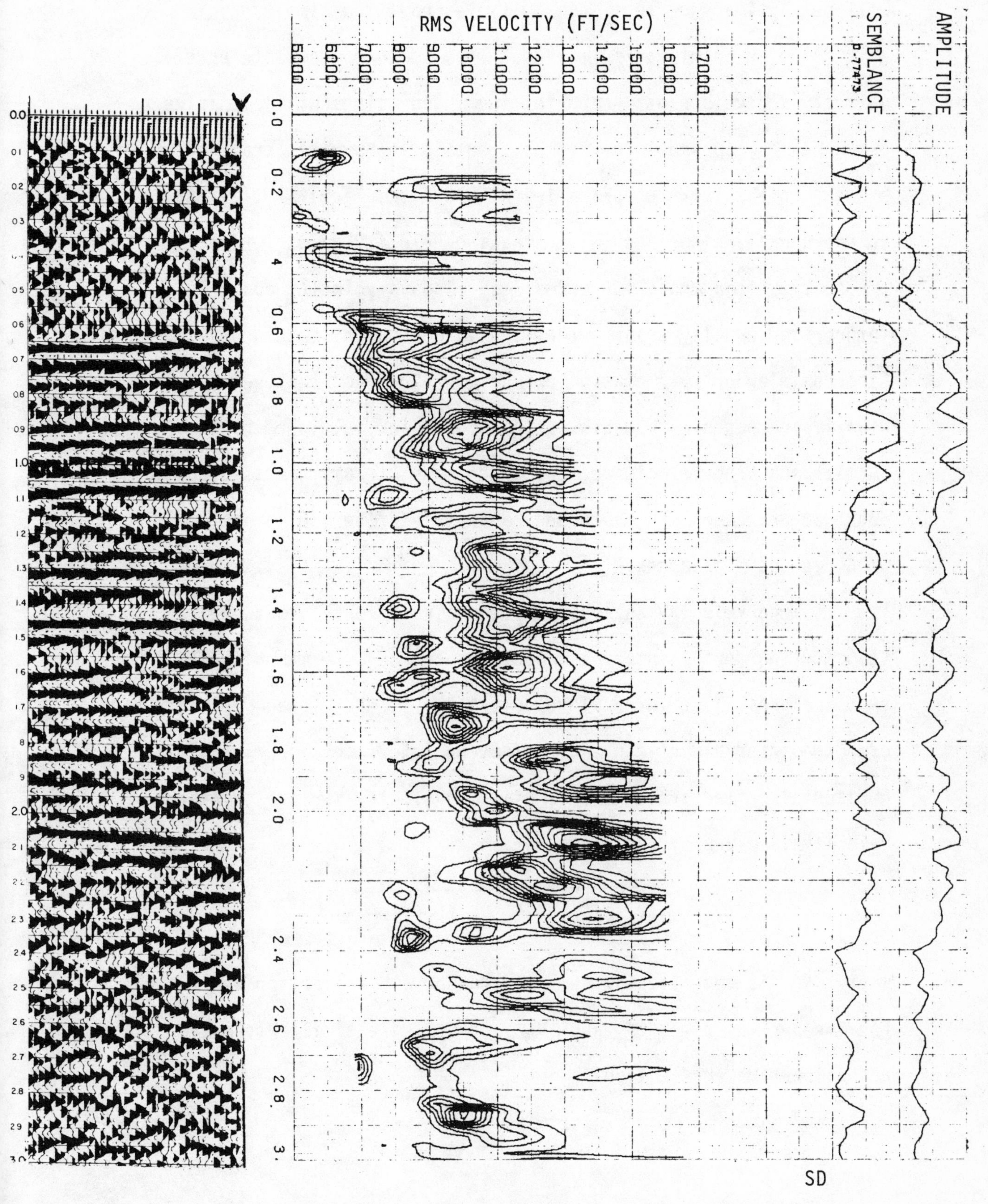
RMS VELOCITY (FT/SEC)
5000
6000
7000
8000
9000
10000
11000
12000
13000
14000
15000
16000
17000
SEMBLANCE
AMPLITUDE
SD

Figure 15.6

A large value of coherence merely means that an event will be brought out by the appropriate assumption, not that the event will be a primary reflection. Multiples have the same type of $X^2 - T^2$ relationship as primaries. Thus someone must interpret velocity analysis plots and determine which values will emphasize primaries as opposed to multiples. Multiples have not traveled as deep in the earth as primaries with the same arrival time and so are usually associated with stacking velocity values which are smaller than those for primaries. Thus a velocity analysis interpreter usually assumes that the larger values should be honored in order to emphasize primaries. The fact that velocity usually increases with depth is the reason for multiples having smaller stacking velocities, but under some circumstances velocity does not increase with depth and hence this criterion may be wrong in such instances.

Since velocity analysis implies a relationship between velocity and depth, interval velocities can be determined from such analyses. The interval velocity V_i is the average velocity over the interval between two reflecting interfaces. For parallel horizontal reflectors and horizontal isovelocity surfaces, interval velocity is given by the Dix equation.

$$V_i = \sqrt{(V_L{}^2 t_L - V_U{}^2 t_U) / (t_L - t_U)} \;,$$

where V_L is the stacking velocity to the Lth reflection which has the arrival time t_L, and V_U and t_U are similar terms for a shallower U^{th} reflection.

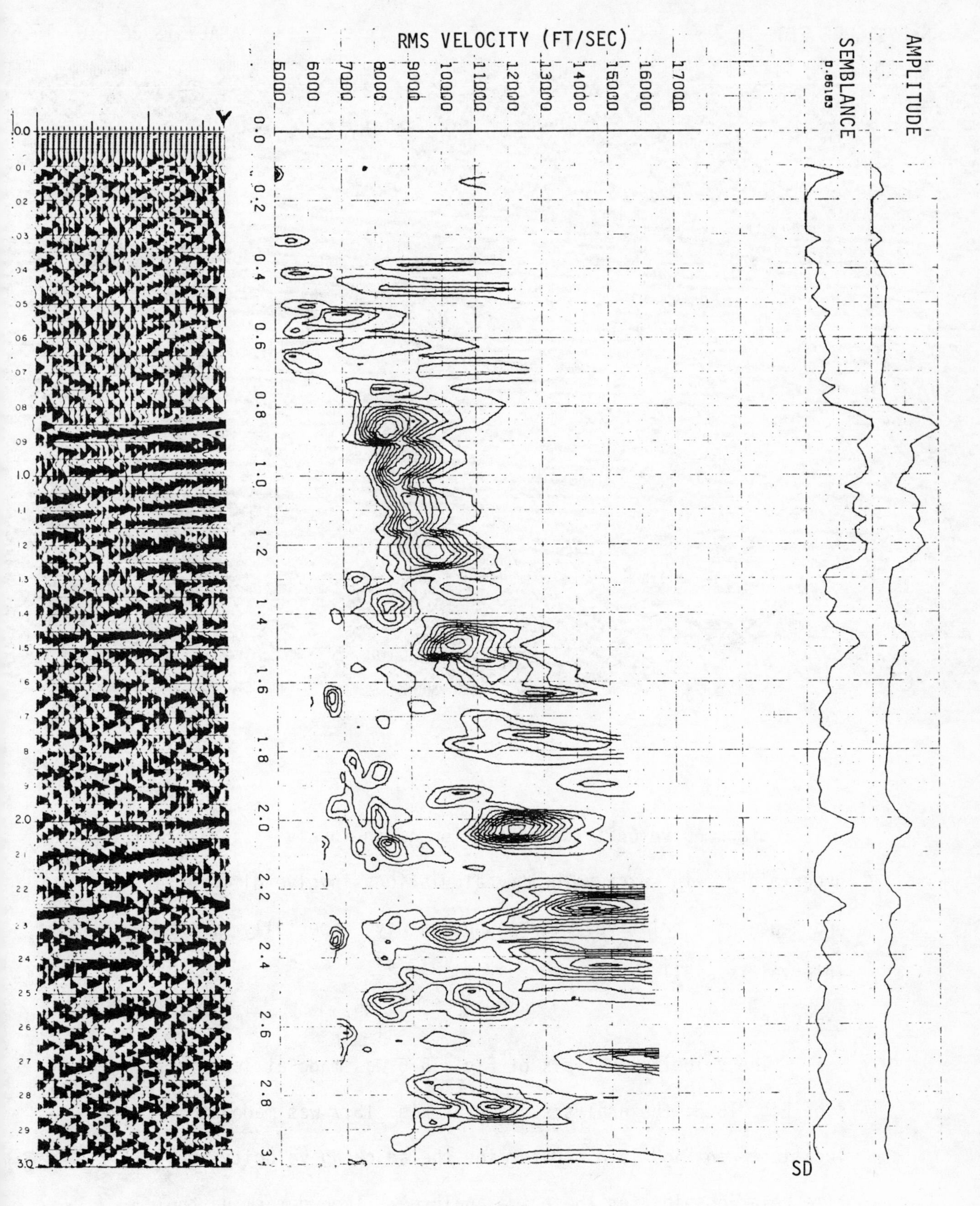

RMS VELOCITY (FT/SEC)
5000
6000
7000
8000
9000
10000
11000
12000
13000
14000
15000
16000
17000
SEMBLANCE
AMPLITUDE
SD

Figure 15.7

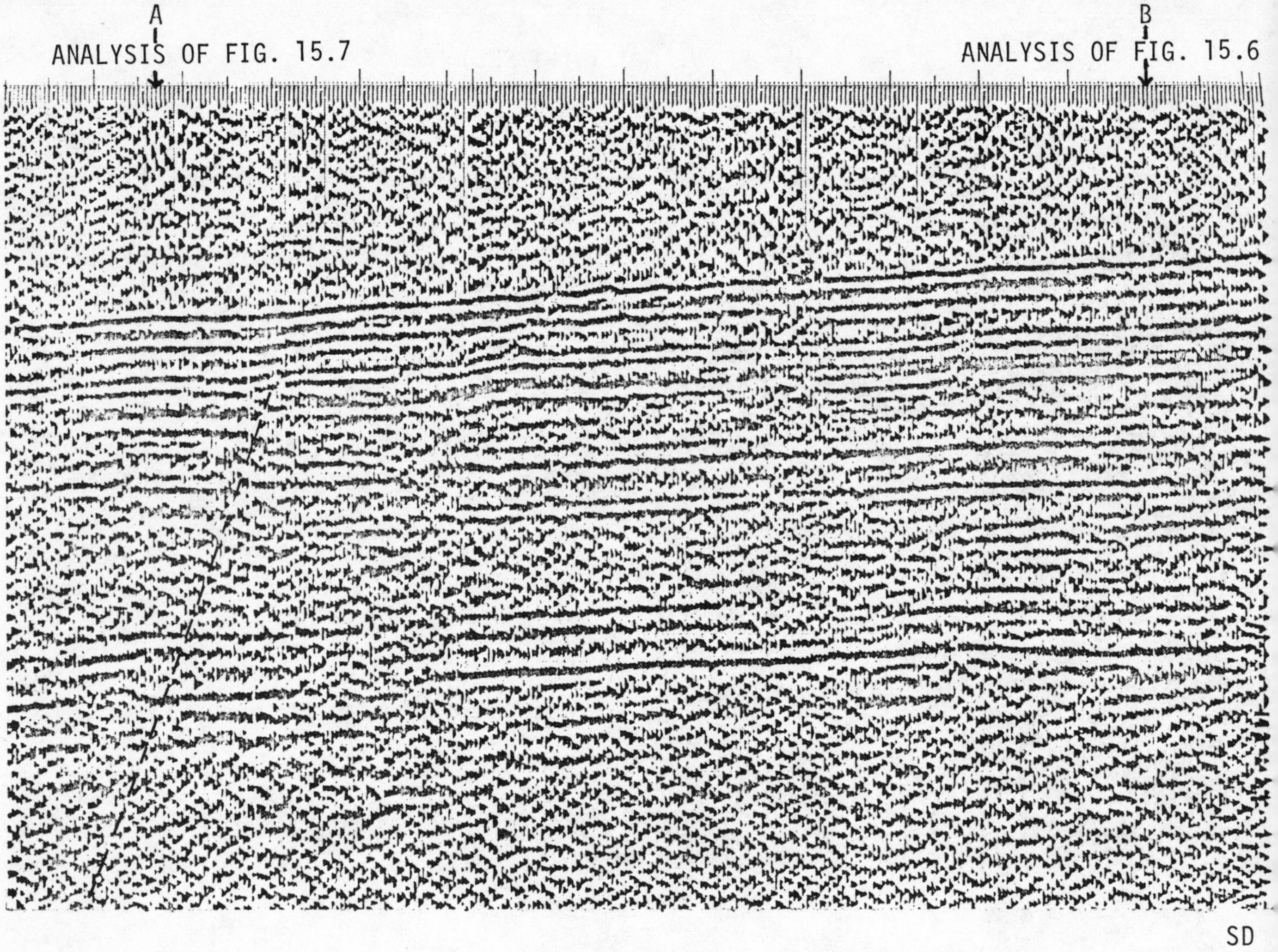

A

B

ANALYSIS OF FIG. 15.7

ANALYSIS OF FIG. 15.6

SD

Figure 15.8

Stacking velocity determination often involves significant uncertainty. Interval velocity calculations involve differences and hence have relatively large uncertainty, especially if the interval is small.

Exercise:

The velocity analysis of Fig. 15.6 was made at location B on Fig. 15.8; the analysis shown in Fig. 15.7 was made at location A on Fig. 15.8. Determine the stacking velocity-arrival time relationships for these two analyses. Then determine the interval velocity for two intervals. What lithologic implications result from these determinations?

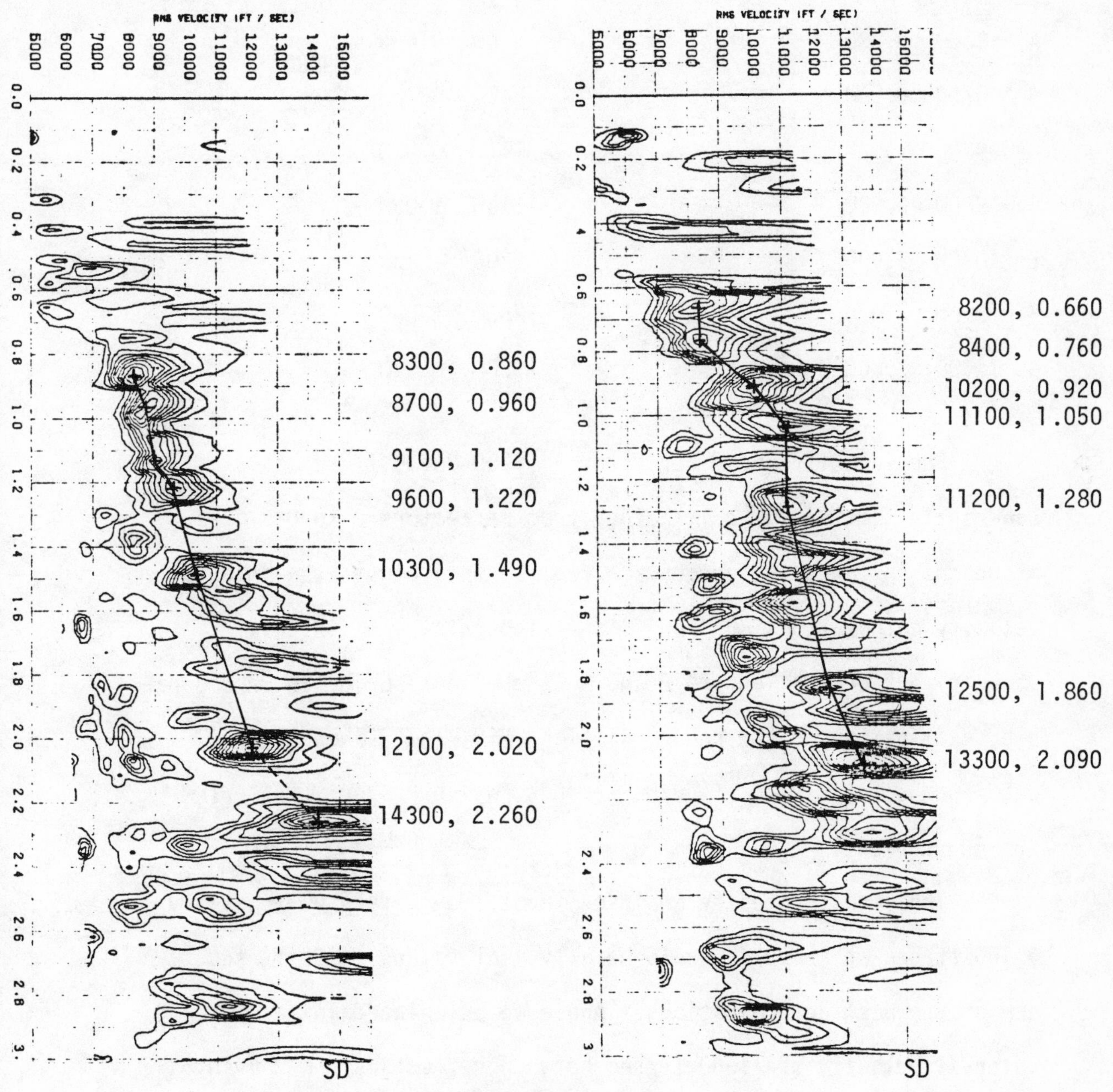

Figure 15.9

Discussion of exercise:

Figure 15.9 shows an interpretation of the velocity analyses in Fig. 15.6 and 15.7. Values of velocity and arrival time corresponding to these picks are tabulated below along with interval velocities calculated by the Dix equation.

	LEFT SURVEY			RIGHT SURVEY		
	V_{stack}	t	$V_{int.}$	V_{stack}	t	$V_{int.}$
A	8300	0.860		8200	0.660	
			11600			9600
B	8700	0.960		8400	0.760	
			11200			16200
C	9100	1.120		10200	0.920	
			14000			16100
D	9600	1.220		11100	1.050	
			13000			11600
E	10300	1.490		11200	1.280	
			14800			15000
F	12100	2.020		12500	1.860	
			26300			17200
G	14300	2.260		13300	2.090	

The velocity above reflector A (8200-8300)ft/sec) is a sand-shale section. The lithology from reflectors B to D increases in carbonate percentage from left to right; the interval velocity calculations are consistant with this. The interval velocity between reflectors F and G in the left survey is unreasonable; examination of the record section (Fig. 15.8) shows a fault cutting the reflection at 2.260 seconds, which probably distorted the measurement.

Individual <u>velocity</u> analyses involve <u>uncertainty</u> - possibly ± 400 ft/sec or about 4% would be a typical value, ± 200 ft/sec one of the best determinations. Above we calculated an interval velocity of 11600 ft/sec between reflectors A and B in the left survey. If the measurement of A were 200 ft/sec too large and that of B 200 ft/sec too low, an interval velocity of 1400 ft/sec would be calculated. Thus 2% error in stacking velocity can produce 20% error in interval velocity.

CHAPTER 16

SEISMIC DATA PROCESSING

The most common objective of data processing is to increase the signal-to-noise ratio. By "signal" we mean the type of data elements we wish to see and by "noise", everything else. In most seismic interpretation, the signal consists of primary reflections, seismic waves which have been reflected only once by bedding underneath the seismic line. We also wish to preserve diffractions because they are essential to locating discontinuities, such as where beds terminate. For certain special objectives, we might also consider other types of energy as signal.

The primary problem arises because different types of events can arrive at the geophone at the same time and so confuse each other. As shown in Figure 16.1, surface waves, shallow refractions, multiples,

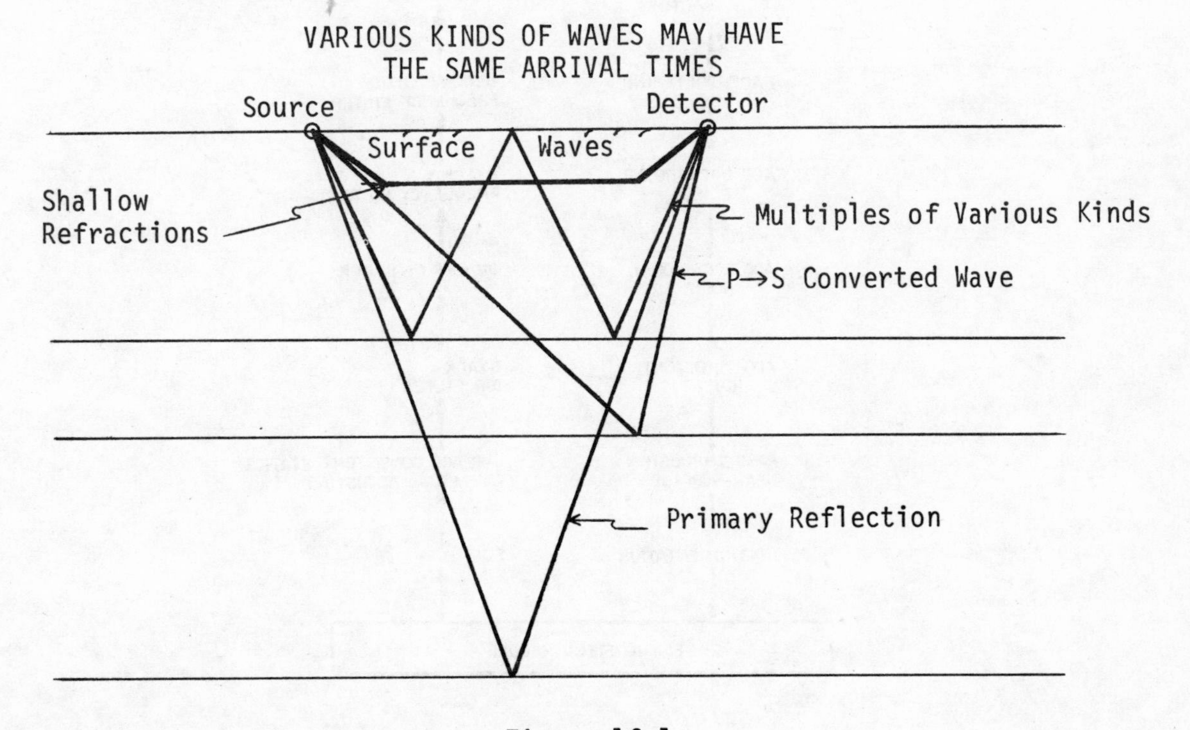

Figure 16.1

converted waves from different horizons, or other types of waves may all arrive at the same time as a primary reflection and be superimposed on the seismic record. The problem, then, is to separate the primary events related to the reflector which we wish to map. To effect this separation, we need a discriminant, i.e., a respect in which the signal and noise elements differ, such as frequency, amplitude, normal moveout, or some other regard.

A number of things can happen to a seismic wave in its passage from generation at the source to recording. Figure 16.2 lists on the left some things which might change the seismic waveshape.

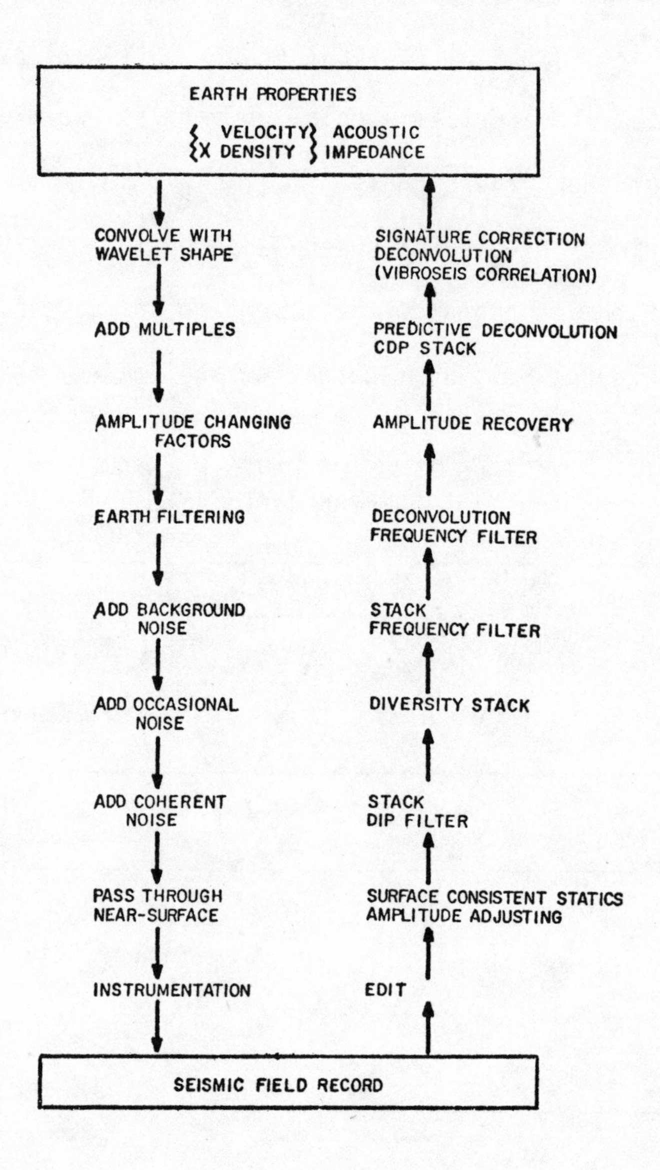

Changing a waveshape is called "filtering" or "convolution". The list on the left thus can be thought of as a series of filters, each of which changes the waveshape. The result is that the waveshape which is recorded does not represent only information about the earth properties we wish to determine, but rather a version distorted because of the various filters which have intervened.

One of the objectives of data processing is to undo the effects of these filters, that is, to remove the distortions which they produced. On the right in Fig. 16.2 are listed processes that help undo these distortions and thus minimize the effects of the respective types of noise. The processes are paired with the respective noises which they are effective in combating, rather than being arranged in the sequence in which operations are usually performed in a data processing center. In data processing we analyze the data to determine how the noise differs from the signal and apply a process which utilizes this difference to attenuate the noise.

Another objective of data processing is to extract information additional to the arrival time of events, such as information about the amplitude, frequency content, velocity or other "attributes" of the data. In processing we also reposition each reflecting event appropriate to its subsurface location, a process called "migration". And lastly, processing has the objective of producing a display of the data which is intelligible to an interpreter from which he can infer the geological significance.

The next sequence of figures shows how processing can improve data. Figure 16.3 shows a section characterized by appreciable near-

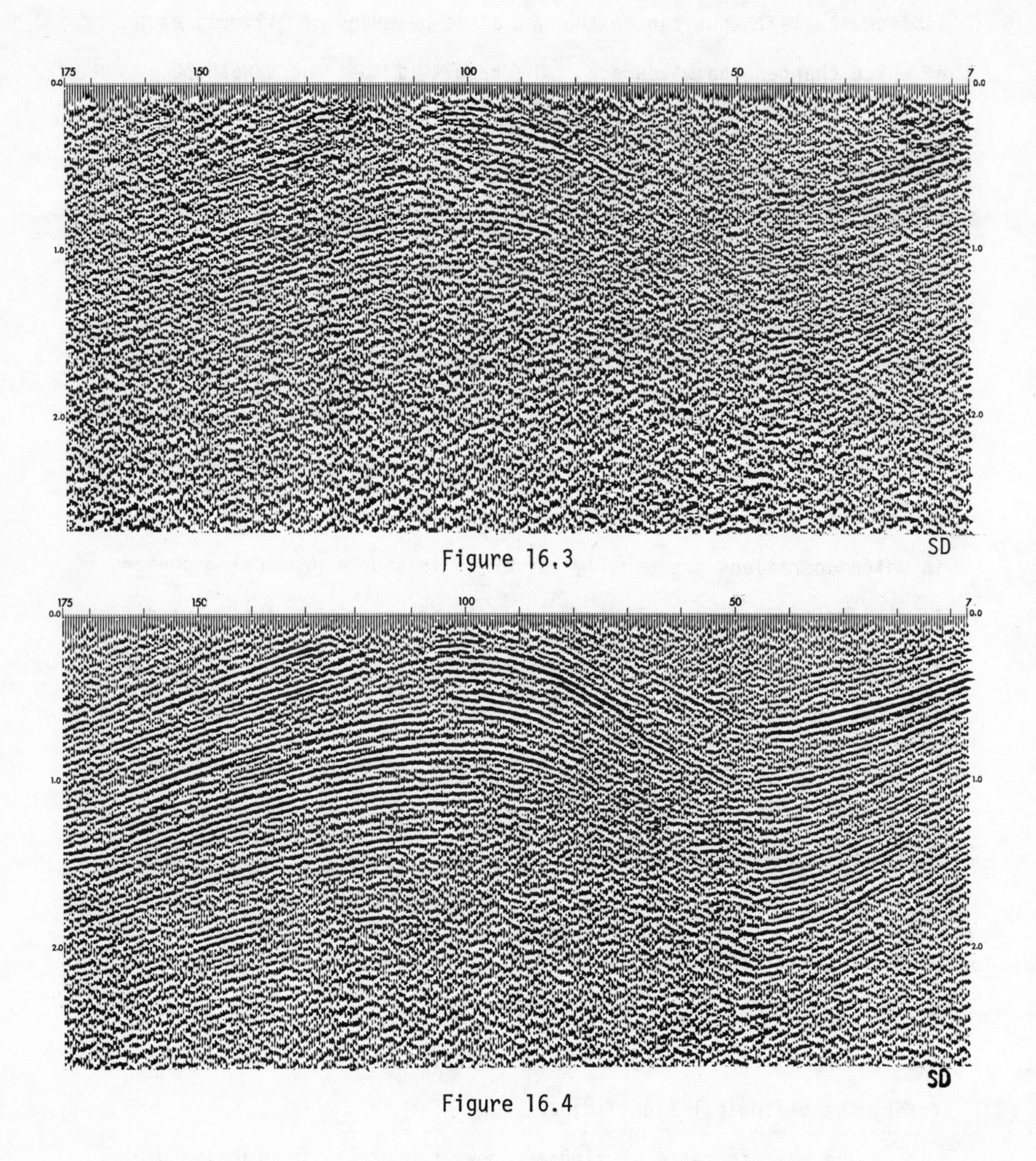

Figure 16.3

Figure 16.4

surface noise, near-surface velocity and/or weathering thickness variations and irregularity in the surface elevation. Variations in the seismic layout because of access or permit difficulties prevented shots being located at regularly spaced points. These data have been reprocessed to correct for these near-surface variations on a surface-consistent basis, resulting in Fig. 16.4. Surface-consistent means associating a filtering effect with a particular portion of the earth and assuming that this filtering effect will affect all waves which pass through this portion. The vicinity of a geophone may have a time delay associated with it because the geophone is at a higher elevation, because the velocity of the earth underneath the geophone is exceptionally low, or because the geophone is poorly coupled to the earth. Likewise a shot may have a delay associated with it because of a slight delay in its firing, variations in the velocity of the earth near it, or variations in the coupling of the shot to the earth. Inasmuch as each geophone is used to record a number of different shot locations and each shot is recorded at a number of different geophones, certain shot and geophone combinations show the same effects and make it possible to determine a consistent set of delays which most nearly account for observations. A statistical procedure for carrying out such an analysis, the "least-squares procedure", minimizes the sum of the squares of the residual errors. The section corrected for the delays determined by surface-consistent statics is shown in Figure 16.4. The result is a section in which we can have more confidence. There is still noise on the section, so the problems are not completely solved, and further processing might achieve an even better section.

Surface-consistent amplitude adjustment is a process which associates attenuation factors with each geophone, each shot, etc., also on a least-squares basis, so that remaining amplitude variations may be associated more directly with subsurface factors.

Figure 16.5 shows data before and Fig. 16.6 after surface consistent statics correction. Velocity analyses were made as shown by the V at the

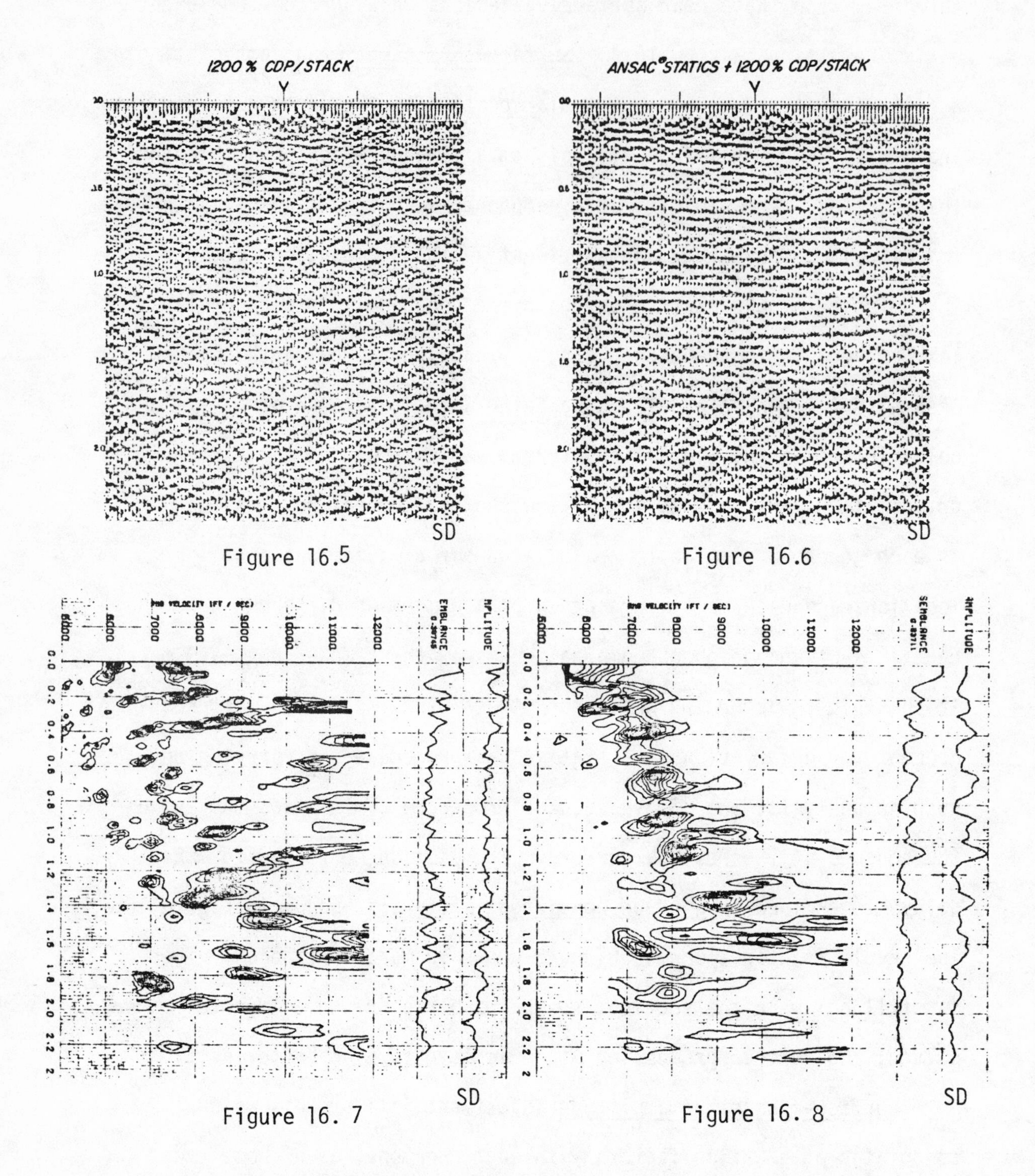

Figure 16.5

Figure 16.6

Figure 16. 7

Figure 16. 8

top. Good static corrections help in achieving a more reliable velocity analysis. Figure 16.7 shows velocity data before statics correction; the values of stacking velocity which will maximize coherence are not clear. Figure 16.8 shows a velocity analysis of the same data after statics correction; the stacking velocity values can now be determined with confidence. We expect interaction of statics corrections and normal-moveout measurements, from which stacking velocities are calculated, because both involve time-shifts. If static time shifts are in error, then the velocity analysis will be misleading. A common procedure is to make an approximate correction for velocity, then calculate statics corrections, use these to achieve a better analysis of the stacking velocity, and finally redetermine the statics corrections using the improved stacking velocity values.

A dramatic example of the effect of stacking velocity on a seismic section is shown in Figs. 16.9 and 16.10. A sharp velocity increase occurs at an angular unconformity at shallow depth; this unconformity is a strong generator of multiples. The stacking velocity used for Fig. 16.9 was that appropriate to the flat multiples, so the steeply dipping primary events were cancalled. The section was restacked using the stacking velocity for the primaries, giving the result shown in Fig. 16.10 and the dipping primary reflections are now clear. Although the effect of stacking velocity is rarely as extreme as in this example, a misleading section can result from a wrong choice of velocity. Stacking velocity must be specified in processing and a knowledge of the local situation is often needed to make this choice intelligently. Sometimes the required background knowledge is not communicated to whomever does the processing and erroneous velocities may be used inadvertently. The end user might not even realize that poor stacking velocity was used.

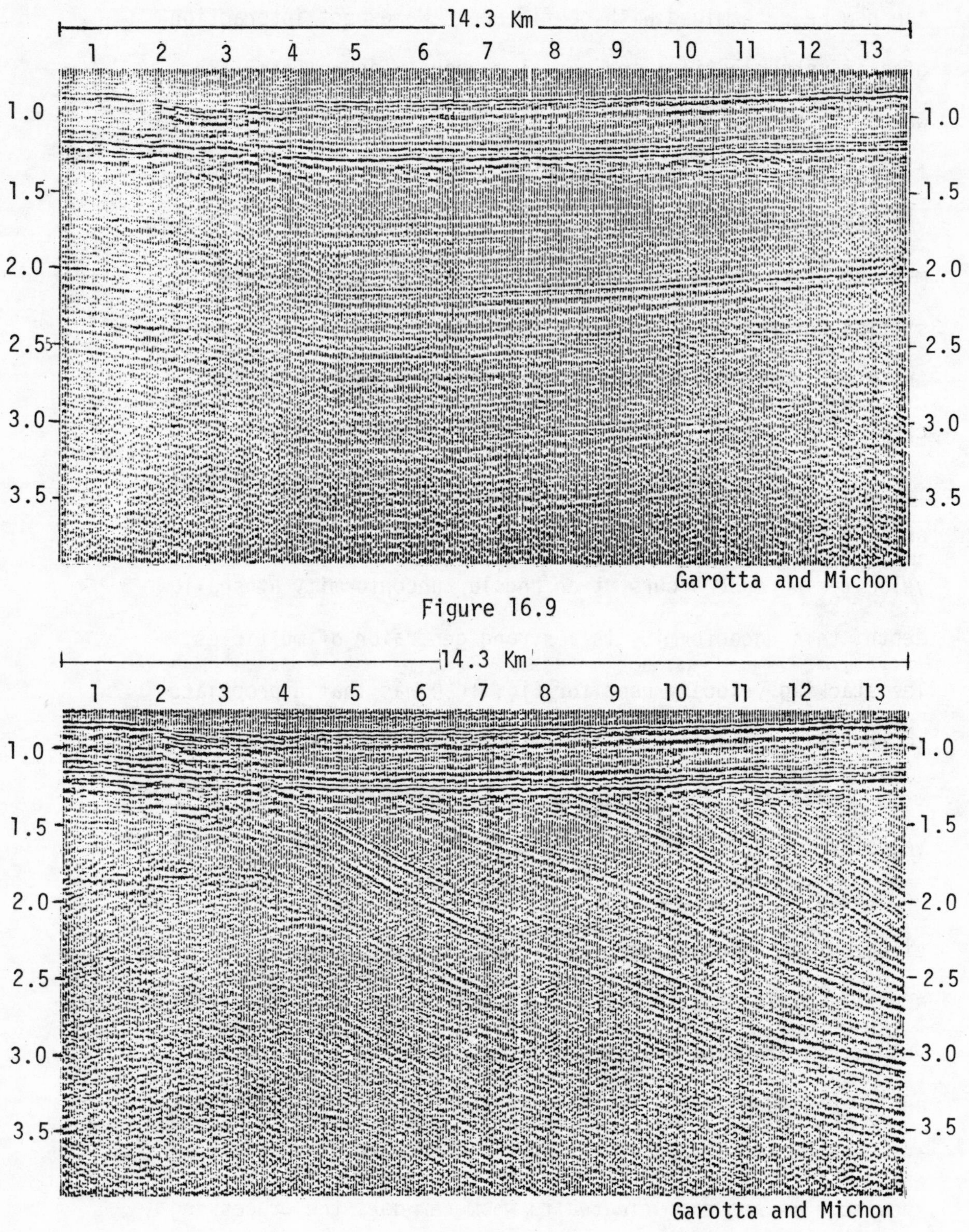

Figure 16.9

Figure 16.10

Figure 16.11 contains strong multiples of the seafloor. An extra bounce in the water layer produces not only multiples of the water bottom but also multiples of other reflections. These data were subjected to a radial multiple attenuation process, a multichannel process which allows for the variations of reflection coefficient with angle of incidence. In the resulting section, Fig. 16.12, the deeper primary data are clearer although there are still residual effects of multiple energy.

One of the earliest applications of data processing was deconvolution for water-layer reverberations. The seismic record of Fig. 16.13a is highly reverberatory due to such energy trapped in the water layer. Whenever new reflection energy arrives, a new train of waves is set up. Figure 16.13b shows the section after deconvolution (filtering) to remove the reverberatory effects. An interpretation is now easier to make.

Wavelet processing to shorten and tailor the seismic wavelet shape is a fairly new deconvolution technique, used especially to improve the high-frequency response of seismic data and hence improve resolution. Figure 16.14 shows land data which were accorded good processing a couple of years ago. Wavelet processing involves finding on a statistical basis a wave form for all the data recorded from a single shot, and then filtering so as to change this wave shape into one which is the same for all of the shots. Thus the high frequency components can be stacked together more nearly in phase so that they will not be cancelled. Since we measure the earth in terms of seismic wavelength, retaining the high frequency (short wavelength) content of the

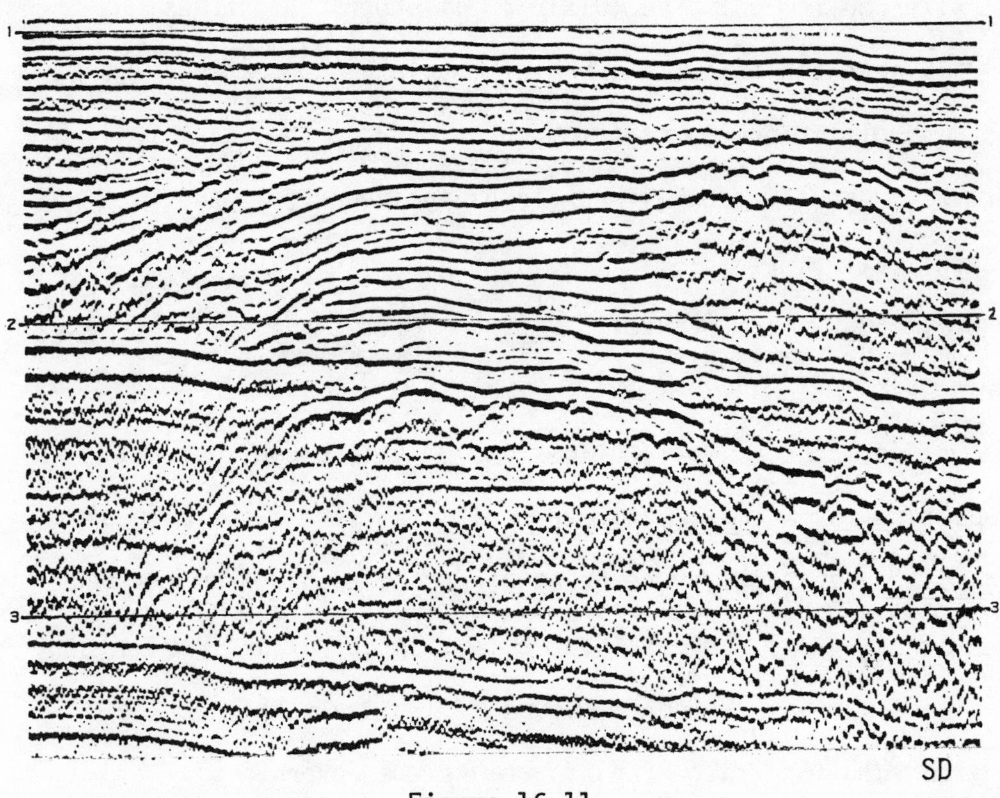

Figure 16.11

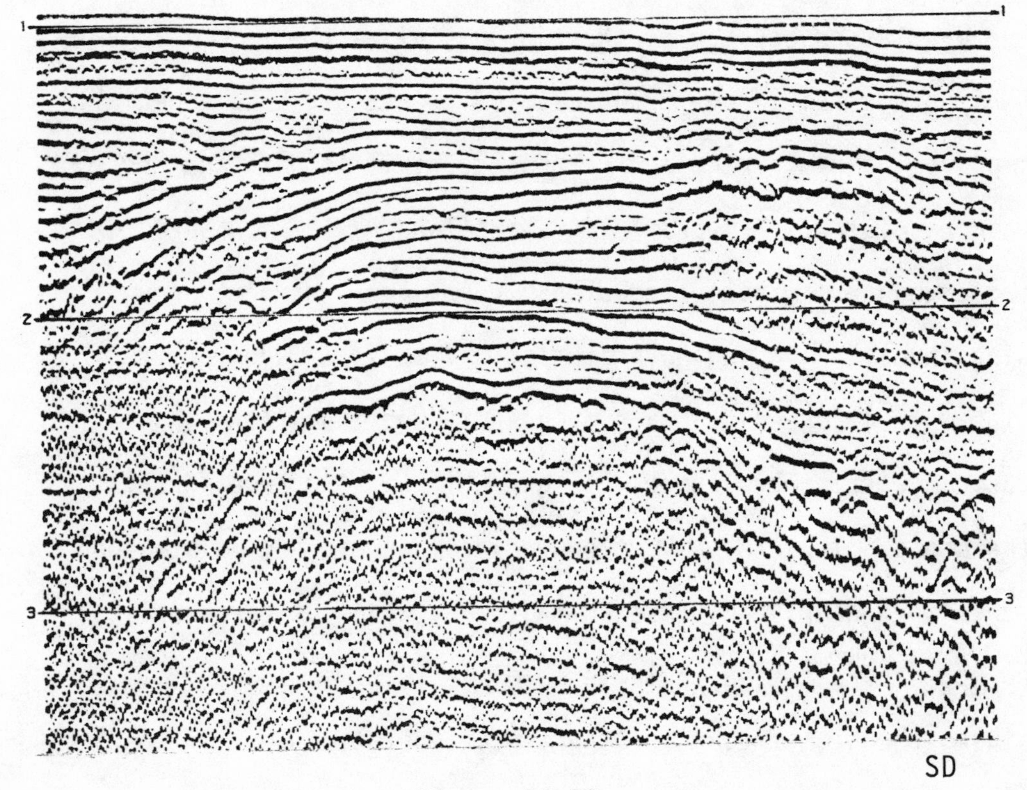

Figure 16.12

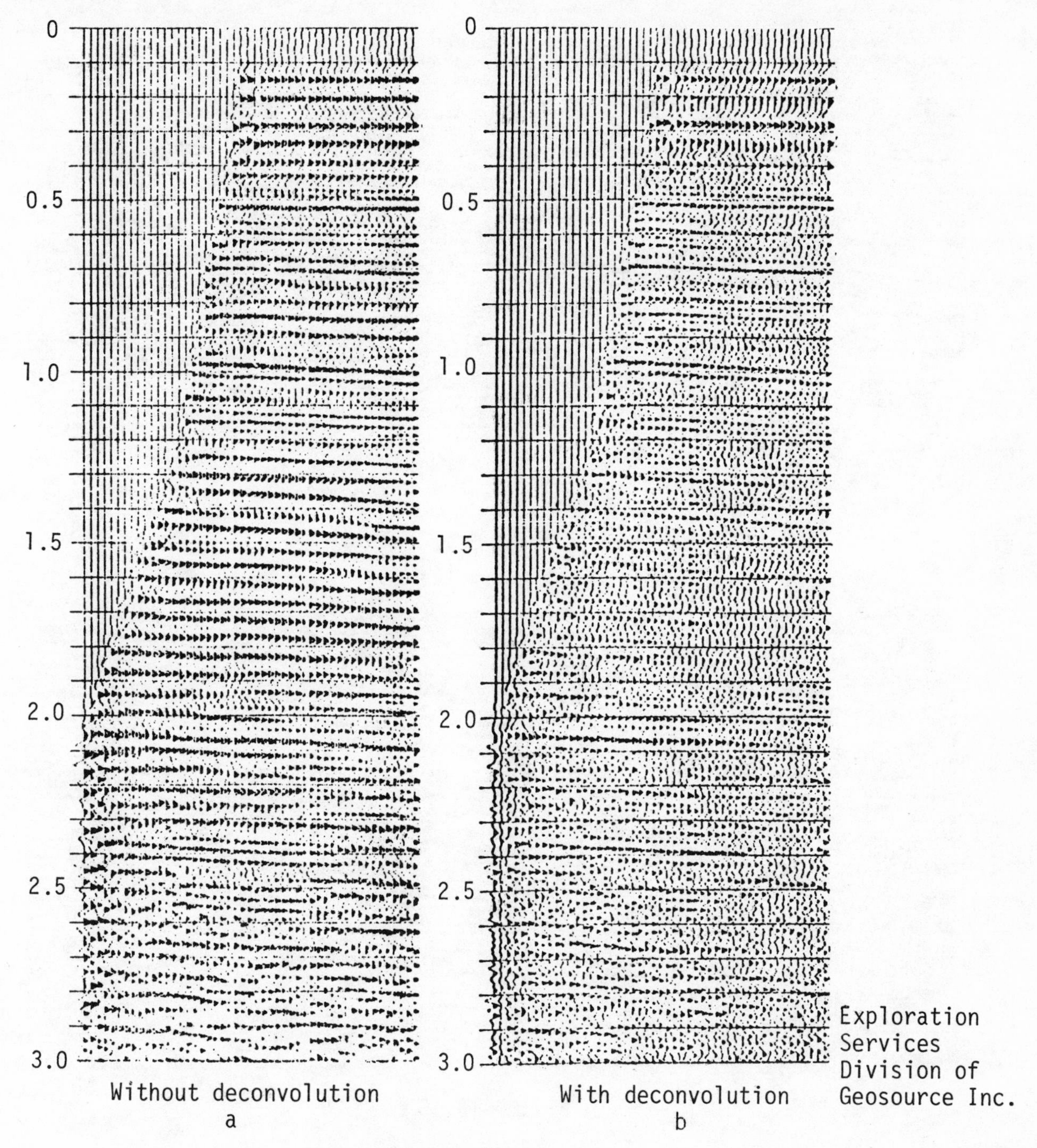

Figure 16.13

data allows us to see features of smaller magnitude. These data after reprocessing to achieve the same equivalent wavelet shape are shown in Fig. 16.15. The frequency content has been broadened, events sharpened and higher resolution achieved. The dipping events can now be followed more closely to their pinchouts.

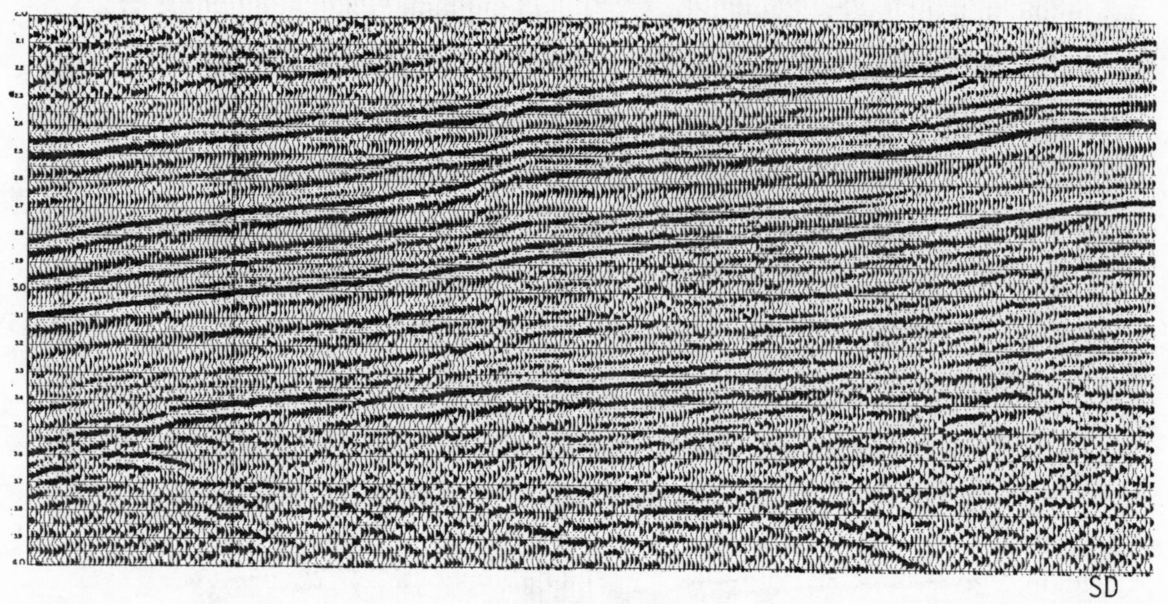

Figure 16.14

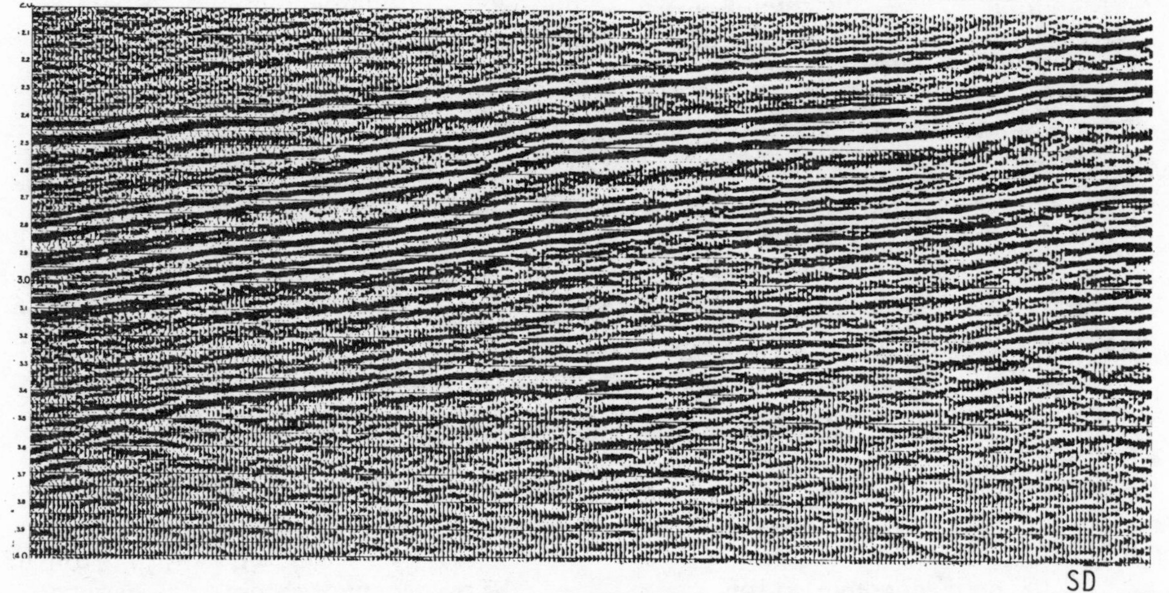

Figure 16.15

Another version of wavelet processing can be used when the wave-shape is known. The waveform of each shot was recorded for the marine data of Fig.16.16, so we know it varied slightly from shot to shot. The data from each shot were filtered to change each shot's waveform into the same effective waveshape. The data were then stacked, giving the result in Fig. 16.17 which shows greater continuity and more stratigraphic detail.

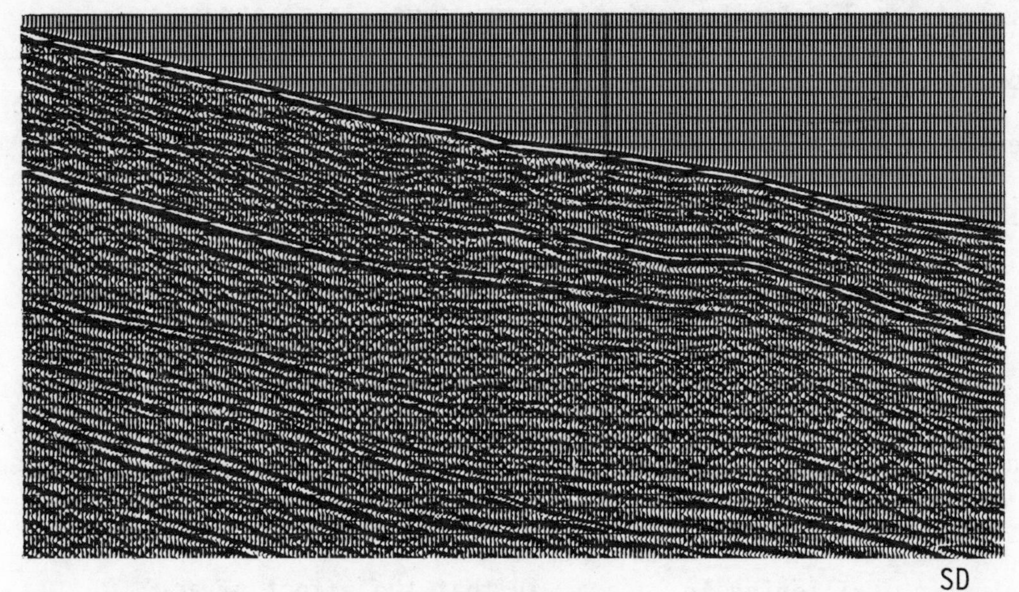

Figure 16.16

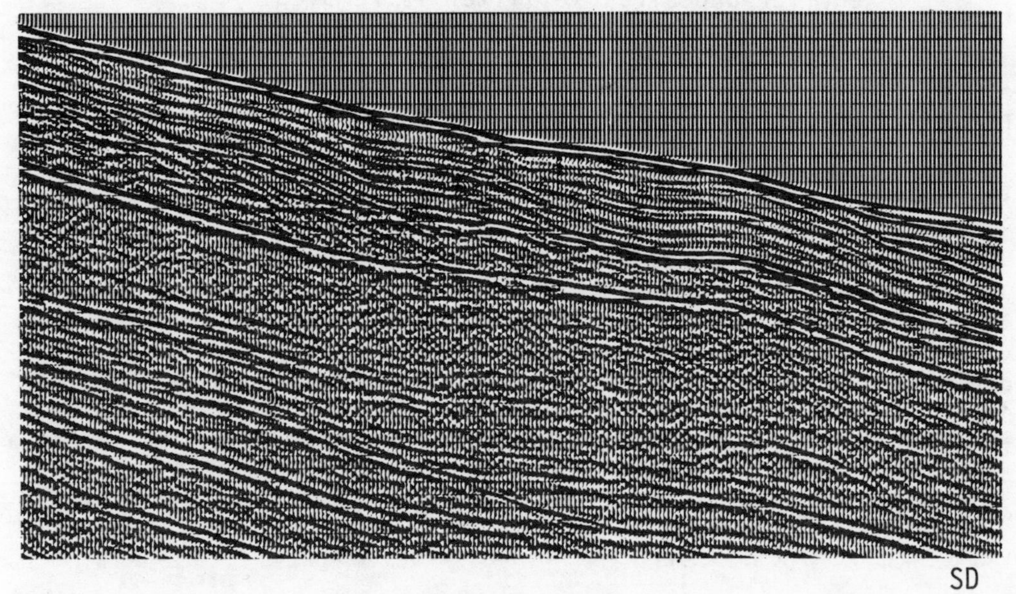

Figure 16.17

It is important that processing be done correctly using <u>optimum choice of parameters</u>. Errors created in processing are sometimes but not always evident on the section used for interpretation. Among processing errors sometimes encountered are misties or labeling errors. Misidentification prior to stack can result in the wrong elements being stacked together. Improper mixing may smear detail. Statics errors may show up as vertical features. Wrong normal moveout may emphasize multiples or other noise. AGC may build up noise. Lateral velocity variations may cause false structure. The wrong design gate may result in emphasizing the wrong features.

We now define a few terms:

<u>operation</u> - doing something to a set of input data to result in a set of output data.

<u>linear system</u> - a system whose effect does not depend on the magnitude of events or the time.

<u>convolution</u> - linear filtering which results in change in wave shape.

<u>deconvolution</u> - filtering in such a way that the effect of an earlier undesireable filter is removed.

Let us look at some ways of manipulating numbers to accomplish the equivalent of some physical processes in the earth or filtering in data processing. We use symbols to indicate entire sets of numbers; for example, a_t might stand for a set of three numbers which occur at successive time intervals. The individual numbers might be $a_0=1$, $a_1=2$, $a_2=-1$. We understand that values are zero unless otherwise specified; thus $a_3=0$, $a_{-1}=0$, etc. We can also show a_t as a bar graph:

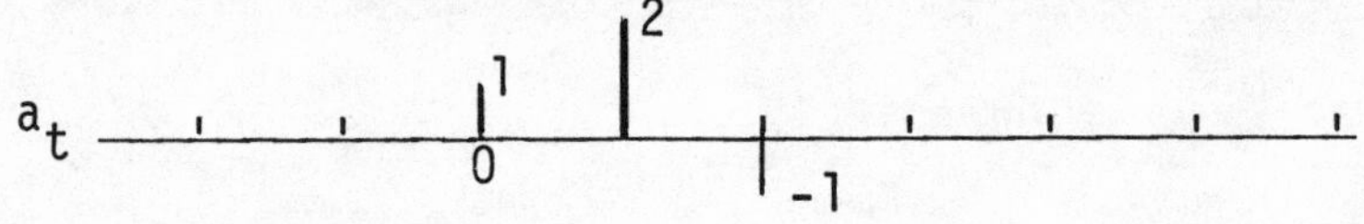

We could have another set, b_t where $b_0=2$, $b_1=1$, $b_2=1$, $b_3=-1$. It's graph would be:

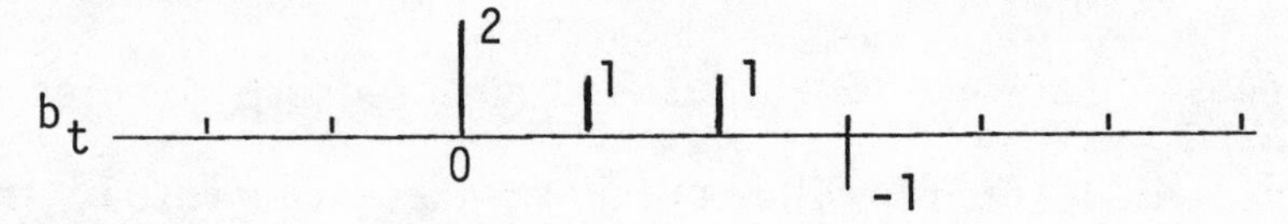

We can add sets, by which we mean adding corresponding members of the sets. Thus if $c_t = a_t + b_t$, $c_0 = a_0 + b_{0_2} = 3$, $c_1 = a_1 + b_{1_2} = 3$, $c_2 = a_2 + b_2 = 0$, $c_3 = a_3 + b_3 = -1$, etc.

We can shift a wavelet to the right by subtracting from the index value. Thus $d_t = a_{t-1}$ gives $d_0 = a_{-1} = 0$, $d_1 = a_0 = 1$, $d_2 = a_1 = 2$, $d_3 = a_2 = -1$.

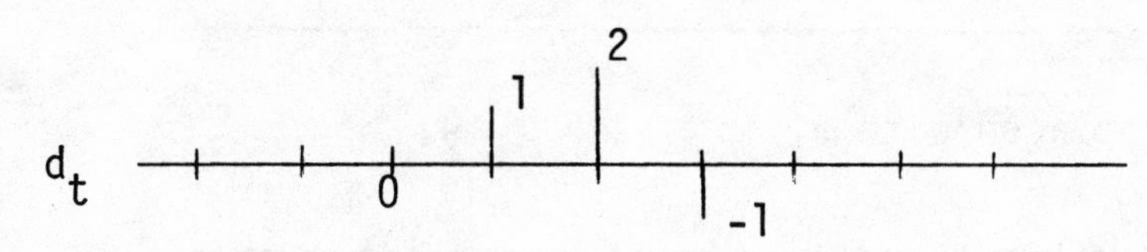

Or $e_t = a_{t+2}$ is a shift of 2 to the left:

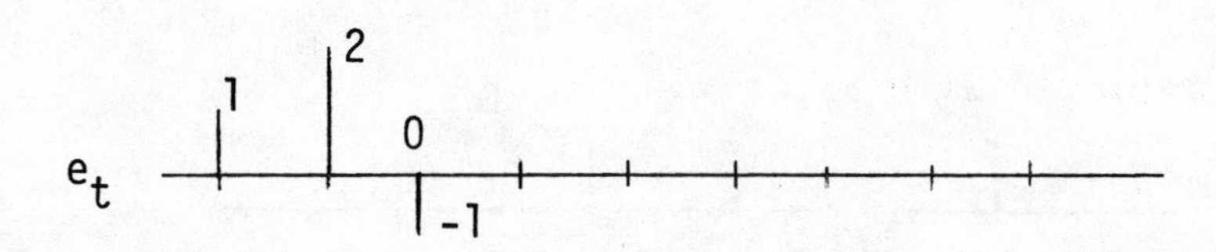

We can simply reverse a trace: $f_t = a_{-t}$, $f_{-2} = a_2 = -1$, $f_{-1} = a_1 = 2$, $f_0 = a_0 = 1$, $f_1 = a_{-1} = 0$, etc.

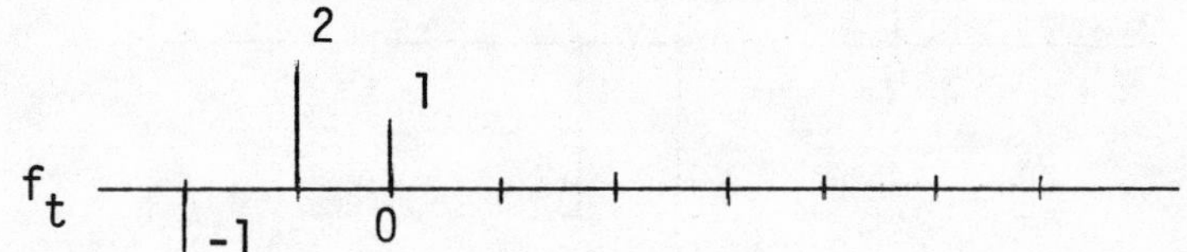

<u>Convolution</u> is the process of replacing each element of one set with a scaled version of another set. Let us call a filter g_t where $g_0 = 3$, $g_1 = 0$, $g_1 = -2$, $g_2 = 1$.

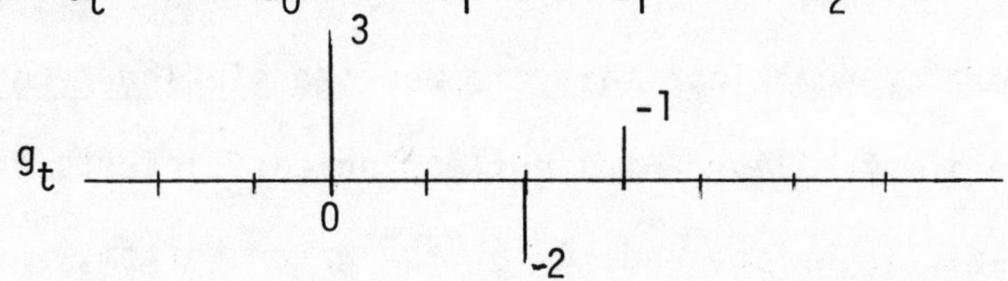

Let us convolve g_t and a_t; we use an asterisk to symbolize convolution. The result is $h_t = g_t * a_t$:

g_o replaced with $3a_t$:

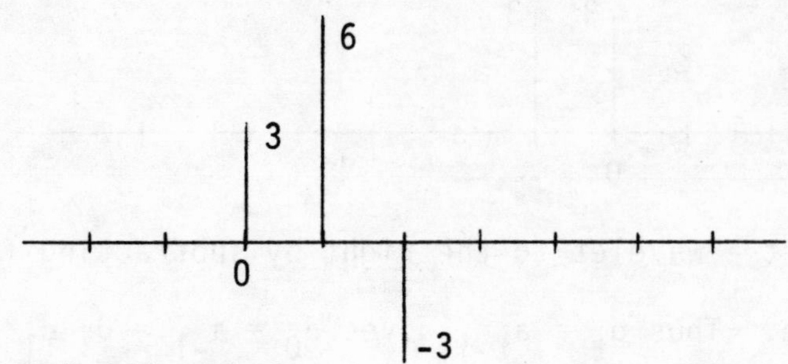

g_1 replaced with 0:

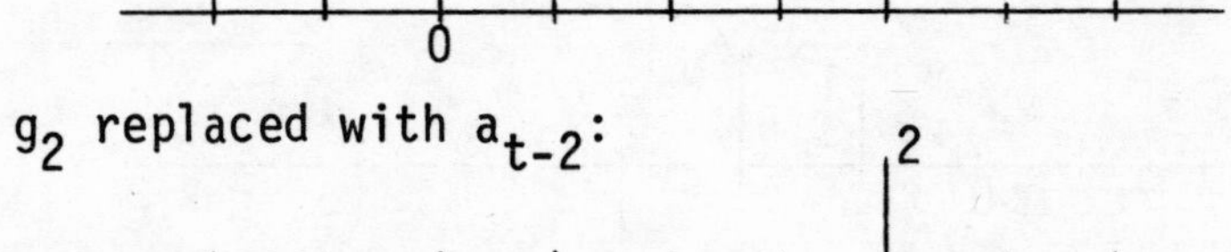

g_2 replaced with a_{t-2}:

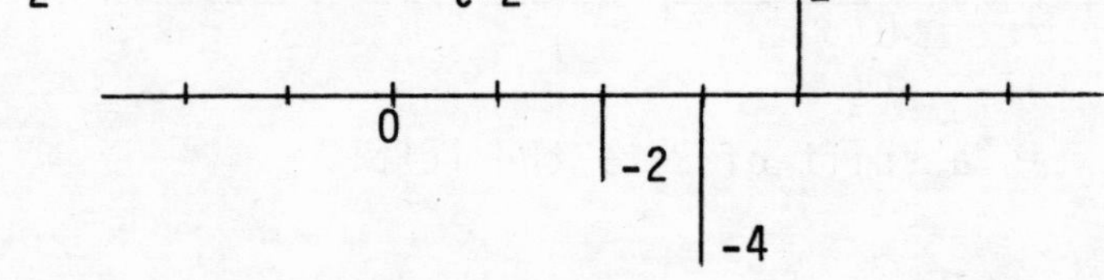

g_3 replaced with a_{t-3}:

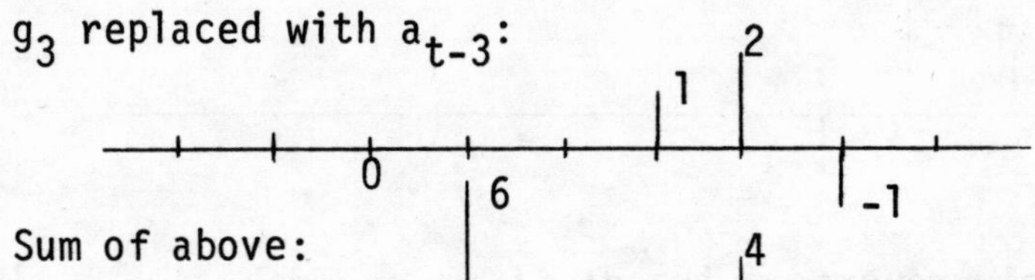

Sum of above:

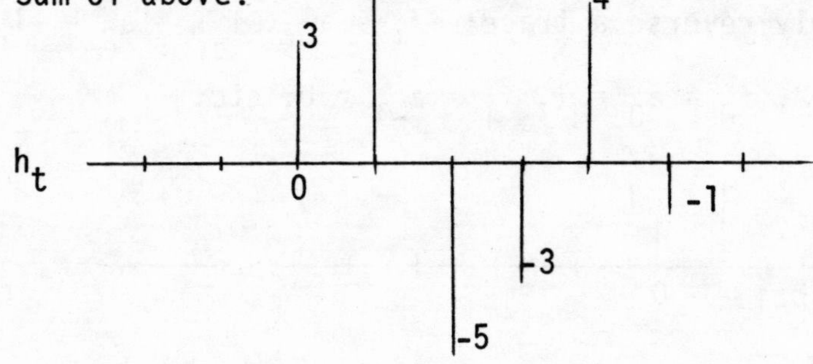

We may note that $h_t = \sum_i g_t a_{i-t}$:

The first important seismic data processing problem solved by this sort of numerical analysis was the <u>singing record problem</u>. We imagine a water layer with reflection coefficients of -1 at the water-air interface and R at the water bottom, and we let the water depth be 20 ft. so that the round trip travel time in the water is 8 msec, or 2 samples at 4 msec sampling.

Energy might bounce once in the water layer before adding to the downgoing wave; such a path would involve an amplitude of -R with respect to the direct wave and would be delayed by 2 samples. Part of the direct wave might bounce at the detector end of its travel path, with an amplitude of -R. Thus two travel paths could arrive with delays of 2.

There are 3 paths for double-bounce energy with delays of 4, each with amplitude of R^2 with respect to the direct wave: a double bounce at the source end, at the receiver end, or a single bounce at each end. Likewise there are 4 triple-bounce paths, each with amplitude $-R^3$; etc. We may therefore express the filtering effect of the water layer as

$k_t = 1, 0, -2R, 0, 3R^2, 0, -4R^3, 0, 5R^4, 0, \ldots..$

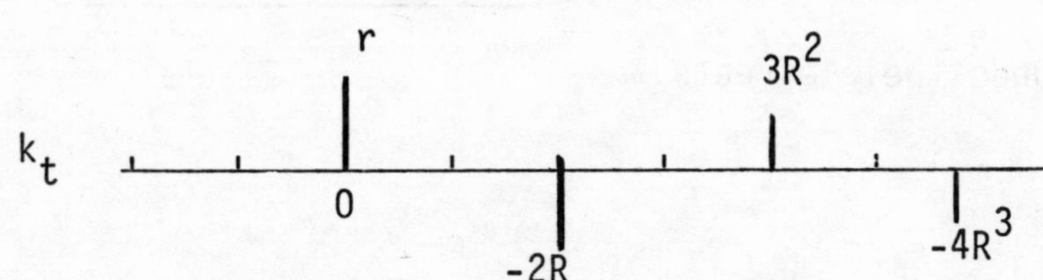

This natural filtering action is responsible for the singing which used to make marine exploration almost impossible in hard water bottom areas where R was sizeable.

This natural convolution which produces singing can be undone in data processing by <u>deconvolution</u> if we can find an appropriate operator. If the appropriate operator m_t, called an inverse filter, is convolved with k_t, the result is a single impulse: $k_t * m_t = 1$

For the specific k_t above,

$$m_t = 1, 0, 2R, 0, R^2.$$

This can be shown by carrying out the convolution $k_t * m_t$:

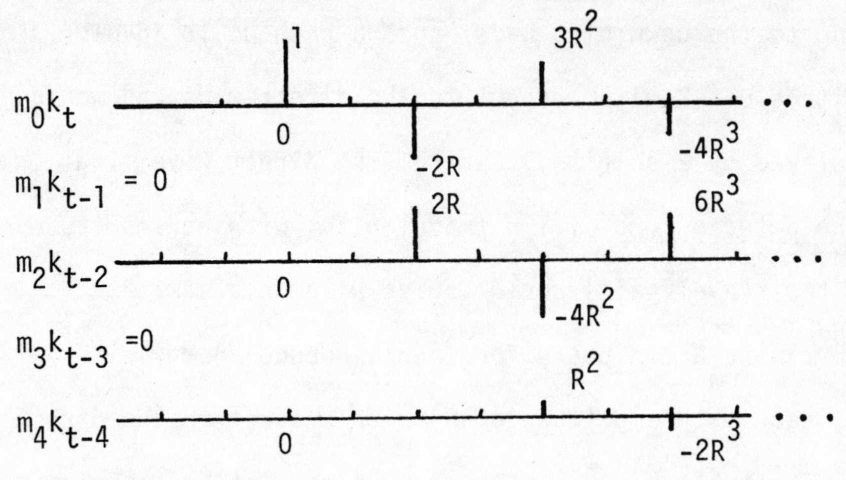

Other types of natural filtering can often be expressed in appropriate filters. In data processing we try to <u>find</u> exactly what <u>the natural filters</u> were from an examination of the data, so that we can then <u>design inverse filters</u> which will undo their effects.

Seismic Data Processing Procedures

Seismic data processing generally follows a basic routine. This basic routine may be varied to tailor processing for specific needs of the data.

The Processing Group

A small group in the processing center has the responsibility for processing a particular set of data. This group prepares the specific instructions, makes processing decisions, chooses processing parameters, and monitors the quality of the results.

This processing group should be well acquainted with the objectives of the processing so that they can make the processing decisions in optimal fashion. The group needs knowledge of the geological section and tectonic framework of the area, including knowledge as to the type of features to be expected (unconformities, faults, reefs, etc.) and where in the section such features might occur. They need to know the extent of the objective section and what constitutes a trap (structure or stratigraphic), and the relationship of the objective section to other portions of the geologic section. This background knowledge includes an appreciation of the types of problems expected, especially near-surface problems and noise wave trains, and expected acoustic impedance variations which may involve multiples.

Processing which is optimal for one set of objectives may not be optimal for another set. For example, large offsets needed to define stacking velocity accurately may not be compatible with muting noise-trains. Processing groups who do not understand the objectives might spend appreciable effort trying to improve the quality of portions which are not of exploration interest, or they might not optimize critical data because doing so does not improve the overall section. On the other hand, the keys to understanding a particular portion of the section are sometimes found elsewhere in the section and overconcentration on limited portions may defeat achieving objectives.

Format Verification

Once data (field tapes) are actually in hand, the processing group verifies how the data are arranged on the magnetic tape. Data are not always arranged as expected. Format verification customarily involves dumping (displaying on a printout the magnetic pattern on the tape) the first few records, possibly the first 10, and comparing the dump with what is expected. If a match does not ensue, detective work is needed to ascertain what data are actually recorded and how they are arranged. With Vibroseis recording, format verification often includes a check on the Vibroseis sweep length and spectrum.

Editing

Editing follows format verification. The first operation usually is to rearrange the data, a process called "demultiplexing". Field seismic data are usually recorded in time-sequential form. If 48 channels are used, the first samples for each of the 48 channels are recorded, then the data for the next moment in time for each of the 48 channels, and so on. Processing requires trace-sequential data, i.e., all the data for the first channel, then all the data for the second channel, and so on.

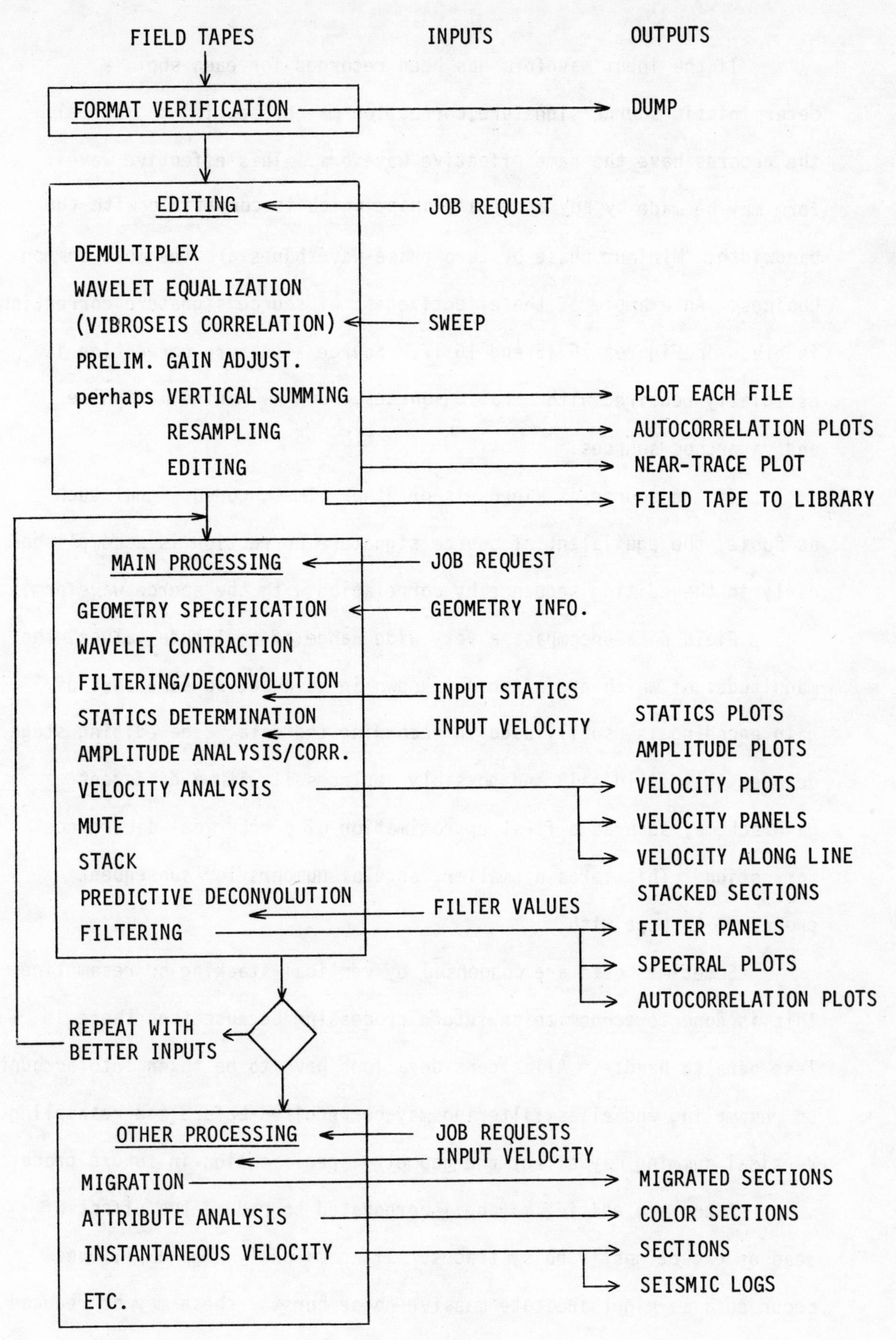
FIELD TAPES
INPUTS
OUTPUTS
FORMAT VERIFICATION
DUMP
EDITING
JOB REQUEST
DEMULTIPLEX
WAVELET EQUALIZATION
(VIBROSEIS CORRELATION)
SWEEP
PRELIM. GAIN ADJUST.
perhaps VERTICAL SUMMING
PLOT EACH FILE
RESAMPLING
AUTOCORRELATION PLOTS
EDITING
NEAR-TRACE PLOT
FIELD TAPE TO LIBRARY
MAIN PROCESSING
JOB REQUEST
GEOMETRY SPECIFICATION
GEOMETRY INFO.
WAVELET CONTRACTION
FILTERING/DECONVOLUTION
INPUT STATICS
STATICS DETERMINATION
INPUT VELOCITY
STATICS PLOTS
AMPLITUDE ANALYSIS/CORR.
AMPLITUDE PLOTS
VELOCITY ANALYSIS
VELOCITY PLOTS
MUTE
VELOCITY PANELS
STACK
VELOCITY ALONG LINE
PREDICTIVE DECONVOLUTION
STACKED SECTIONS
FILTER VALUES
FILTERING
FILTER PANELS
SPECTRAL PLOTS
AUTOCORRELATION PLOTS
REPEAT WITH
BETTER INPUTS
OTHER PROCESSING
JOB REQUESTS
INPUT VELOCITY
MIGRATION
MIGRATED SECTIONS
ATTRIBUTE ANALYSIS
COLOR SECTIONS
INSTANTANEOUS VELOCITY
SECTIONS
SEISMIC LOGS
ETC.

Figure 16.18

If the input waveform has been recorded for each shot, a deterministic source signature correction may be applied to make all the records have the same effective waveform. This effective waveform may be made by any arbitrary shape which is compatible with the bandwidth. Minimum phase or zero-phase waveshapes are the most common choices. An example of the effectiveness of source signature correction is given in Figures 16.16 and 16.17. Source signature correction is especially required with erratic sources, such as marine maxipulse and vaporchoc sources.

If the source is Vibroseis or other time-encoded signal such as Sosie, the equivalent of source signature correction is accomplished early in the editing sequence by correlation with the source waveform.

Field data encompass a very wide range of amplitude values, the magnitudes of which are often not known in advance, so some sort of gain encoding is usually used in recording the data. The editing stage decodes this field gain and possibly replaces it with a different gain scheme, such as a first approximation of a spherical divergence correction. This makes a smaller range of numbers for subsequent processes to cope with.

Sometimes data are condensed by vertical stacking or resampling. This is done to economize on future processing because then there is less data to handle. Alias considerations have to be taken into account on resampling and alias filtering may be required before the resampling. Vertical summing may affect the geometry specification in future processing.

Automatic editing may be incorporated to reduce the effect of dead or exceptionally noisy traces. If anomalously high amplitudes occur such as might indicate massive noise bursts, these may be reduced to the level of the surrounding data or possibly to zero.

Outputs of this editing pass usually include:

(1) a plot of each file, so that one can see what data need editing and the types of noise attentuation processes required;

(2) a near-trace plot, a plot of the shortest offset trace from each record; such a plot may be used to give a quick look at structural complexities, and decisions as to where velocity analysis studies are to be made may be based on it;

(3) near-trace autocorrelation plots; these plots indicate the degree to which multiples are a problem and deconvolution decisions will be based in part on them.

The output of this editing pass is a trace-sequential tape which will be used for subsequent processing. The field tape is usually returned to the library after its use as the input for this pass.

Main Processing - Preliminary Pass

The object of the preliminary pass is to ascertain processing parameters which are data-dependent. Static time shifts, amplitude adjustments, normal-moveout values and frequency content have to be ascertained from analysis of the actual data.

Information about the field geometry is input so the computer will be able to determine which data involve common geophone locations or have other factors in common, the offset for each trace for normal-moveout determinations, etc.

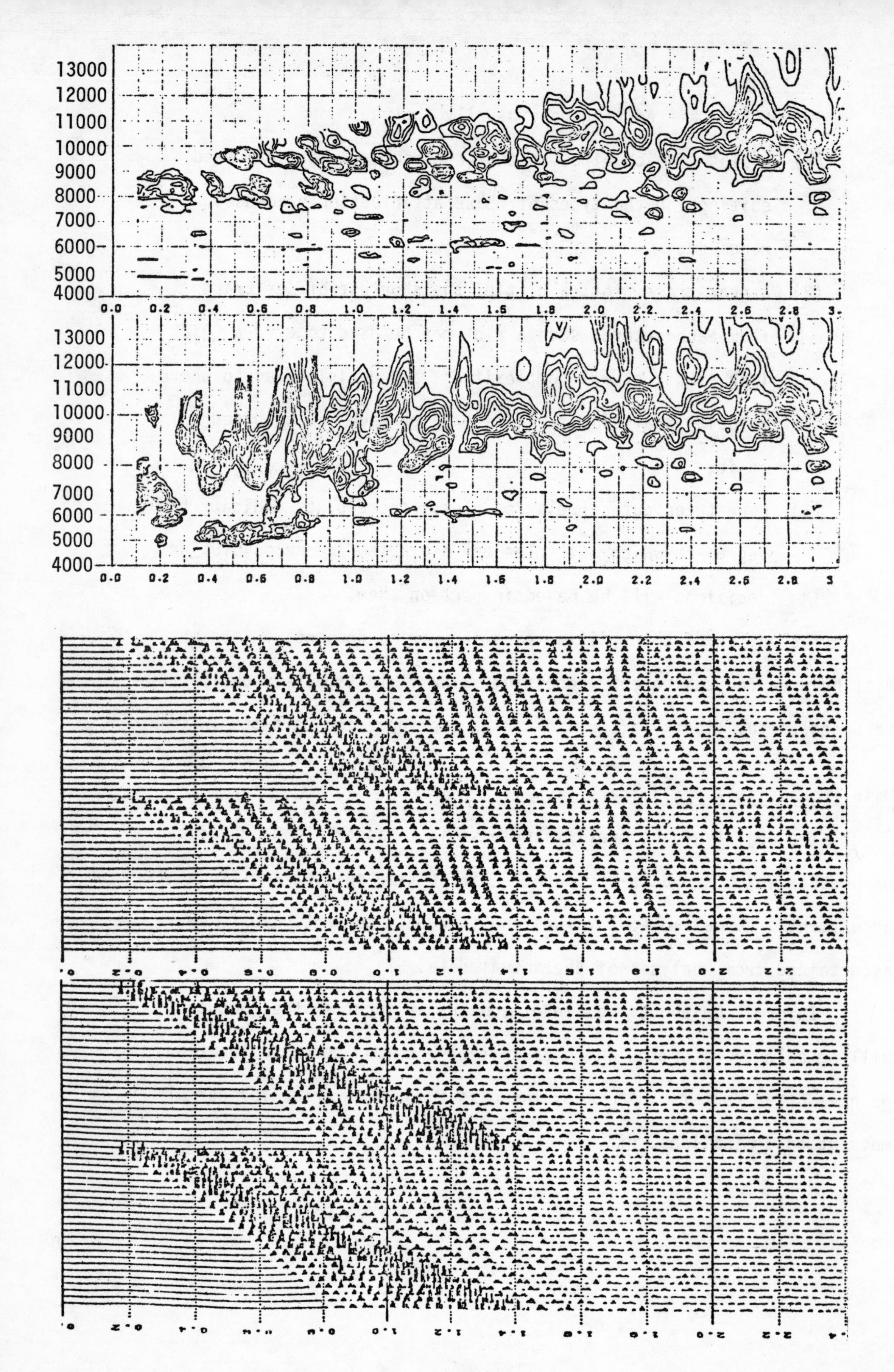

Figure 16.20

Figure 16.19

If the near-trace plot shows excessive wave trains, such as might be associated with horizontal traveling energy like surface waves or shallow refractions, apparent-velocity filtering (also called 2-D filtering) may be applied to remove them. Figure 16.19 shows gathers before and after such 2-D filtering and Figure 16.20 shows velocity analyses without and with such 2-D filtering. The accuracy of velocity analyses is greatly enhanced by this filtering. Other filtering or muting operations may also be applied to remove other types of noise.

Statistically based wavelet contraction may be applied where the shape of the source wavelet was not recorded. The effectiveness of such is shown by Figures 16.14 and 16.15.

The analysis of the statistical characteristics of the data for wavelet contraction and other filtering and for statics and velocity analysis should be based on data which exclude known bad traces and zones of non-reflection energy such as the first-breaks region. A wide window is usually chosen so as to provide appreciable statics, but the window should exclude zones where primary reflections are not expected, such as below the basement.

The statics analysis program looks for systems such as would be expected if time shifts were associated with particular shots, particular geophones, etc. Preliminary statics, as determined in the field office from first-break information and from the elevation of geophone stations, is usually input before the statics analysis so that the statics analysis determines the residual statics errors. The effectiveness of surface-consistent statics is shown in Figures 16.3 and

16.4 and in Figures 16.5 and 16.6, and the improvement in velocity analysis which results is shown in Figures 16.7 and 16.8. A control plot from a statics program is shown in Figure 16.21.

A type of analysis similar to the statics analysis is carried out for amplitude, to determine system to the amplitude. Thus effects associated with a weak shot, a poor geophone plant, etc., are isolated so that they can be allowed for. A control plot from such an amplitude analysis program is shown in Figure 16.22.

Velocity analysis is usually run at locations approximately 1-2 km apart but with locations selected so as to be relatively free of structural complications. The locations for these analyses are based on the near-trace section output by the prior editing pass. A first guess as to velocity is input prior to the velocity analysis so that the velocity analysis looks for residual normal-moveout. If the data dip, as ascertained by the near-trace plots, the dip information at various times is input because velocity values depend on dip.

Compromises have to be made in velocity analysis specification, as in most processing decisions, as to how much data should be included in the analysis. The more data, the better the statistics but then the determinations do not apply at specific points. The compromise is thus between determining a more accurate average and a value which is less accurate but applicable to a specific location.

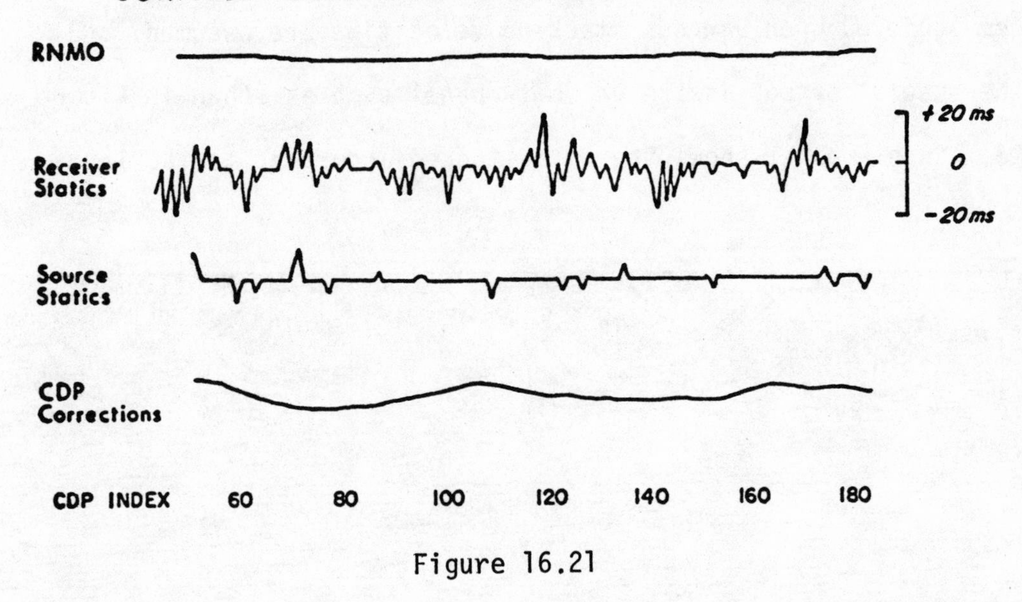

Figure 16.21

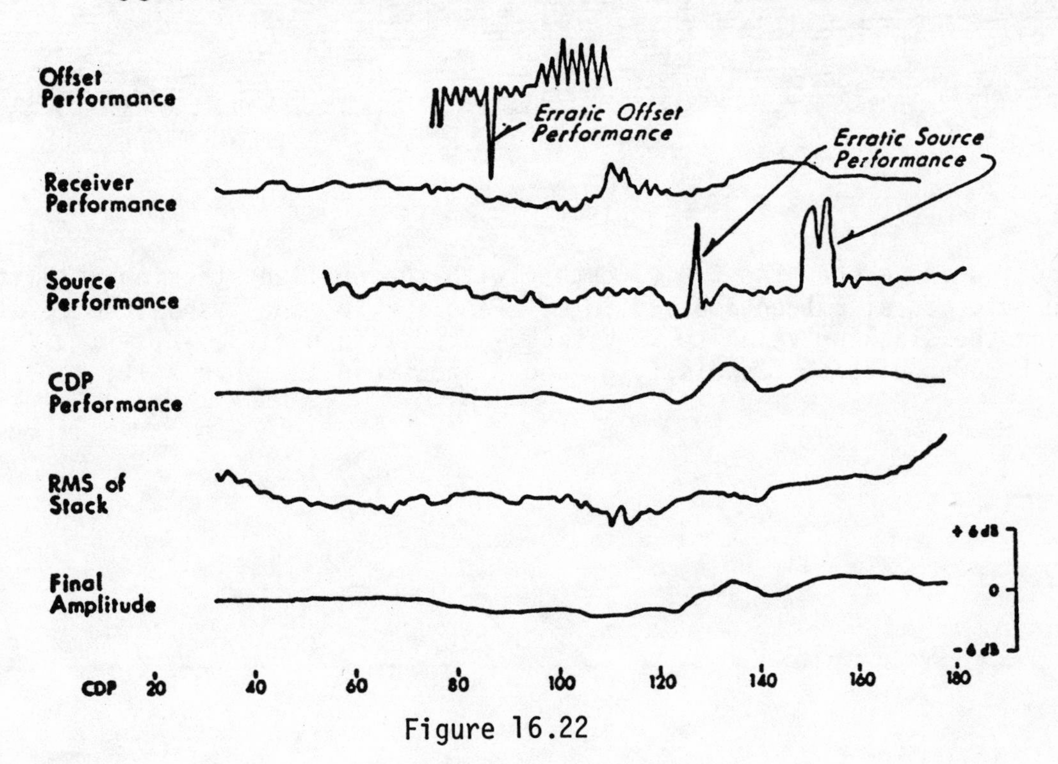

Figure 16.22

At least three types of outputs may result from velocity analysis: (1) the most common output is simply a velocity spectral plot, such as shown in Figures 15.6 and 15.7. These show contours of the semblance achieved when various stacking velocities are assumed. (2) Another useful output is the velocity panel such as shown in Figure 16.23. Such a panel shows the data stacked according to the input

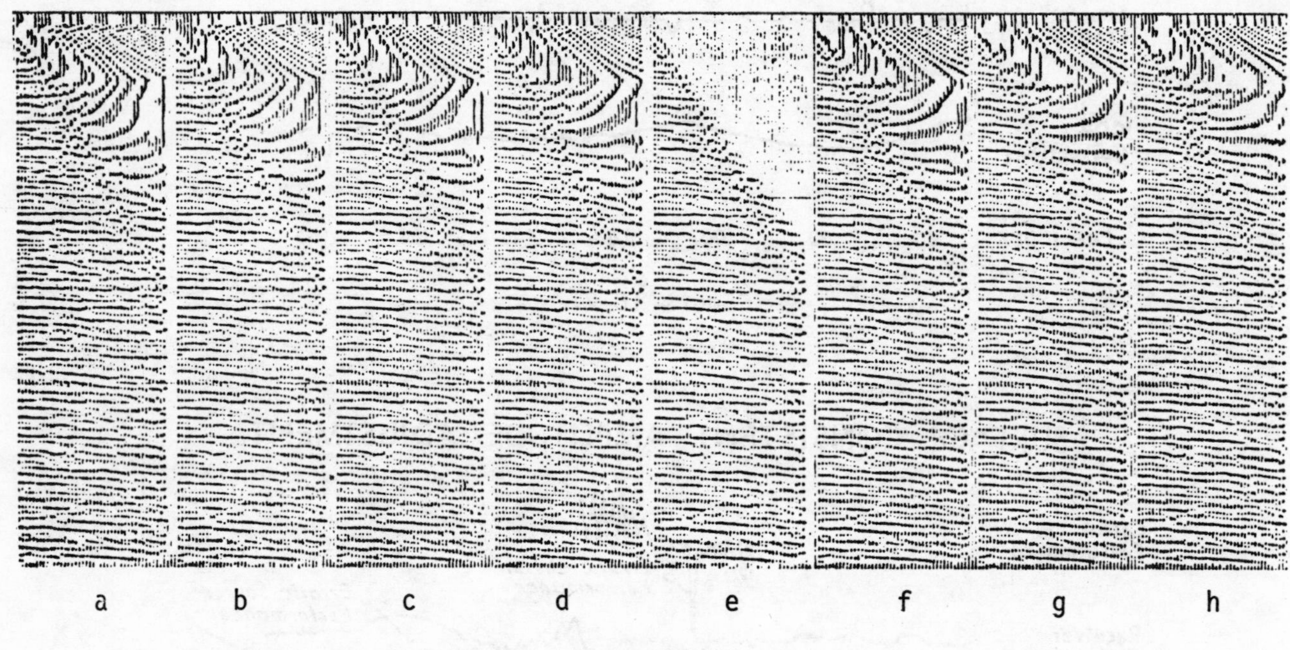

Figure 16.23

Panels d and e show the C.D.P. gather with the applied stacking velocity; the mute has also been applied in e. Panels a, b, and c show the results where the stacking velocity is Vstack - $n \Delta V$ with n = -3, -2, -1; ΔV is often 200-500 ft/s. Panels f, g, and h show results for n = +1, 2, 3.

velocity information and according to velocities which are slightly below and slightly above the input velocity. Such velocity panels allow one to see if certain events require different stacking velocity than other events. Stacking velocity is not necessarily a single-valued function, especially when events approach the spread from different directions. Such panels also allow one to see how sensitive the stack is to velocity assumptions. (3) Another output may show how velocity determinations at different locations along the lines relate to each other. Such a plot is shown in Figure 16.24.

Finally, a preliminary stack may be made as a check on the effectiveness of the various process steps and as an aid in diagnosing additional problems, selecting additional velocity analysis locations,

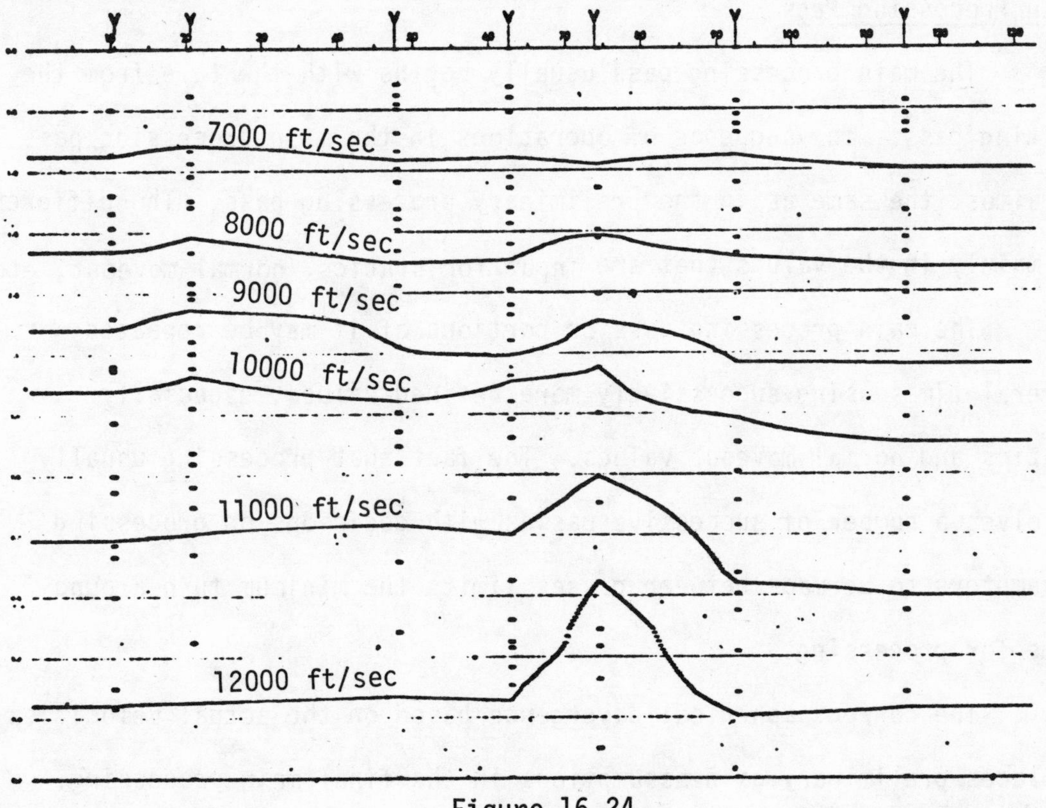

Figure 16.24

Stacking velocity variations along seismic line. Velocity analyses are usually run at only selected points and then stacking velocity in between is interpolated. This plot shows the interpolations the computer is making.

determining the windows over which additional analyses are to be carried out, etc.

The data may also be filtered by a sequence of narrow bandpass filters yielding a filter panel which is used to ascertain the frequency content as the arrival time varies. A filter panel is used to determine subsequent filtering parameters. Autocorrelations and spectral plots of various sorts may also be output.

The important outputs from the preliminary processing pass thus are the values of the parameters to be used in the subsequent main processing, although tapes may be pulled off at various stages throughout the preliminary processing for use as the input to subsequent processing.

Main Processing Pass

The main processing pass usually begins with the tape from the editing pass. The sequence of operations in the main processing pass is almost the same as in the preliminary processing pass. The difference is mainly in the values that are input for statics, normal moveout, etc.

The main processing pass or portions of it may be repeated several times using successively more refined values, especially statics and normal moveout values. The fact that processing usually involves a number of successive passes with decisions on processing parameters to be made between passes limits the minimum turn-around time for processing.

The correct spherical divergence based on the actual velocities replaces preliminary gain assumptions in the final main processing.

The final output from the main processing pass will be one or more stacked sections. Often the output consists of several stacked sections which differ in the processes which have affected the data or in the choices of display parameters, for example in amplitude, polarity, filtering, etc.

Migration and Other Processing

The stacked data may then be used as input to various other processes, such as migration and attribute analysis.

Velocity information is required to accomplish migration and so has to be input. This velocity information may be based on the stacking velocity analyses, especially where dips are not extreme, but in general the optimum velocity for migration differs from the optimum velocity for stacking.

Either the stacked or the migrated data may then be analyzed for amplitude, frequency content, apparent polarity, etc. in attribute analysis and displayed as color sections or in other ways. Other types of processing such as iterative modeling may also be done on selected data.

CHAPTER 17

STRUCTURAL INTERPRETATION

A well-processed seismic section often bears a striking resemblance to a diagramatic cross-section of the earth. However, it is not such a cross-section and the objective of this unit is to show some of the ways in which the seismic response of the earth differs from such a cross-section.

The primary objective of interpretation usually is to map the bedding, that is, to determine the geological structure. A secondary objective is to determine the nature of the rocks, and a third to determine the nature of fluids in the pore spaces of the rocks. This unit concerns the first of these objectives.

Figure 17.1 is a structure map of one formation in a hypothetical area. It shows a plunging anticlinal nose and a parallel plunging syncline, both cut by a normal fault. The locations of 3 seismic lines are also indicated. While these 3 lines are not everywhere perpendicular to the strike, the angle they make with the strike is usually not large and in the following exercise we shall neglect cross-dip effects. In this exercise we shall deal with the relationship between seismic data and geologic maps backwards of the normal sense. That is, given a geological structure, what will the seismic sections look like.

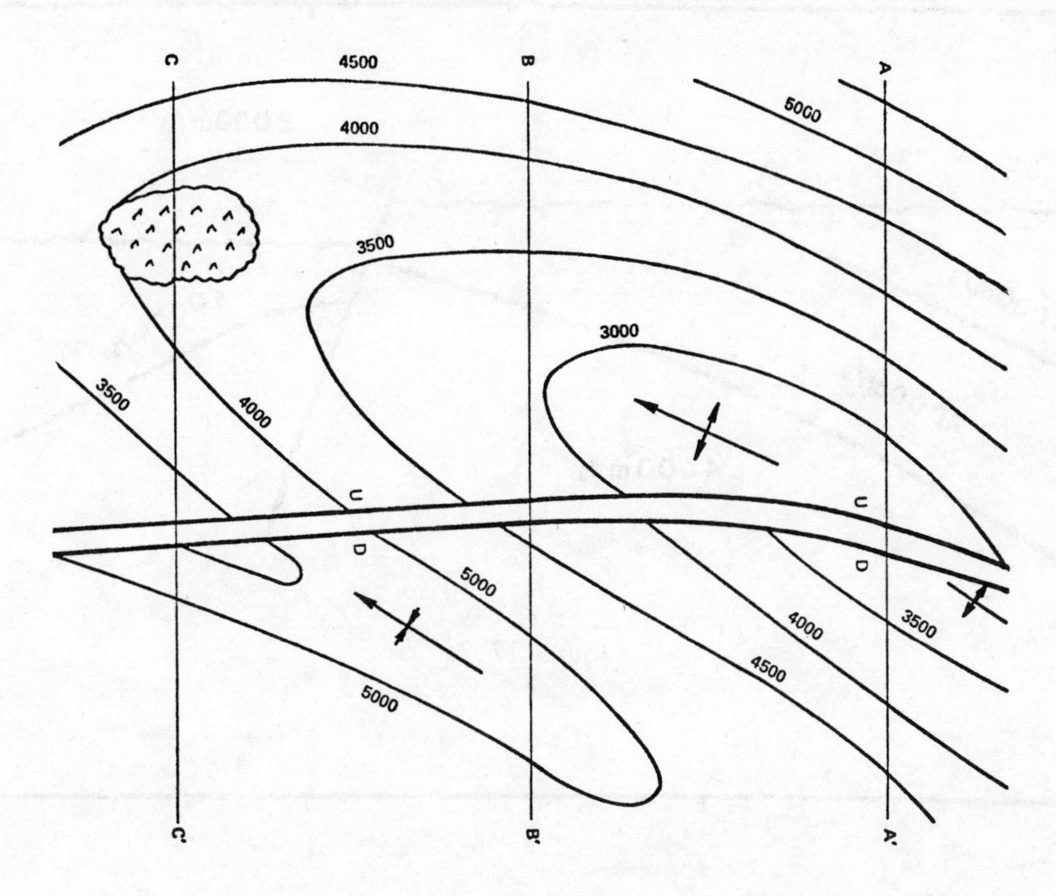

Figure 17.1

Exercise:

The top of Fig. 17.2 shows the geological structure along the profile line AA' and the bottom half is a blank seismic section on which you are asked to draw the arrival time of the various types of seismic events you would expect. This section will then be discussed prior to the similar exercises on the other two seismic lines. The geologic section has the same horizontal as vertical scale. Note that the scale of the seismic section has been selected to facilitate calculations.

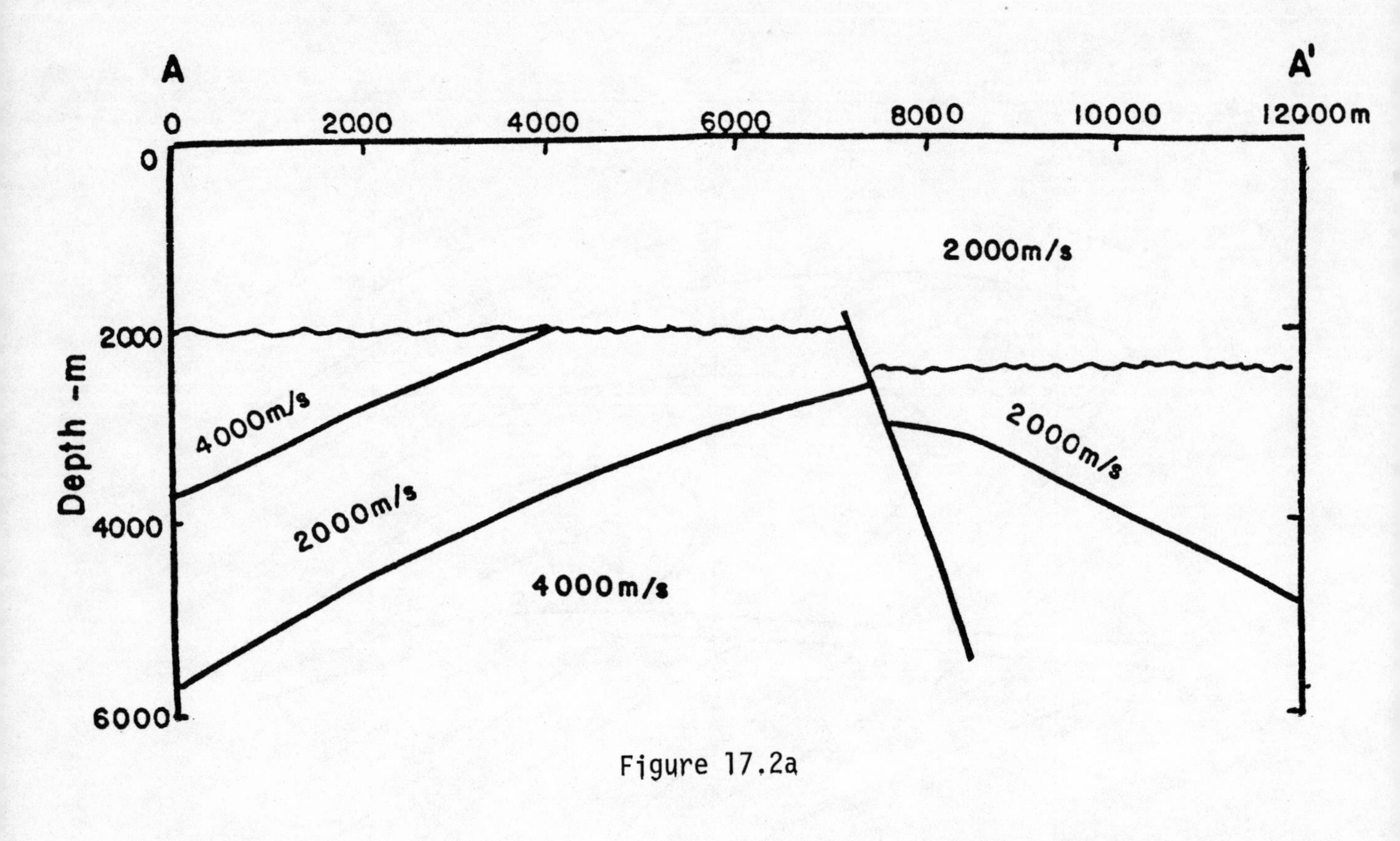

Figure 17.2a

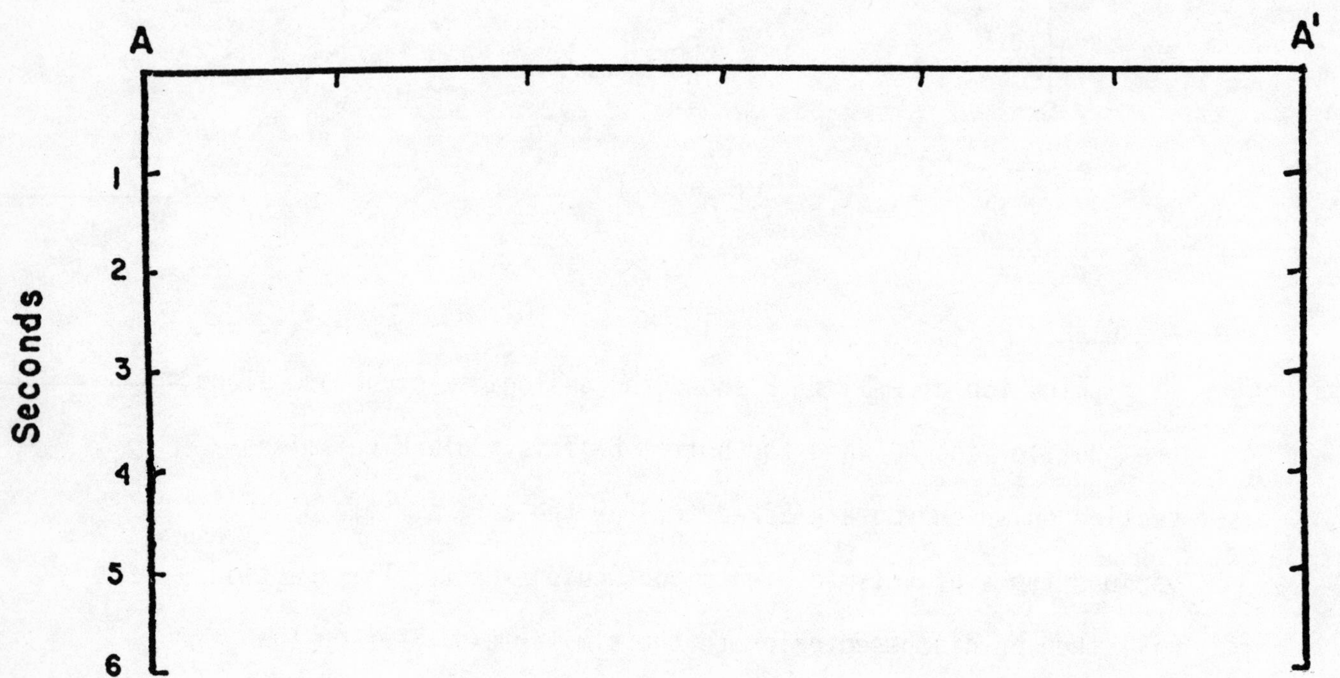

Figure 17.2b

Discussion of section AA':

The top diagram of Fig. 17.3 shows some construction lines and the bottom diagram some of the seismic events which would be expected. The reflectors and diffraction points are indicated by letters which have also been placed on the corresponding events on the seismic diagram.

The reflection from the unconformity A will travel 2000 m down to A and 2000 m back; at a velocity of 2000 m/s this 4000 m of travel path will take 2 seconds. Since reflector A is flat and the velocity above it is constant, reflection A will be flat at 2 sec. The fact that A is drawn as an unconformity probably would not affect the seismic event because the relief on the unconformity is probably very small compared to seismic wave length.

The portion of the unconformity labeled B separates 2 materials of the same velocity. If the density were also the same on both sides of the unconformity, there would be no reflection from it. However, the density could change even though the velocity does not change, in which case there would be a reflection from it. Since not enough information has been given to determine whether B will generate a reflection, we fall back on our general experience; usually some change does occur at unconformities and unconformities are often among the best reflectors. We therefore assume that we will get a very weak reflection from B; it has been dashed in in the lower part of Fig. 17.3. The same considerations apply to C.

The base of the upper 4000 m/sec section labeled D will generate a reflection. The travel path from the shotpoint at

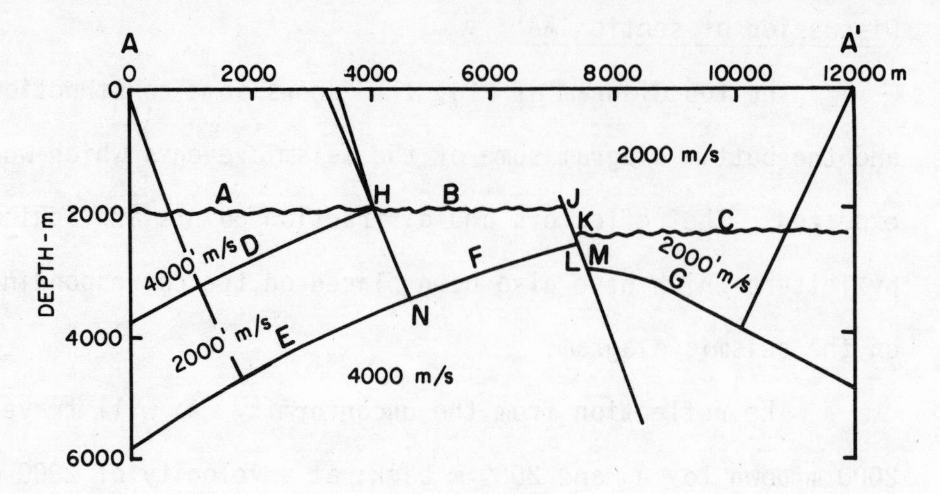

Figure 17.3a

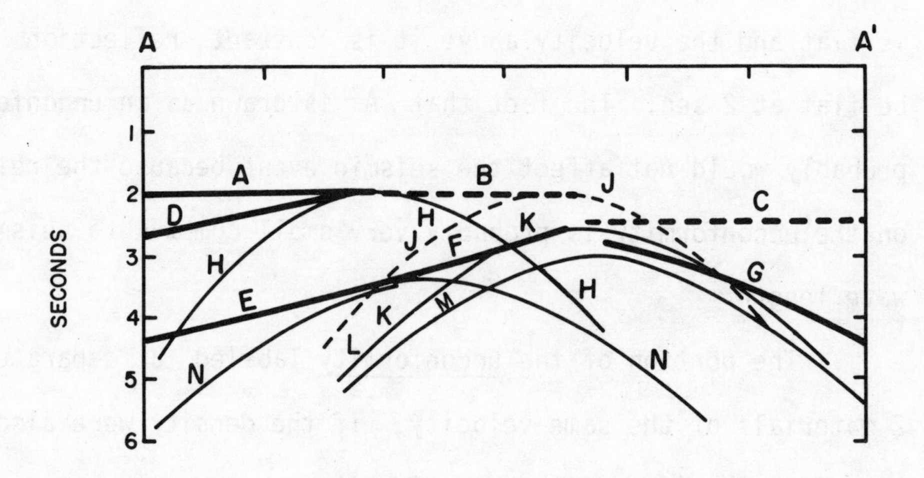

Figure 17.3b

location 0 would bend slightly at interface A according to Snell's Law. The travel time to reflector segment D at location 0 is approximately 2.7 sec. An essentially parallel travel path from location 3200 would strike the other end of reflector segment D, giving a reflection at about 2.1 sec. Connecting these two points gives reflection d, which has a different attitude than reflector D because of the effect of the wedge of 4000 ft/sec material. The reflection from D will be of opposite polarity to the reflection from A and nearly as large, diminished only by the slight loss of transmission loss in passing through interface A. The portion of the reflector to the left of D will not give any observable effects on this portion of the seismic line because the seismic line stopped at 0.

Some of the energy incident on D will pass through it and be reflected from E. By placing rays perpendicular to the portion of this interface labeled F we determine that reflections F and E are not colinear and in fact reflections F and E slightly overlap. The apparent bed is a velocity effect caused by the upper wedge of 4000 m/sec material.

Reflector G on the opposite side of the fault will give rise to reflection G, as indicated. The length of the reflection will be longer than the length of the corresponding reflector, because of the convex upward curvatue of the reflector.

A reflection from the <u>fault plane</u> will not occur on the portion of the section seen here. A contrast at the fault plane would give rise to a fault-plane reflection, but usually one has to get an appreciable distance away from the fault in order to obtain the fault plane reflection and then the arrival time is so long that the event is apt to be disregarded as merely noise.

The point H will give rise to a <u>diffraction</u>. The diffraction curvature can be calculated easily by ray-tracing as shown in Fig. 17.4, where rays to the 0, 2000, 4000, 6000 and 8000 meter locations are indicated and used to locate the diffractions at these points. The diffraction H is tangent to both reflectors A and D because both of these reflectors terminate

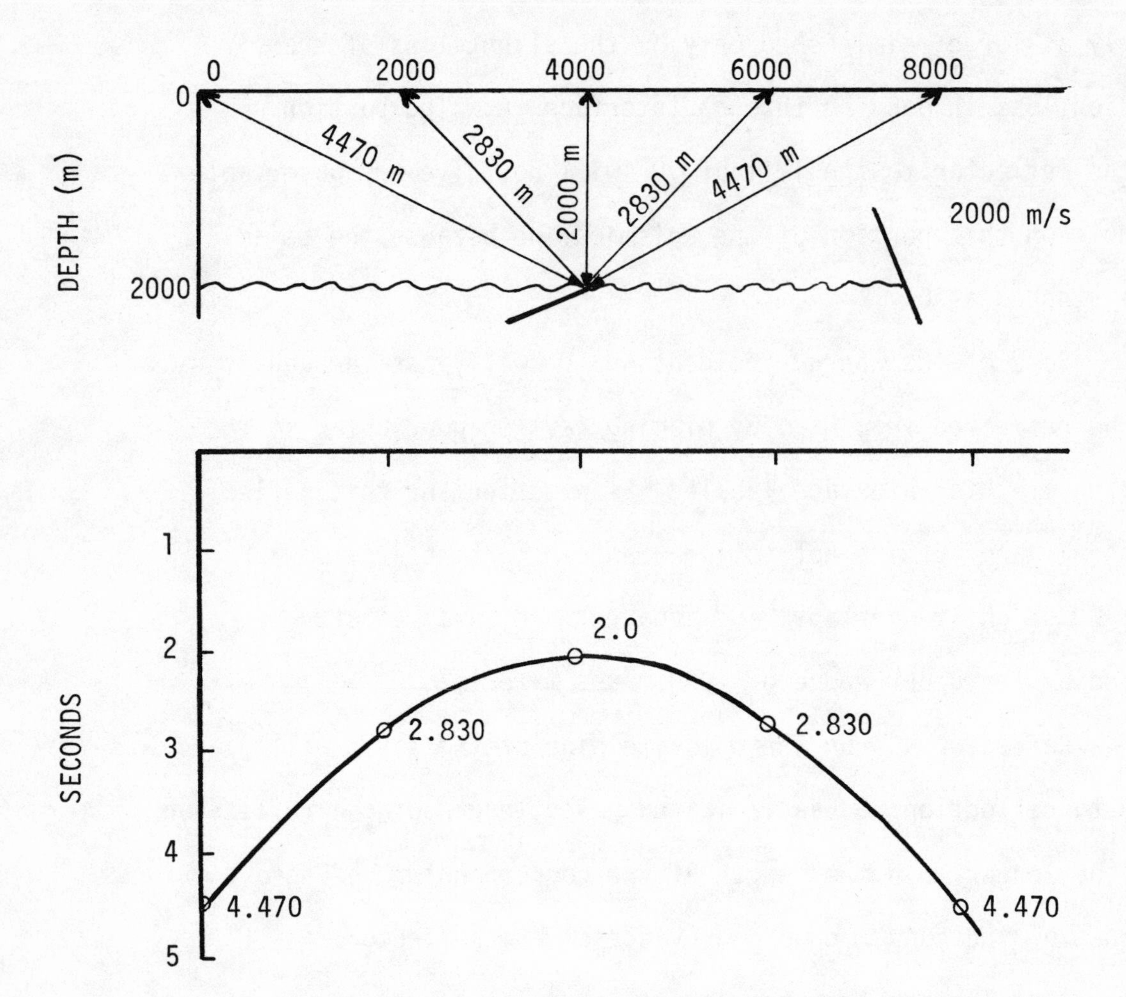

Figure 17.4

at point H. A very weak diffraction of similar curvature would be expected at point J. Diffractions with slightly less curvature might also be expected from points K, L, and M.

Point N will generate a phantom diffraction. Although the reflector is smooth and continuous through point N, a phantom diffraction results from the variation in the velocity on opposite sides of the raypath to the point N. Diffraction N will be tangent to both reflectors E and F (which slightly overlap) and a portion of N will appear as a reverse branch of the phantom buried focus caused by the function of E and F.

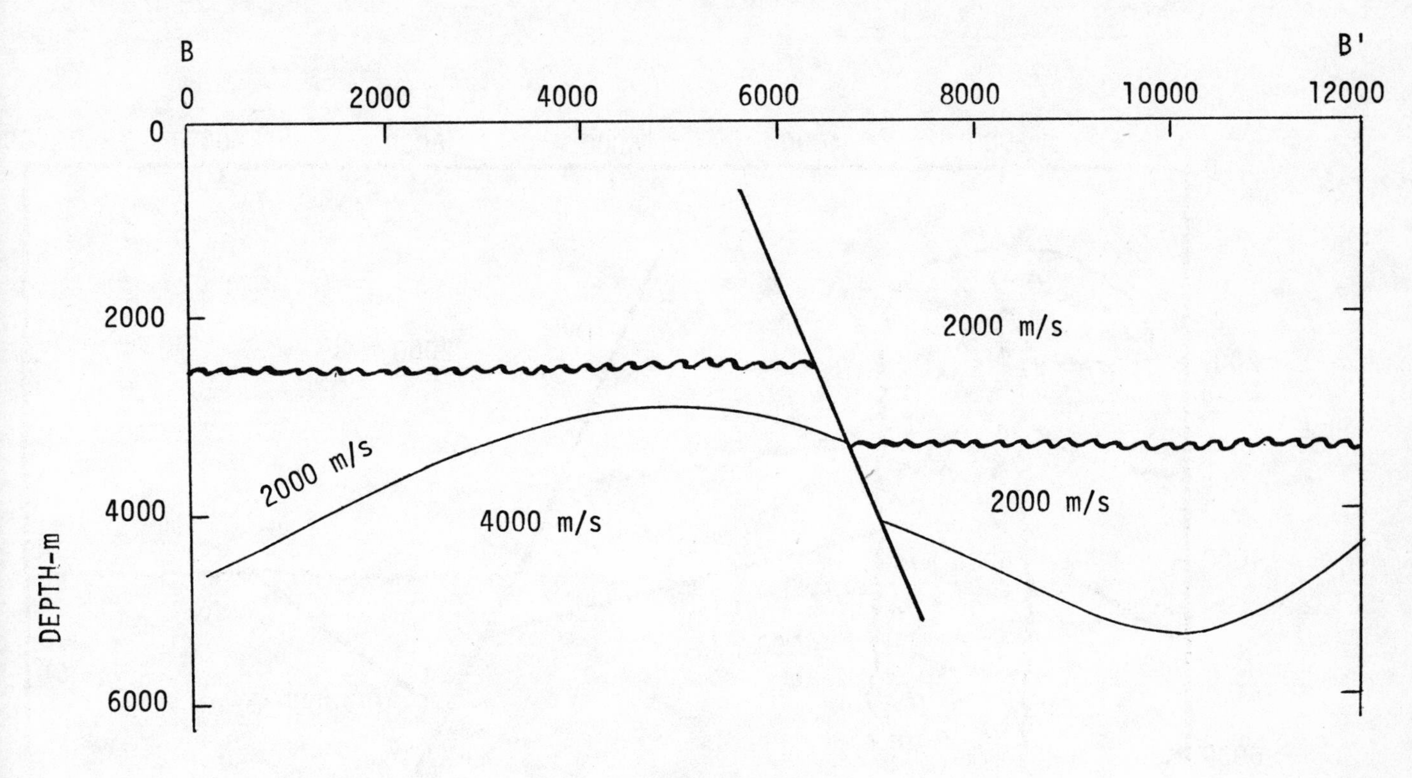

Figure 17.5a

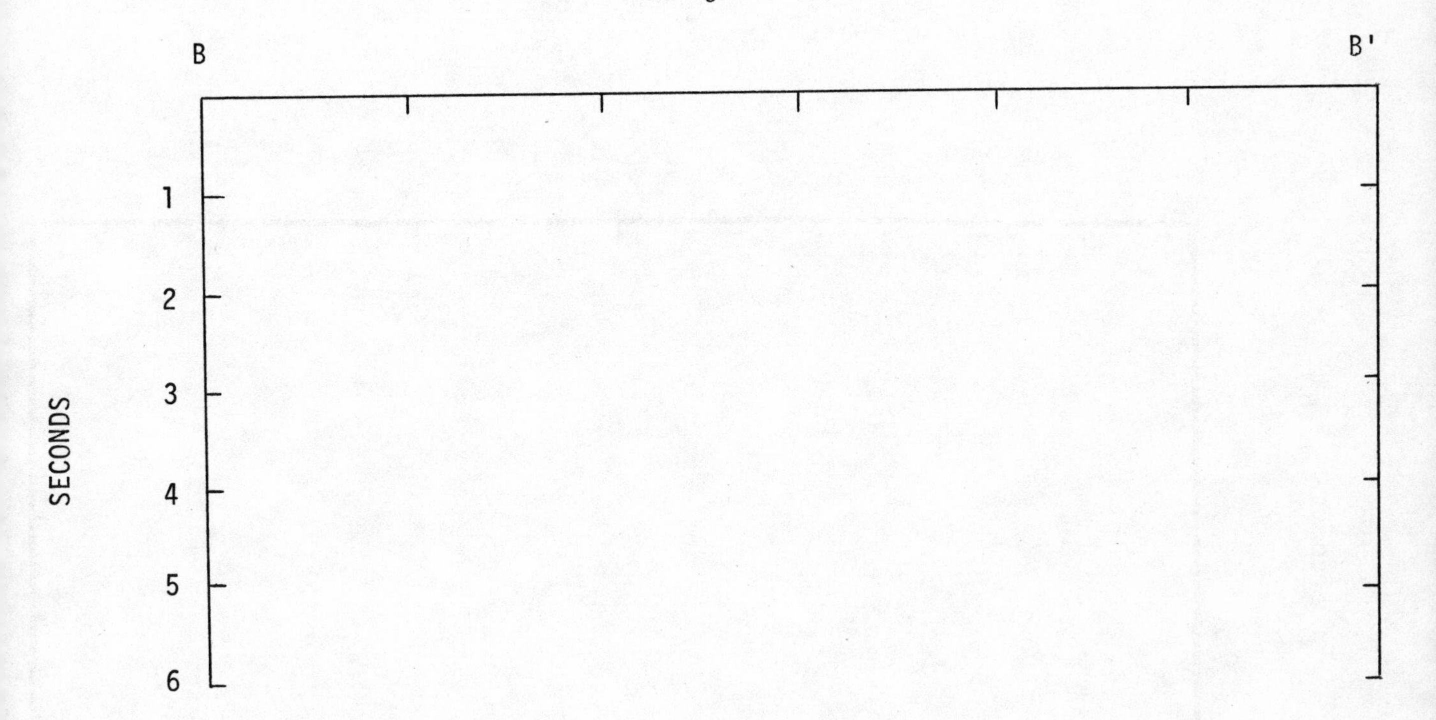

Figure 17.5b

<u>Exercise</u>:

Show on Fig. 17.5b the events which would be expected along seismic line BB'. The results are discussed on Page 187.

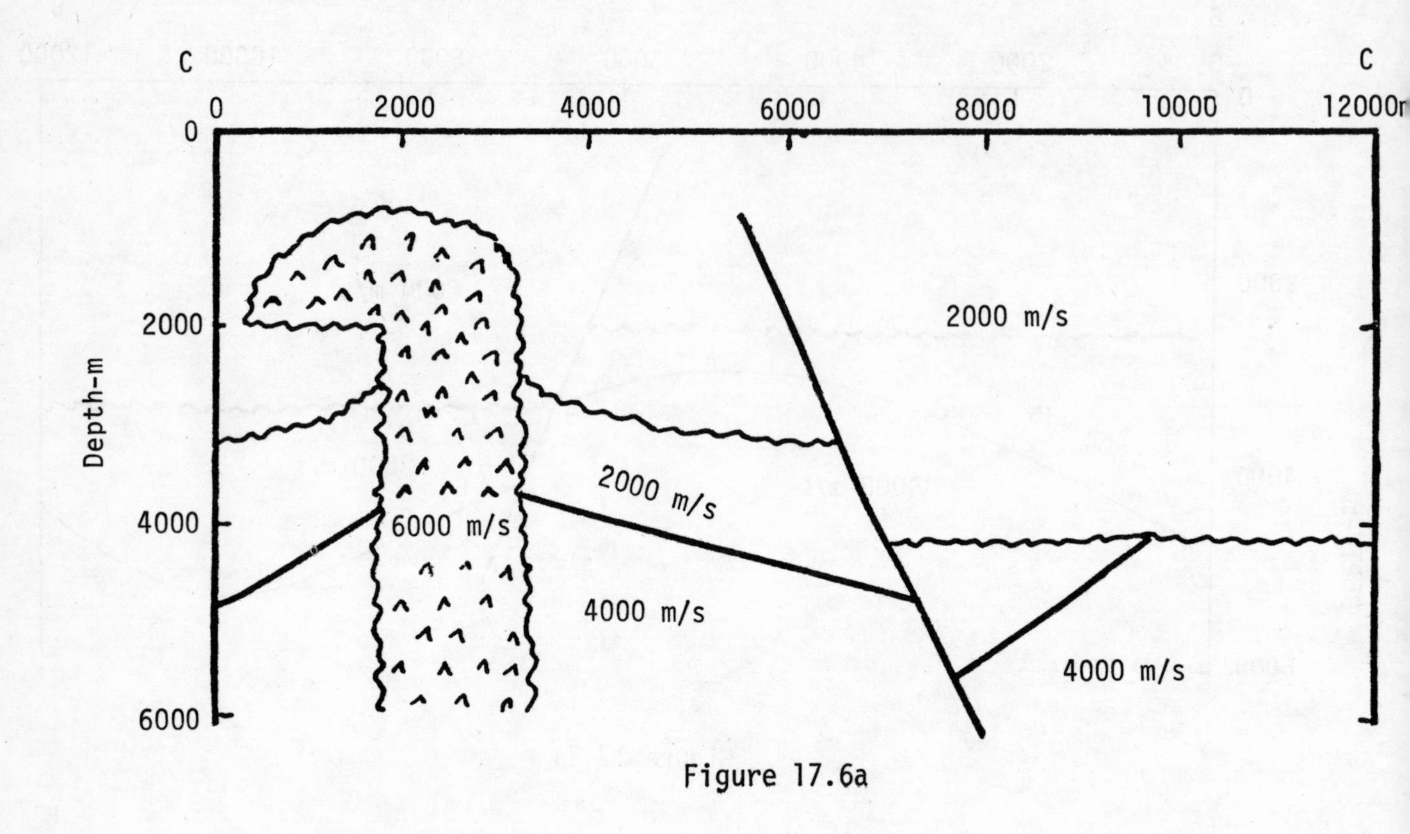

Figure 17.6a

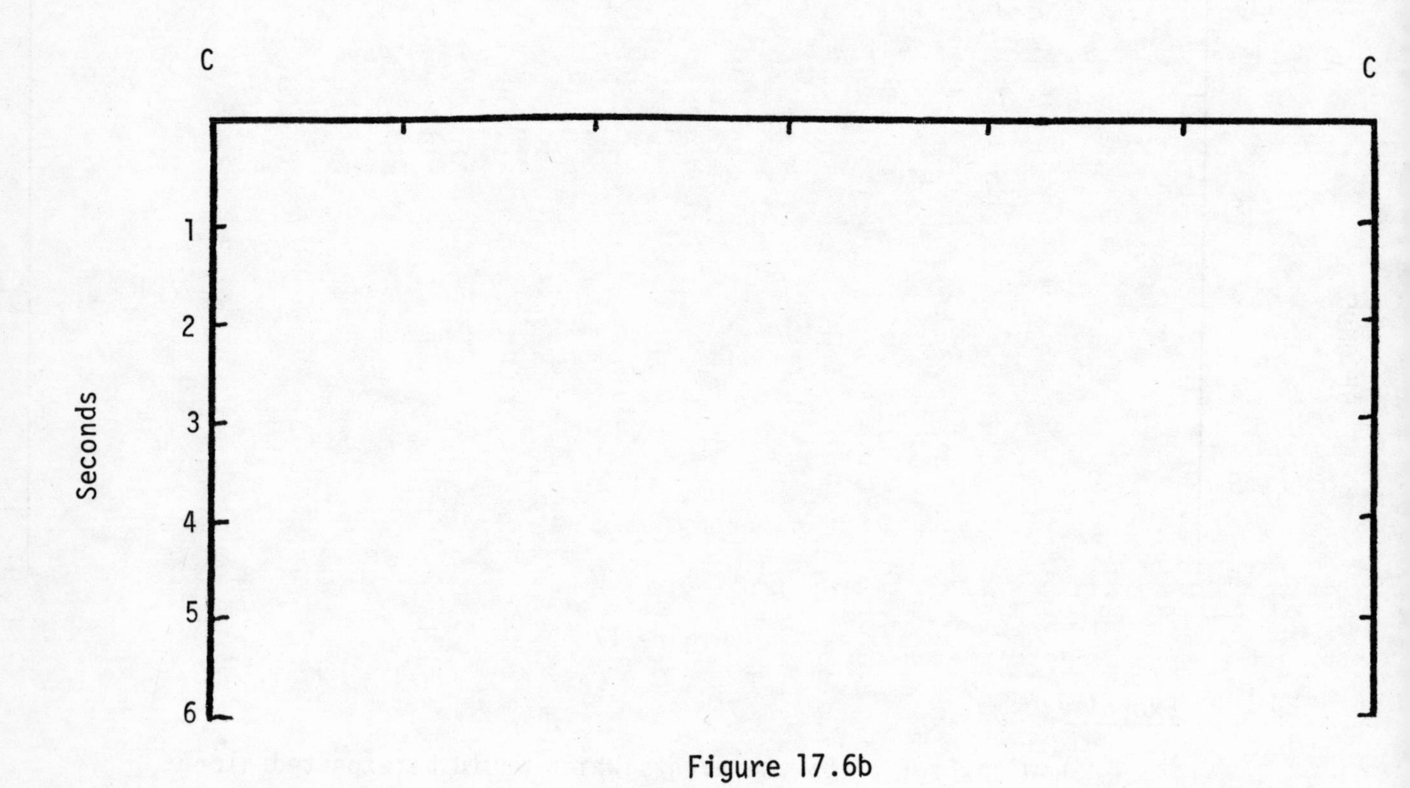

Figure 17.6b

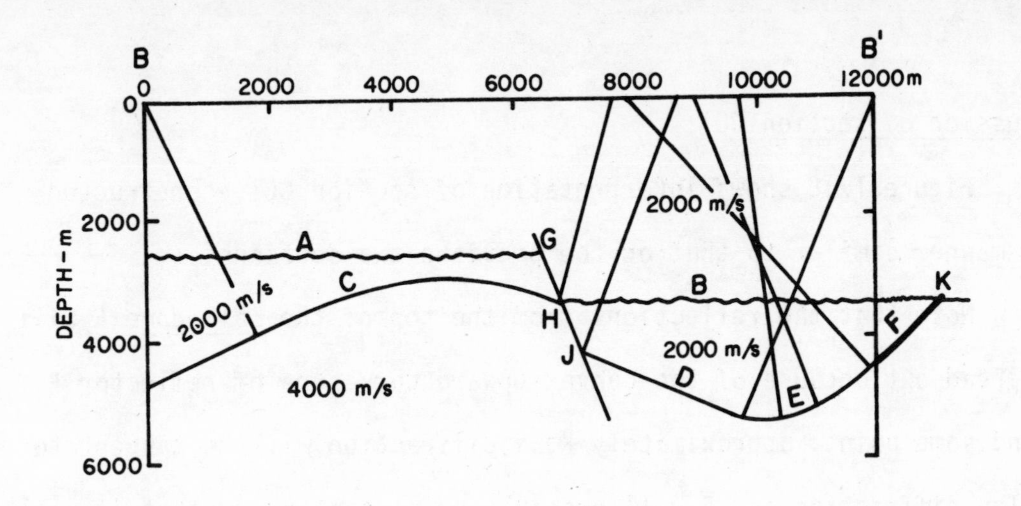

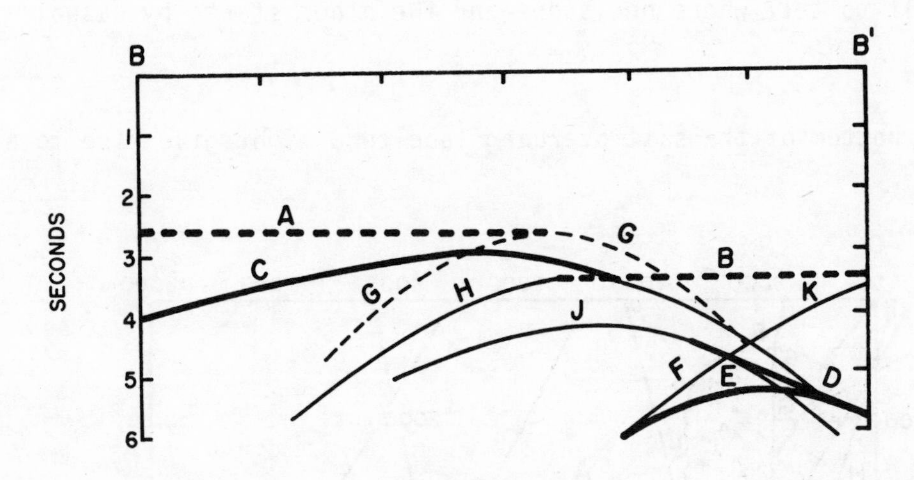

Figure 17.7

Discussion of section BB":

Interpretation of section BB' is shown in Fig. 17.7.

The reflection from the portion of the anticline labeled C appears to be appreciably longer on the seismic section than the corresponding portion of the reflector in the earth because of the anticlinal curvature.

The portion of the reflector D will give rise to a more-or-less straight reflection and the curved synclinal portion E will give rise to a reverse branch. What happens to this reflector beyond the right end of the section will also be seen; for example, reflection F and the diffraction K from the truncation of F by the unconformity.

Points H and J will give rise to diffractions.

Discussion of section CC':

Figure 17.8 shows interpretation of section CC', constructed in a manner similar to that on the previous two sections.

Note that the reflections from the top of the salt dome F will be spread out because of the convex-upward curvature of reflector F. Beyond some point, approximately K, a diffraction will be tangent to F. The diffraction and F will probably be so continuous that it will be difficult to tell where one stops and the other starts by casual inspection.

The bottom of the salt overhang labeled G might give rise to a

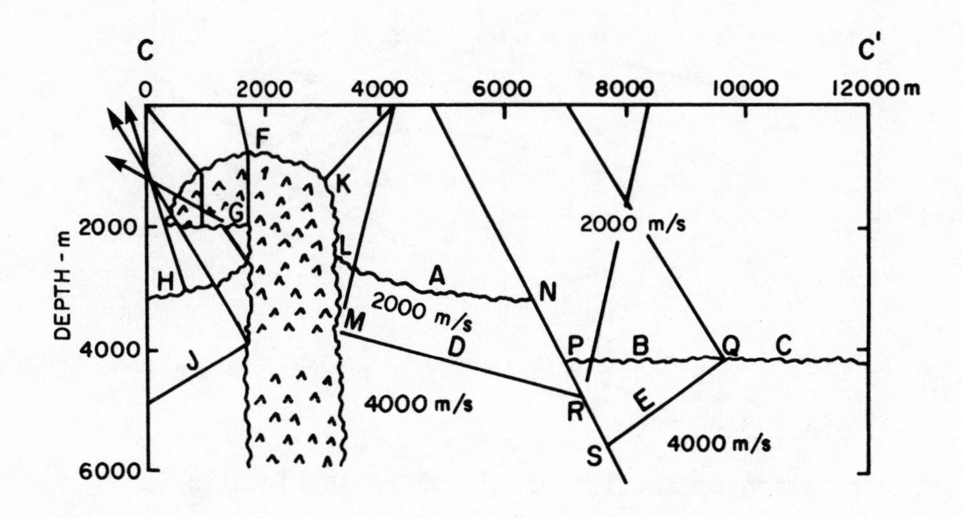

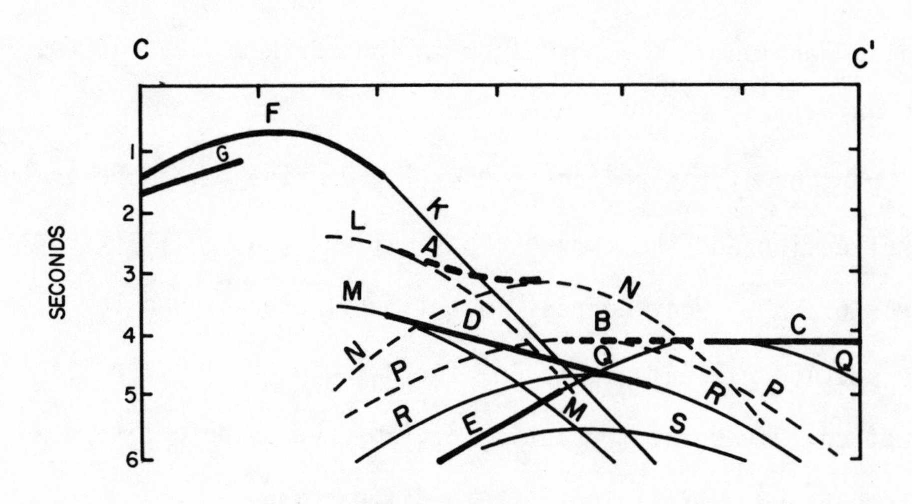

Figure 17.8

reflection, as shown; this reflection will appear to dip even though G is horizontal, because of refraction at the top of the salt. G will also follow so closely after the reflection from the top of the salt that it might not be recognized as a reflection from the bottom of the overhang.

Reflections from H and J will not be seen on the CC' portion of the seismic line.

The right half of diffractions L and M will probably be evident but the left half of these diffractions may be so confused by travel path through the salt and refraction effects at the top of the salt that they may not be distinguishable events. At any rate, the left half would arrive appreciably earlier than the right hand branches because of travel through the high-velocity salt and so the complete diffraction pattern will probably not be recognizable.

CHAPTER 18

MIGRATION

Migration is required because reflections on seismic sections appear at different locations than directly above the reflectors. Migration involves moving each reflection element to a location appropriate to the reflector location. In the following discussion, the data are assumed to be in the same vertical plane as the seismic line. In a subsequent section, we shall examine the errors involved where this assumption is not correct and see how such situations can be handled.

The wavefront chart in Fig. 18.1 is plotted vertically in depth and horizontally in distance. The more-or-less circular curves whose centers are near the shotpoints are wavefronts. Wavefronts show the location at particular times of the downgoing energy from the shot. The wavefronts are labeled according to two-way travel time. If it takes one second for a wave disturbance to reach a particular wavefront location and if the disturbance should be reflected at some point along the wavefront, then it will take another second for the data to return to the shotpoint so that it will be observed at 2 seconds of arrival time, which is how the wavefront is labeled.

The curved lines which radiate from the shotpoint are called ray paths. Ray paths are labeled in terms of the apparent dip at the surface. We shall examine in a moment how this apparent dip is measured.

One particular wavefront in Fig. 18.1, the 2.5 second wavefront, has been specially marked. It shows the location of the downgoing disturbance from the shot 1.25 seconds after the shot. One ray path

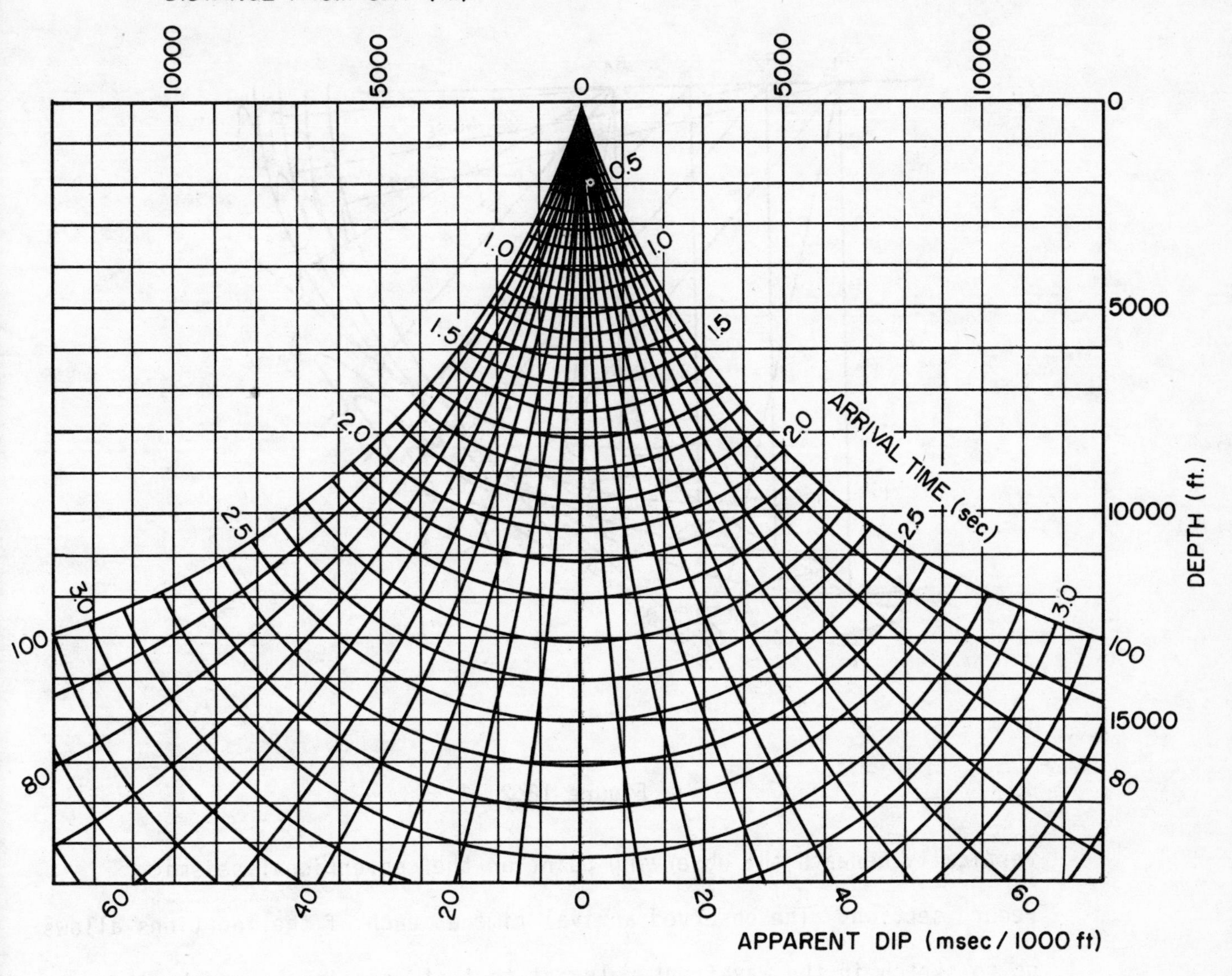

Figure 18.1

has also been specially marked, the one which has apparent dip of 30 milliseconds per 1000 feet. If we can measure from our seismic data the direction of the ray path associated with a given reflector and note the arrival time of reflection, then we can follow down along that ray path until we reach the wavefront appropriate to the arrival time, and thus find the reflector tangent to the wavefront at this point. Following a ray path to the proper arrival time is the raypath method of migration.

Figure 18.2 shows a reflection which might be seen at four different observing stations along a seismic line. Data are plotted

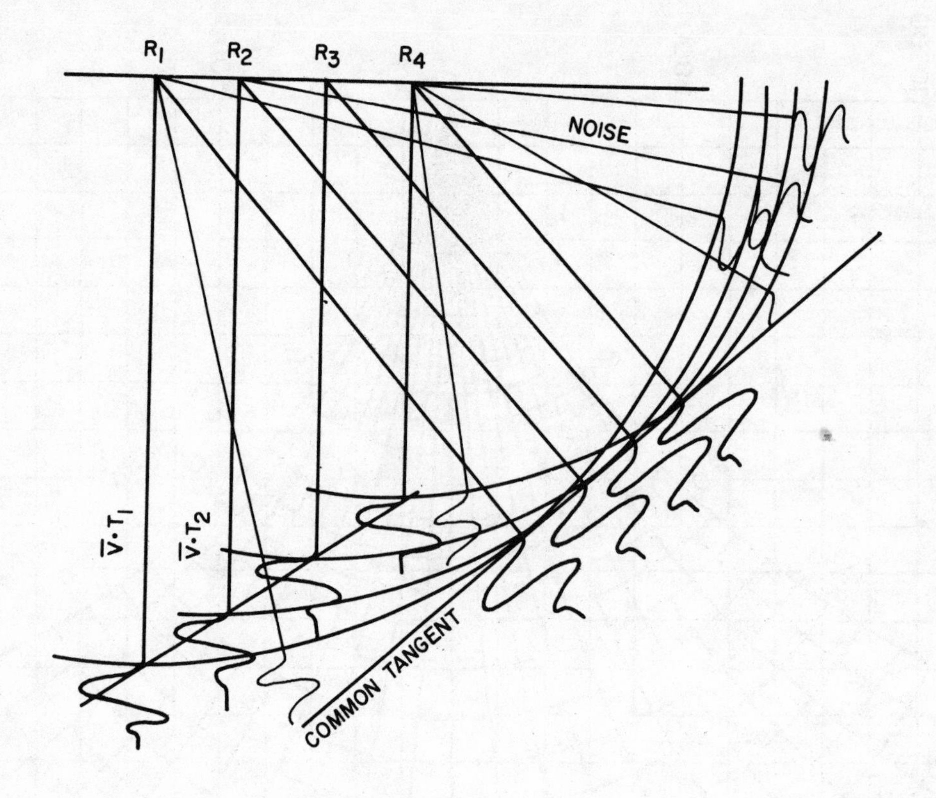

Figure 18.2

vertically beneath the observing point on the conventional seismic record section. The observed arrival time at each of the locations allows us to sketch in the wavefront relevant to that location. To make the illustration simpler, we assume a constant velocity so that the wavefronts are arcs of circles. The common tangent to the various wavefronts for the different observation points locates the reflector. This method of migrating data is sometimes called the common-tangent method of

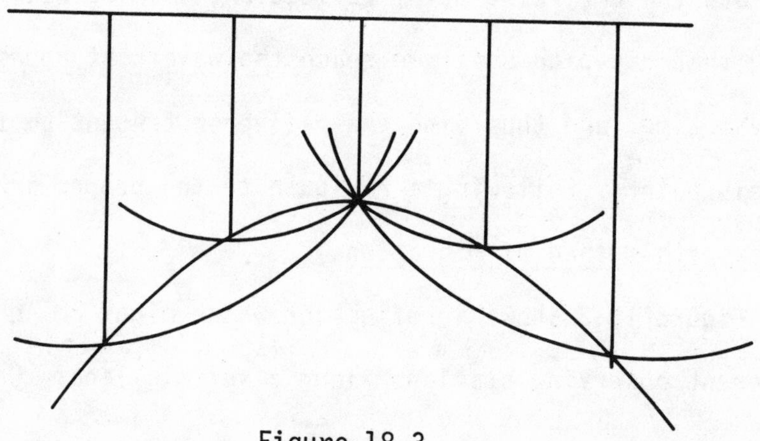

Figure 18.3

APPARENT DIP SPECIFICATION

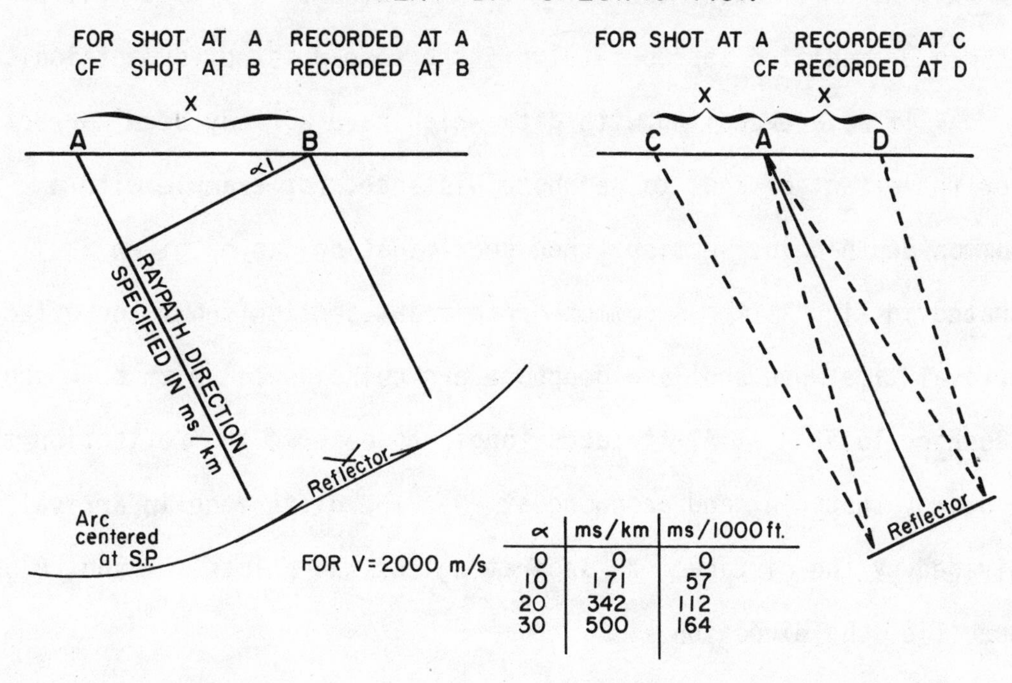

α	ms/km	ms/1000 ft.
0	0	0
10	171	57
20	342	112
30	500	164

Figure 18.4

Figure 18.5

migration.

If we assume a diffraction such as shown in Fig. 18.3, the wavefronts for arrival times at different surface locations all intersect at the crest of the diffraction curve and the common tangent becomes merely a point. Such an intersecting cluster of wavefronts is sometimes called a "diffraction knot".

Now let's examine the methods of ray path specification. Figure 18.5 is drawn for a split configuration of geophones, that is, the geophones are distributed from C to D on the surface of the earth symmetrically about the shotpoint at A. Let's assume a ray path as indicated by the direction of the arrow. The energy will take longer to be received by a geophone at C than by a geophone at D. The difference between the arrival time of the energy at C and at D (in milliseconds) divided by the distance X is the method by which we specify the direction with which the energy reaches the spread. It's usually expressed as "milliseconds per thousand feet" or "milliseconds per kilometer". While the definition relates to a symmetrical spread, it can be modified appropriately if the spread is not symmetrical.

If we are dealing with data which have already been corrected for the effect of shot to geophone distance, for example with a common-depth-point section, then specification can be given as illustrated in Fig. 18.4. A common-depth-point section shows the effective arrival time when shot and geophone are coincident. With both shot and geophone located at A, it takes longer to observe the reflection than at a nearby shotpoint and geophone at B. The difference in arrival time divided by the distance X separating the two points A and B again specifies the direction.

A wavefront chart, which shows how wavefronts spread out, is specific for a certain velocity distribution. Sometimes the vertical scale is arrival time rather than depth and the vertical and horizontal scales are chosen to match a seismic section. The chart can be overlaid

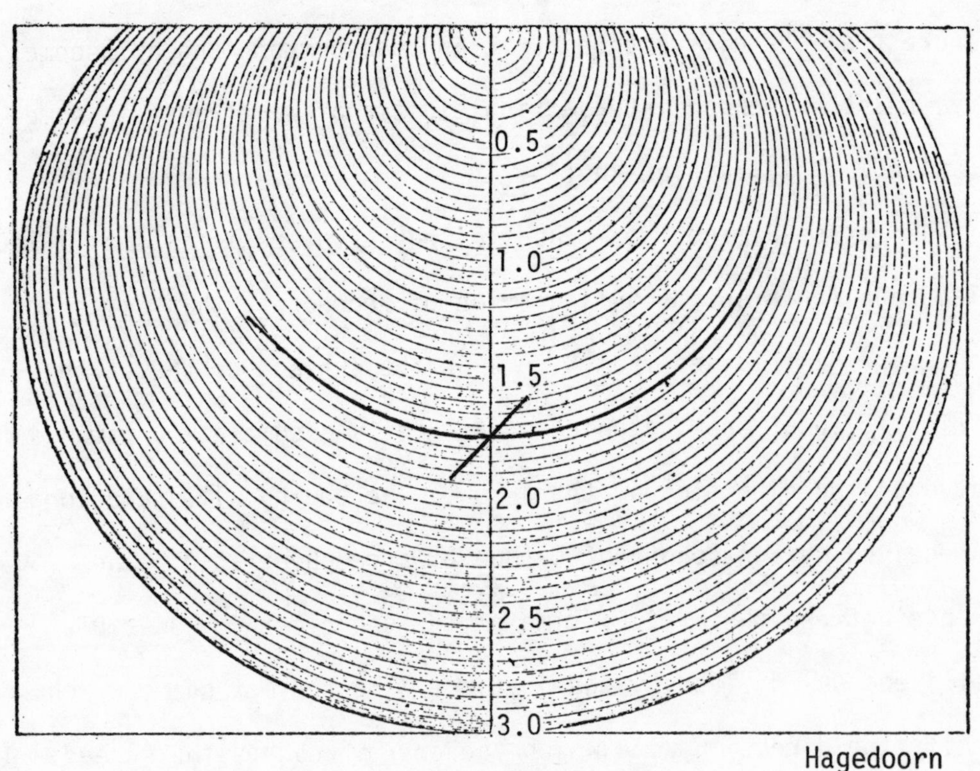

Hagedoorn

Figure 18.6

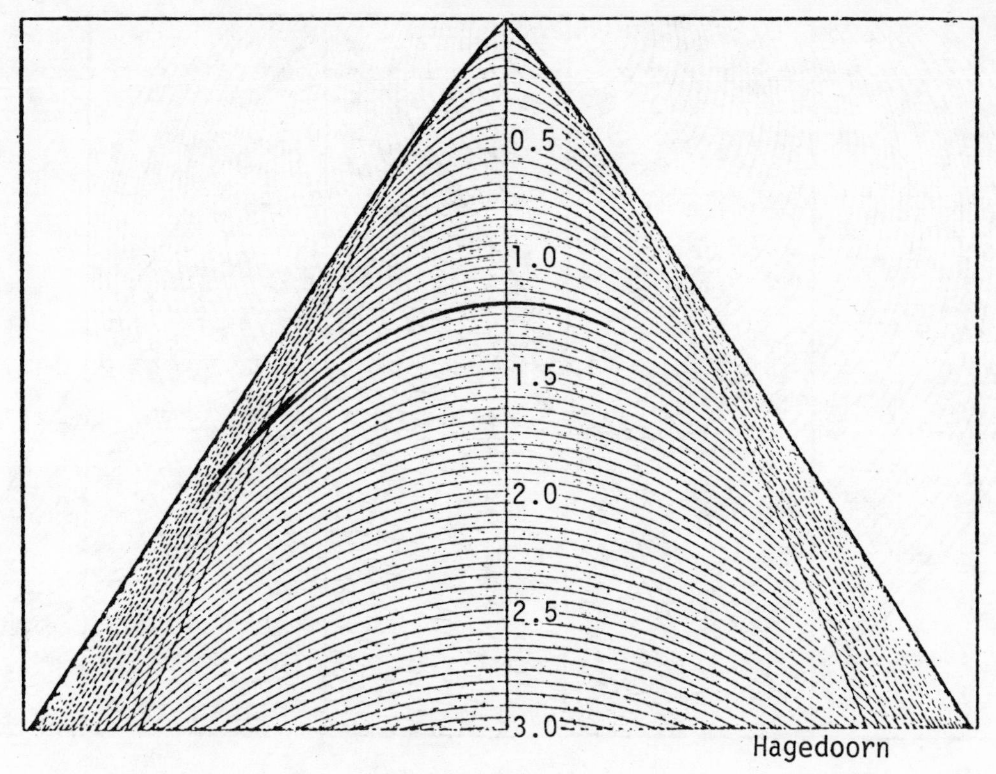

Figure 18.7

on a seismic section centered at a station where a reflection event is observed. The reflection event can be traced onto the wavefront chart, as in Fig. 18.6 where it is indicated by the bar plotted vertically beneath the shotpoint. It has the travel time indicated by the wavefront which is marked on the chart. The reflector location associated with the reflection lies in the updip direction, as shown by the ray path.

A diffraction chart is shown in Fig. 18.7. Like the wavefront chart, a diffraction chart is made for a particular velocity distribution. This chart is made for the same velocity distribution as the wavefront chart and also has the same scale. Any reflector can be thought of as a sequence of diffracting points so the reflection can be thought of as a diffraction fragment. A property of a diffraction chart is that the diffracting point is at the crest of the diffraction curve. If we slide the diffraction chart along the seismic section, at

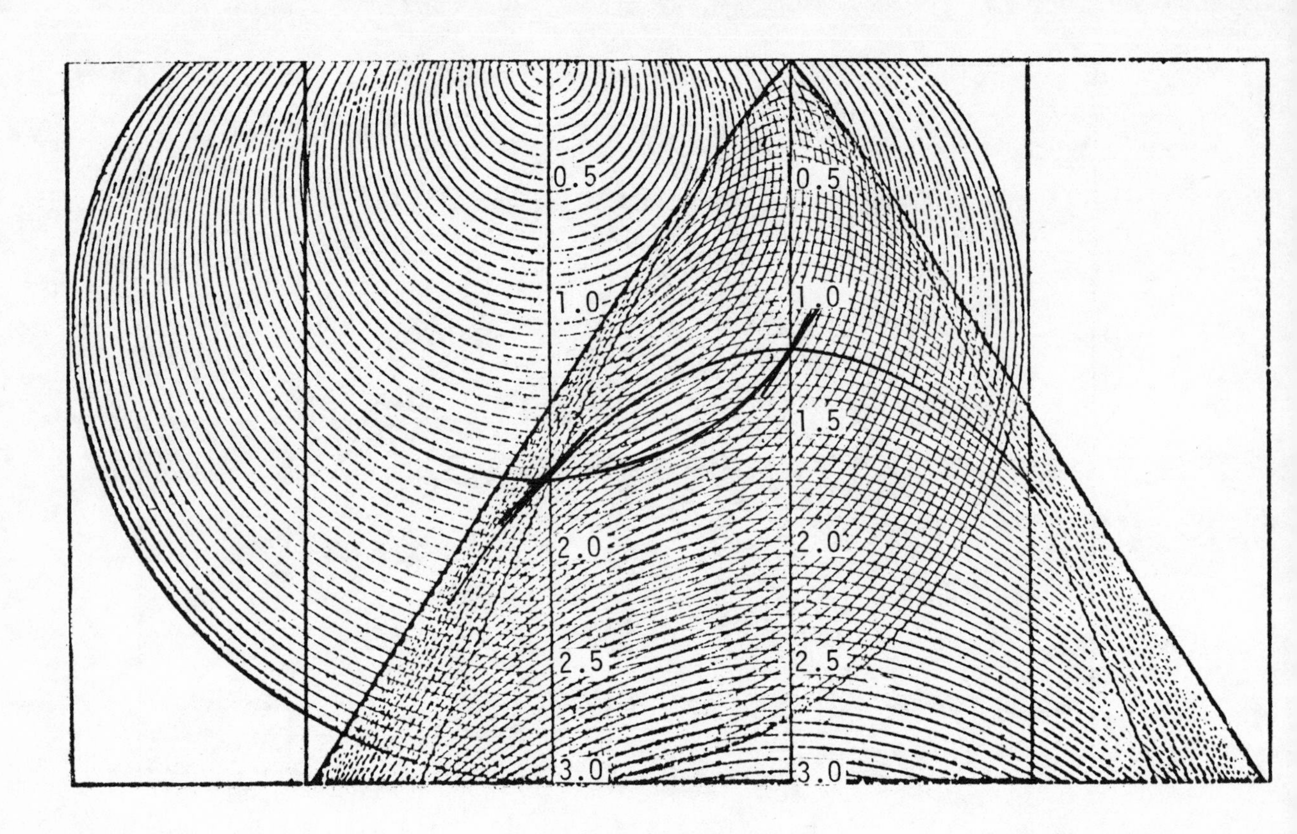

Hagedoorn

Figure 18.8

some point a diffraction curve will be tangent to the reflection element. The reflecting surface is located at the crest of the diffraction curve to which the reflection is tangent. Since we are considering the same reflection element shown in Fig. 18.6, we have the same elements fitting two different curves. If the two curves are superimposed as shown in Fig. 18.8, we see a way to accomplish migration. Both reflection and reflector lie on both the wavefront and the diffraction curve, at the intersection of these curves. The lower intersection is where the reflection appears on the seismic record section, tangent to the diffraction curve; the upper intersection is the location of the migrated reflecting surface, tangent to the wavefront chart. This forms the basis for the diffraction method of migration. This method can be used by digital computers to migrate data automatically. The data are searched along all possible diffraction curves, and the data elements so found are summed and positioned at the crests of the respective curves; the result is a migrated section. The basis of this migration method was worked out by Hagedoorn and the method is also called the Huygens, Fresnel or Kirchoff method of migration.

Migration methods (when the intrinsic assumptions are satisfied) result in repositioning reflections appropriate to the surface locations which give rise to the reflections, but the vertical scale is usually vertical travel time rather than depth (see Fig. 18.9). Knowledge of the velocity permits converting from a vertical time scale to a depth scale.

To summarize methods of migration:

(1) Swinging wavefronts (common-tangent method);

(2) Intersection of wavefront and ray path (including use of plotting arms, timing along ray path, etc.);

(3) Search along diffraction curves;

(4) Wave-equation methods (downward continuation), which will be discussed in Chapter 19.

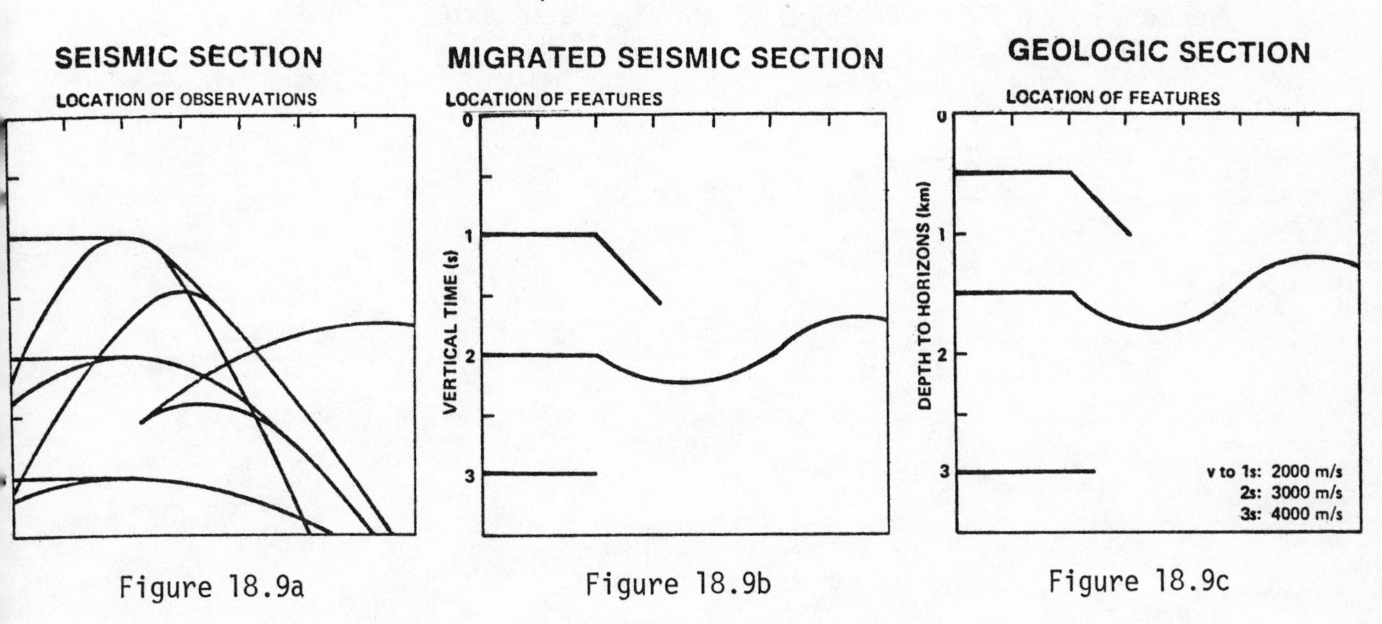

Figure 18.9a Figure 18.9b Figure 18.9c

All of these methods make the same three basic assumptions:

(1) all the events are primary reflections or diffractions

(2) the events are in the plane of the section, that is, we have only the two dimensions of depth and in-line horizontal distance to be concerned with, and

(3) the velocities are known both in the vertical and the lateral sense.

The migrated result will be in error to the extent that these assumptions are not satisfied and subsequent examples will show the effects of such errors. Conceptually true three-dimensional migration can be accomplished by the same methods, but this is not commonly done. Usually the input data required for true three-dimensional migration are not available since data are customarily acquired along lines rather than over areas, and the amount of computer work for a three-dimensional solution is so large as to burden most computers. Further, two-dimensional migration produces a very useful result in most instances. Even where the three assumptions are not satisfied, a two-dimensional migrated section is usually considerably superior to an unmigrated section as an aid in seeing structural relationships.

CHAPTER 19

MIGRATION II: WAVE EQUATION MIGRATION

A general principle of geophysics is that features are seen more sharply if observations are made nearer to them. The wave equation method of migration utilizes this principle.

The seismic wave field varies from location to location and also as a function of time. Ordinarily we make observations as to how the field varies as a function of time only at the surface of the earth. The seismic wave field obeys certain continuity properties which allow us to calculate what the field would be on some other surface based on knowledge of the field over one surface. The surface for which we calculate the field is lower than the surface on which the field is known, so this technique is called downward continuation. The concept is similar to the downward (or upward) continuation of gravity or magnetic fields. Wave equation migration involves downward continuation.

In this migration method we determine the wave field at some depth of the earth, perhaps 250 feet, from measurements of the field on the surface. Then, knowing the wave field at 250 feet, we calculate what it would be at 500 feet, and then 750 feet, and so on. The processing proceeds downward one step at a time. As we approach the particular depth of some feature, the seismic effect of that feature becomes very sharp. Thus we successively sharpen the expressions of features at successive depths.

The top of Fig. 19.1 shows a seismic trace observed at the surface with time increasing to the right. We might see three reflections,

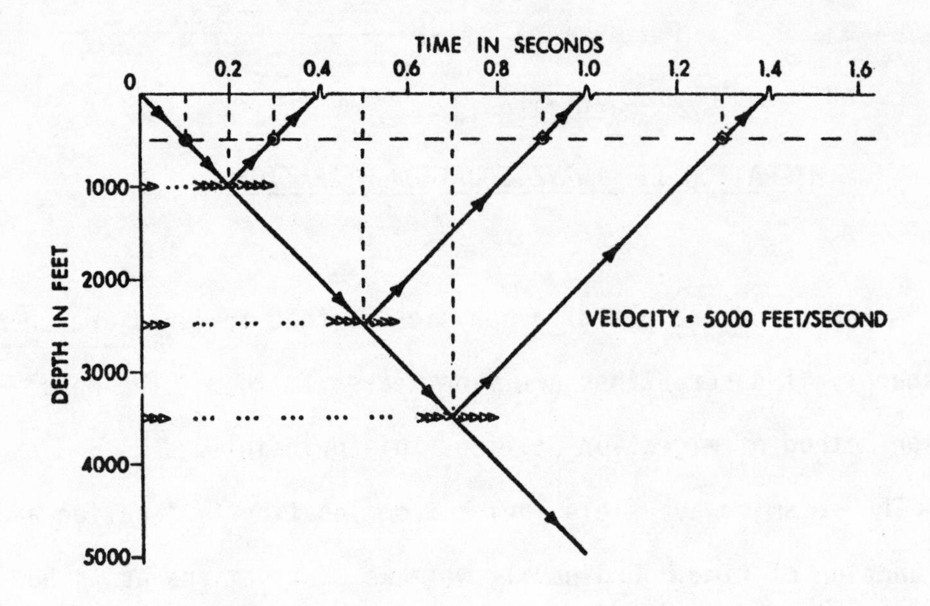

Figure 19.1

at 0.4 seconds, 1 second and 1.4 second. If we observed 500 feet deep in the earth, the downgoing wave from the shot would not arrive until 0.1 second (assuming a velocity of 5000 feet/second) and the reflections would be observed at 0.3 sec, 0.9 sec, and 1.3 sec; the downgoing wave from the source would arrive at 0.1 sec. If we observed at 1000 feet, the first reflection would coincide with the arrival of the downgoing wave because our observation is at the depth of the reflector; the reflection expression of a feature at this depth has its maximum sharpness at this depth. We continue this procedure until we have continued through the entire section. Wave equation migration involves horizontal filtering of data elements to calculate the next set of elements. The procedure is carried out recursively, step by step, down through the section.

This procedure is illustrated in Figs. 19.3 to 19.8 which show successive <u>steps in this continuation process</u> (only a few of the steps are shown). The model in Fig. 19.2 was designed to illustrate different

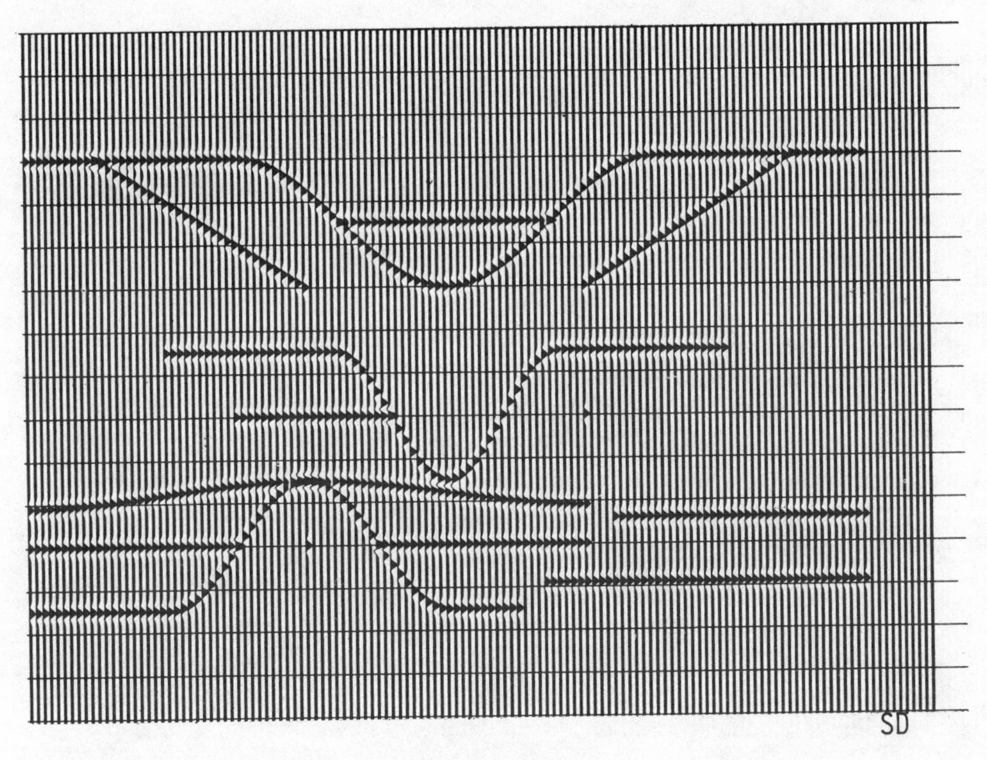

Figure 19.2

kinds of structural features: broad and sharp synclines, broad and sharp anticlines, faulting, truncation of beds. The seismic section which would be obtained over such a model is shown in Fig. 19.3. The general forms of the features show because there is no noise, but the features are not very sharp. All events terminate in diffractions. The anticlines appear broader, the fault locations are fuzzy and the shape of the synclines is turned inside out.

Figure 19.4 shows the result of continuation to the level indicated by the arrow at the side of the figure. Since we are observing at the level of the first reflection, it is now sharp as are the truncation of the dipping beds against it. However, the deeper data are still somewhat confused. In Fig. 19.5 we have continued the data to a deeper

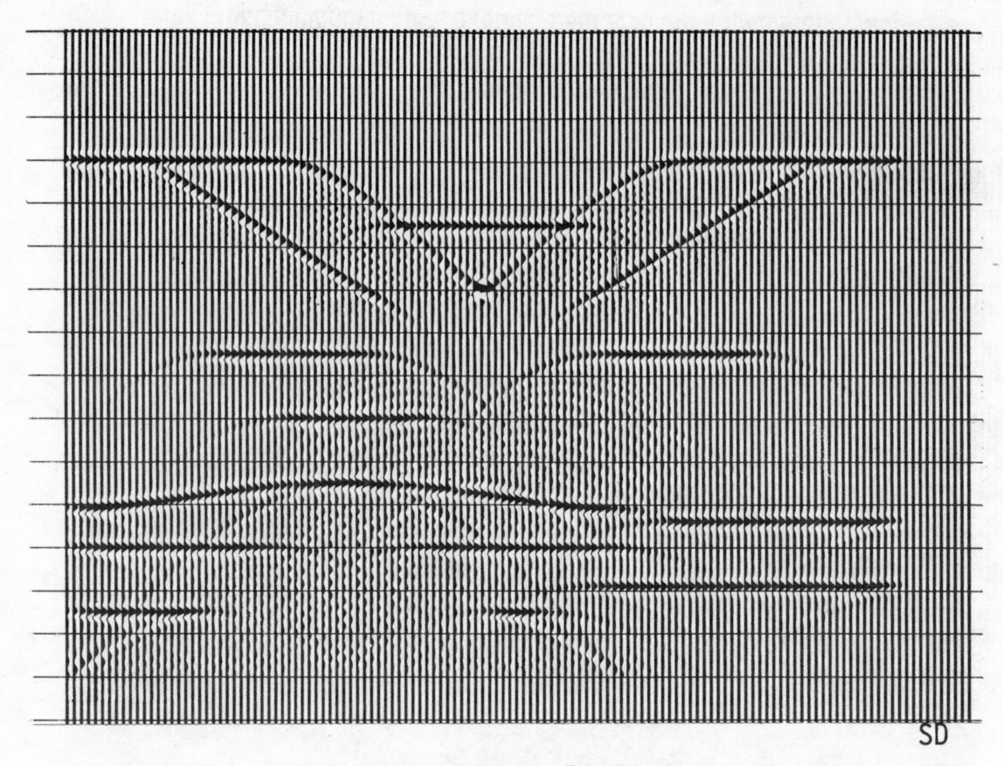

Figure 19.3

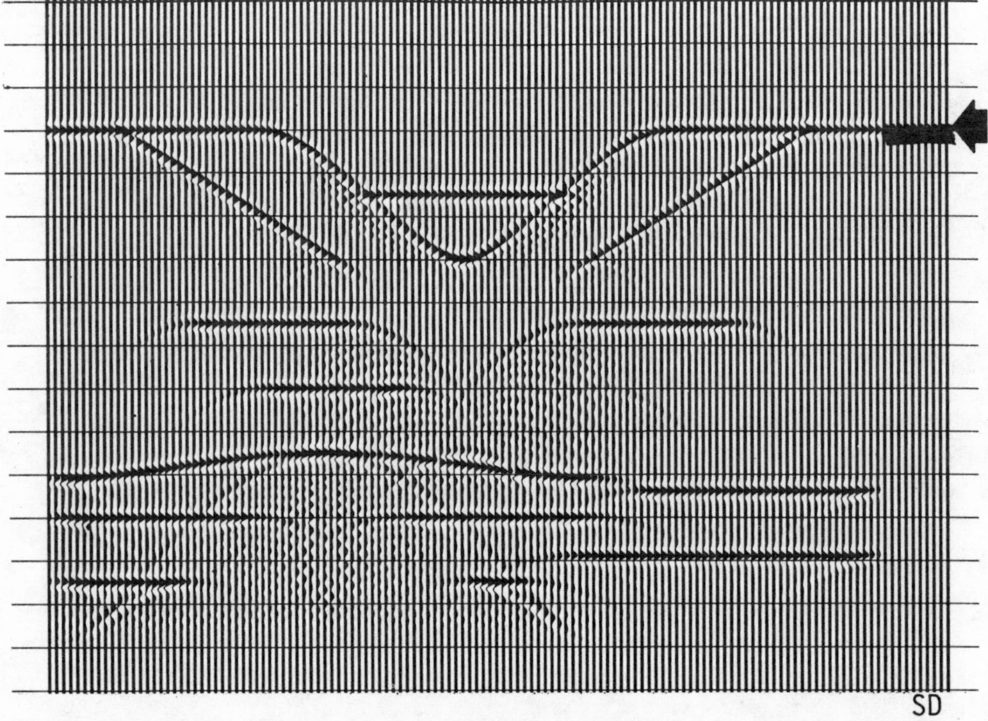

Figure 19.4

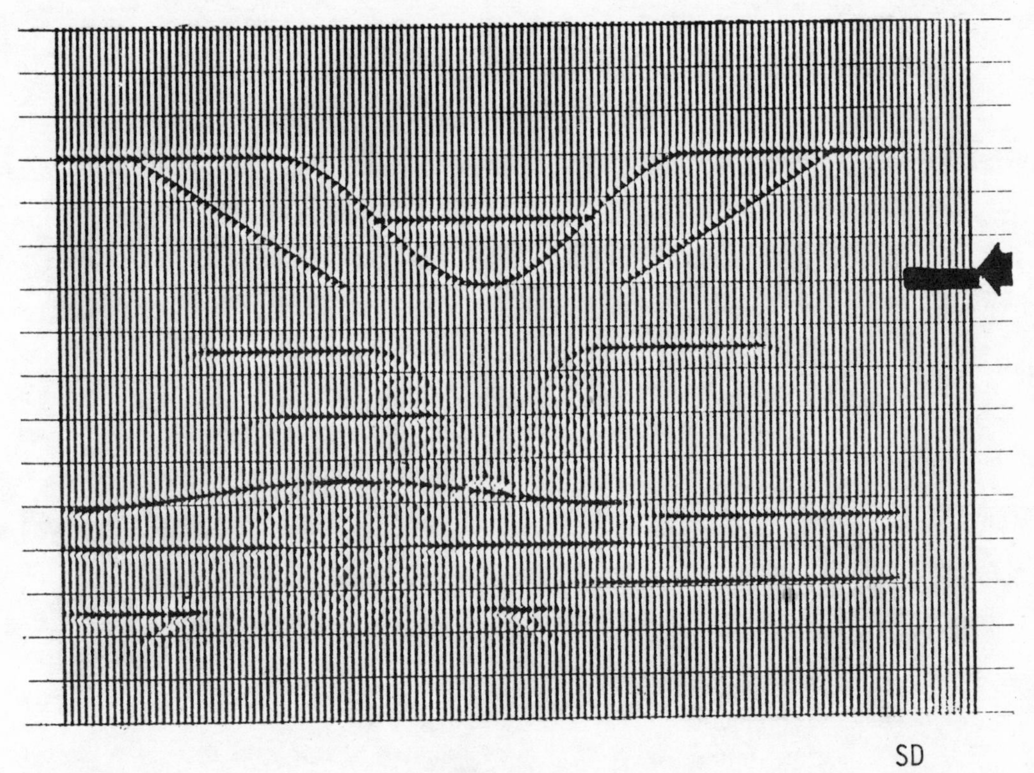

Figure 19.5

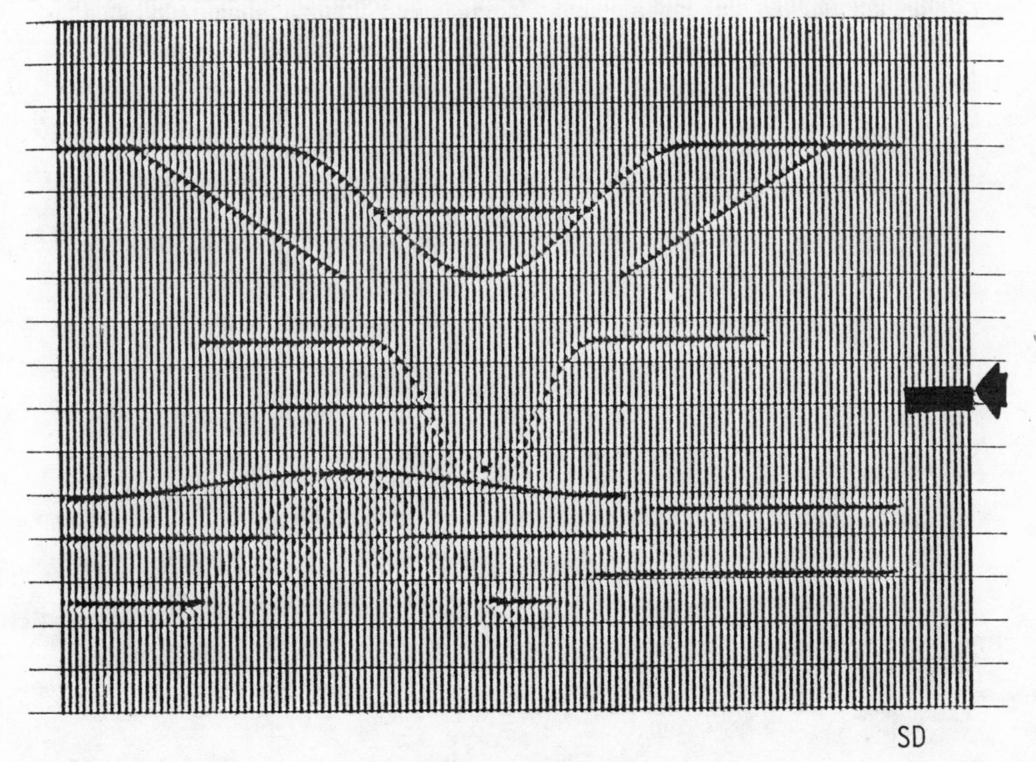

Figure 19.6

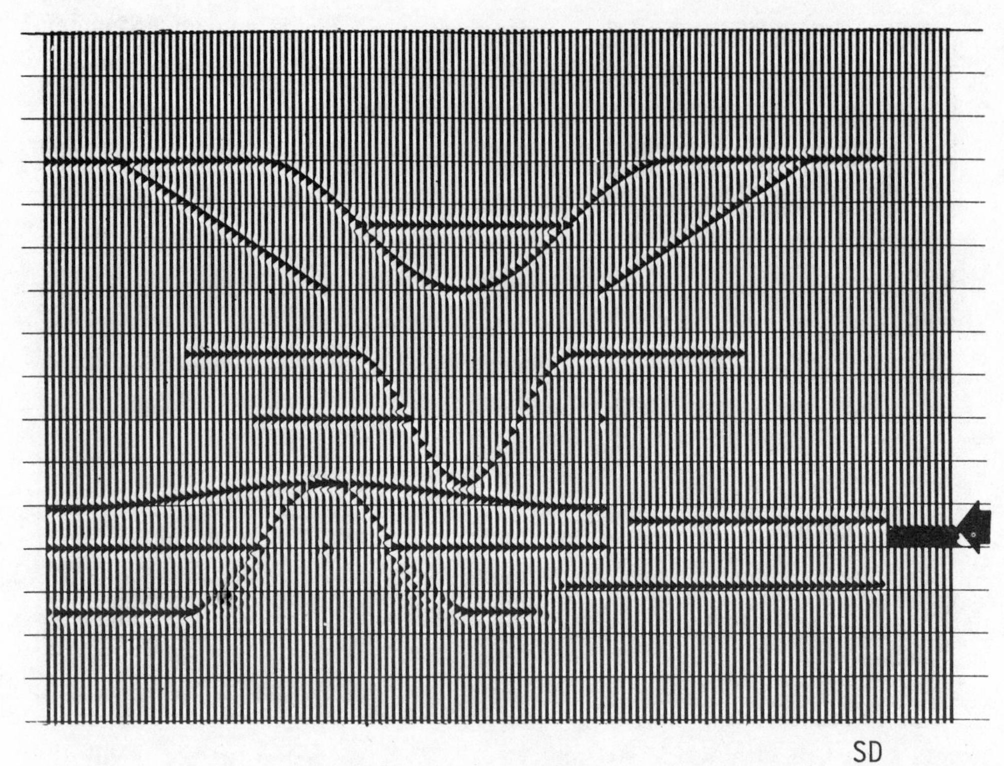

Figure 19.7

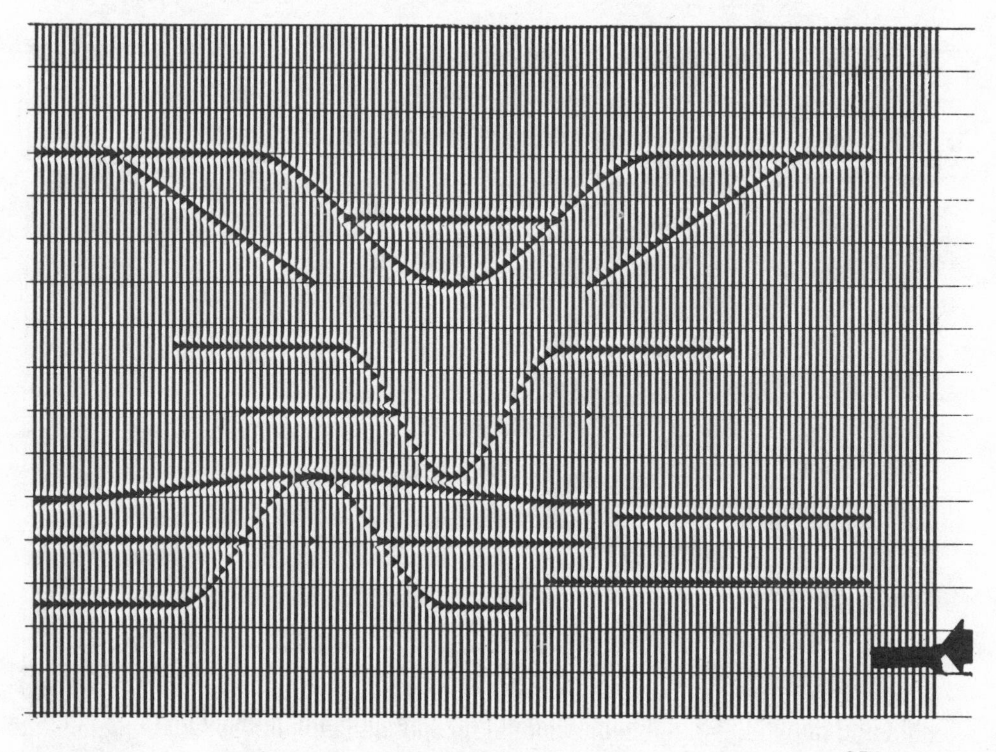

Figure 19.8

level. As the geophones are effectively lowered, we preserve the data above the continuation depth, because we have already achieved the maximum sharpness of such data, and we change only the data below this level. In Fig. 19.6 the geophone level is below the upper syncline but not yet to the lower syncline. The upper syncline has been sharpened and put in approximate location. Figure 19.7 shows the result of observing at the level of the center of the lower syncline. We have passed the center of curvature as the buried-focus effect disappears. In Fig. 19.7, the observing plane is at the top of the lower anticline and the faults are sharper. Finally, in Fig. 19.8 we have continued through all the data and we have a correct image of the original mode. Wave equation migration thus involves calculating the field that would be observed at successive depths in the earth and in so doing sharpens events and puts them in correct position.

Figure 19.9 shows events with various degrees of dip before and after migration. The data are in correct migrated position and the

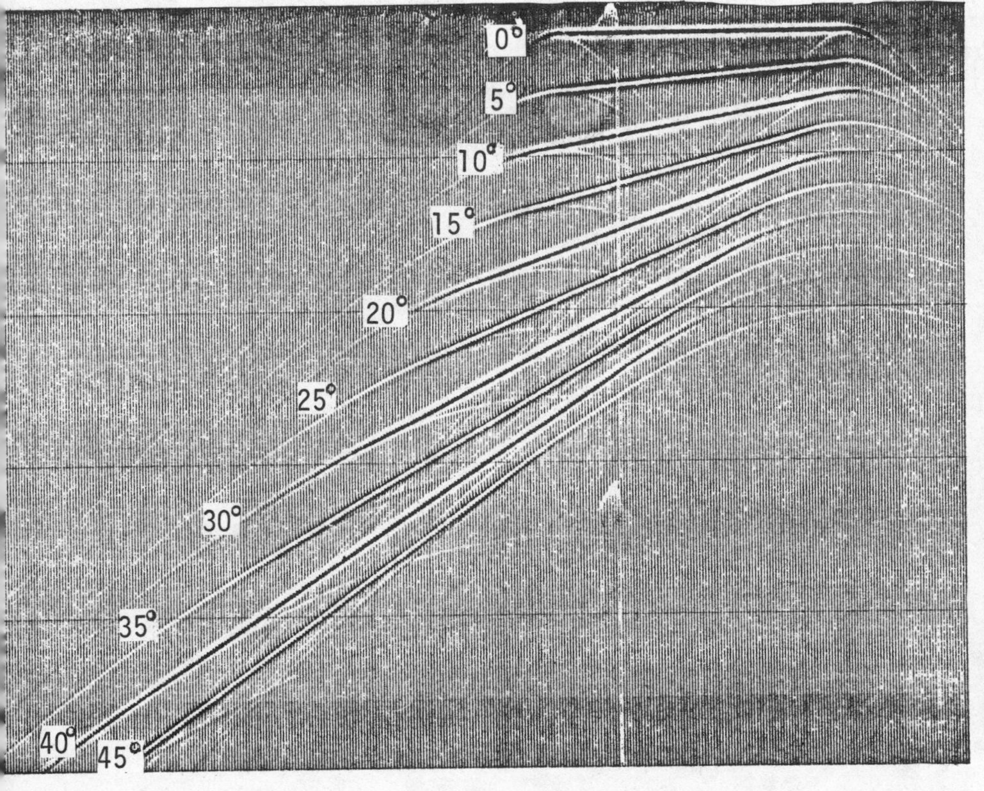

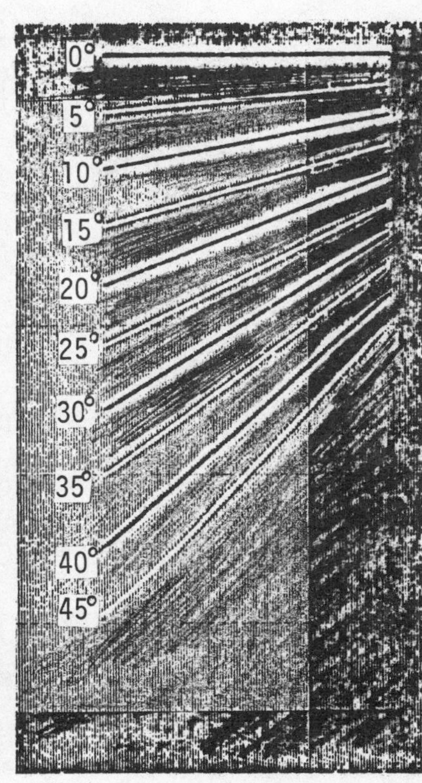

SD

Figure 19.9

terminations of events are sharp. The most steeply dipping event shows some distortion. The input data were cut off at a certain time and location, and some data elements for this steep event are missing so this distortion is a boundary effect, at least in part. Seismic lines do not go on to infinity, and we cannot recover data that aren't given.

Another model seismic section is shown in Fig. 19.10 with the corresponding migrated section shown in Fig. 19.11. The sharp little anticline on the second reflector might be a reef sitting on a platform with which it has no acoustic impedance contrast. The curvature of the top of the reef is so great that the reflection from it appears almost as a diffraction and the reflections from the platform appear to be almost continuous through the reef base. The sharp synclinal curvature at the base of this sharp anticline gives rise to buried focus effects. Clearly a small reef will be difficult to see in seismic data and its effects are apt to be confused with other types of features.

Figure 19.12 shows real data involving buried foci. After migration, Fig. 19.13, events are much sharper. The strong reflection below 1 sec which migration makes into a continuous surface with appreciable relief is the basement surface. The big arcs at the bottom of Fig. 19.13 are multiples which have been migrated as if they were primary events; they do not satisfy the assumption that all data are primary events and hence, are not migrated properly.

Data from the North Sea are shown in Fig. 19.14. The more or less continuous reflection at the bottom is the base of a salt section and the anticline is formed by flowage of the salt. The complex of diffractions near the center are typical of massive mobile salt; they probably

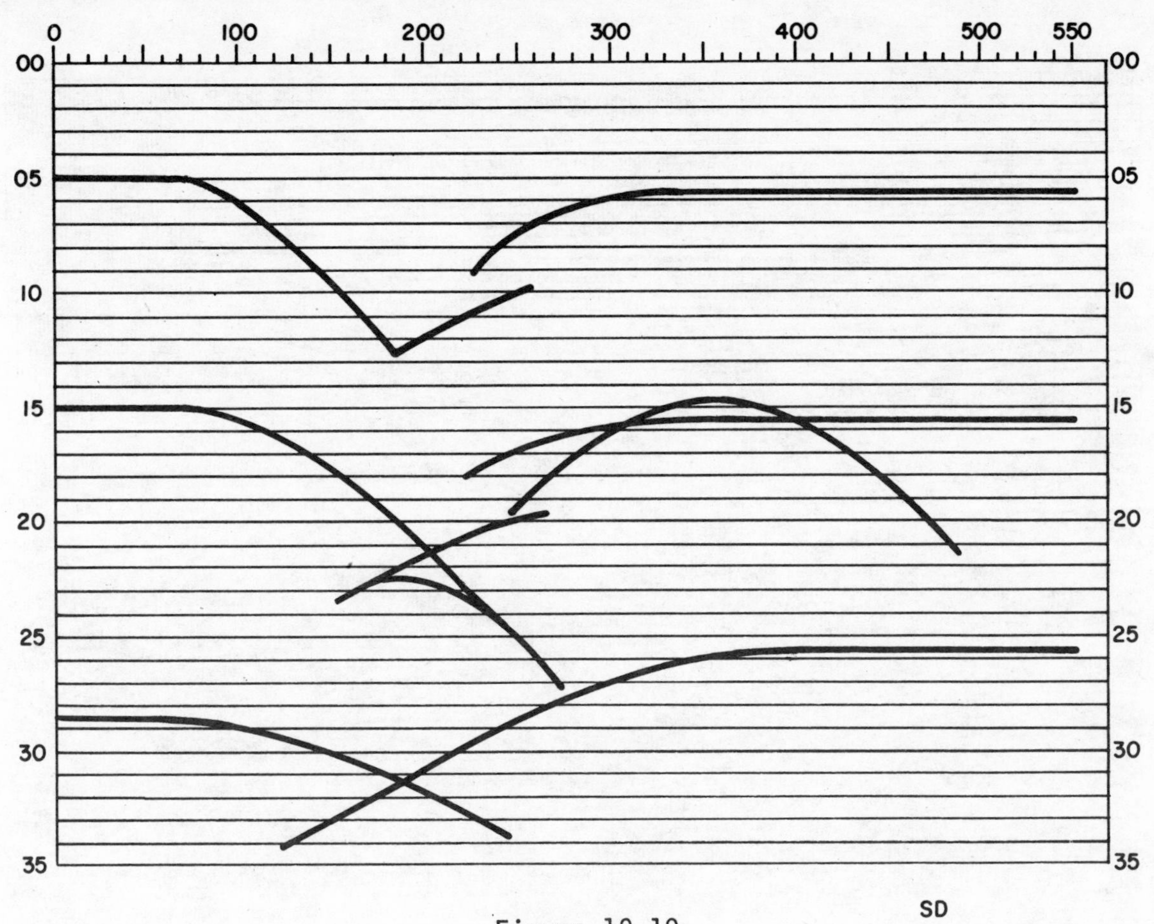

Figure 19.10

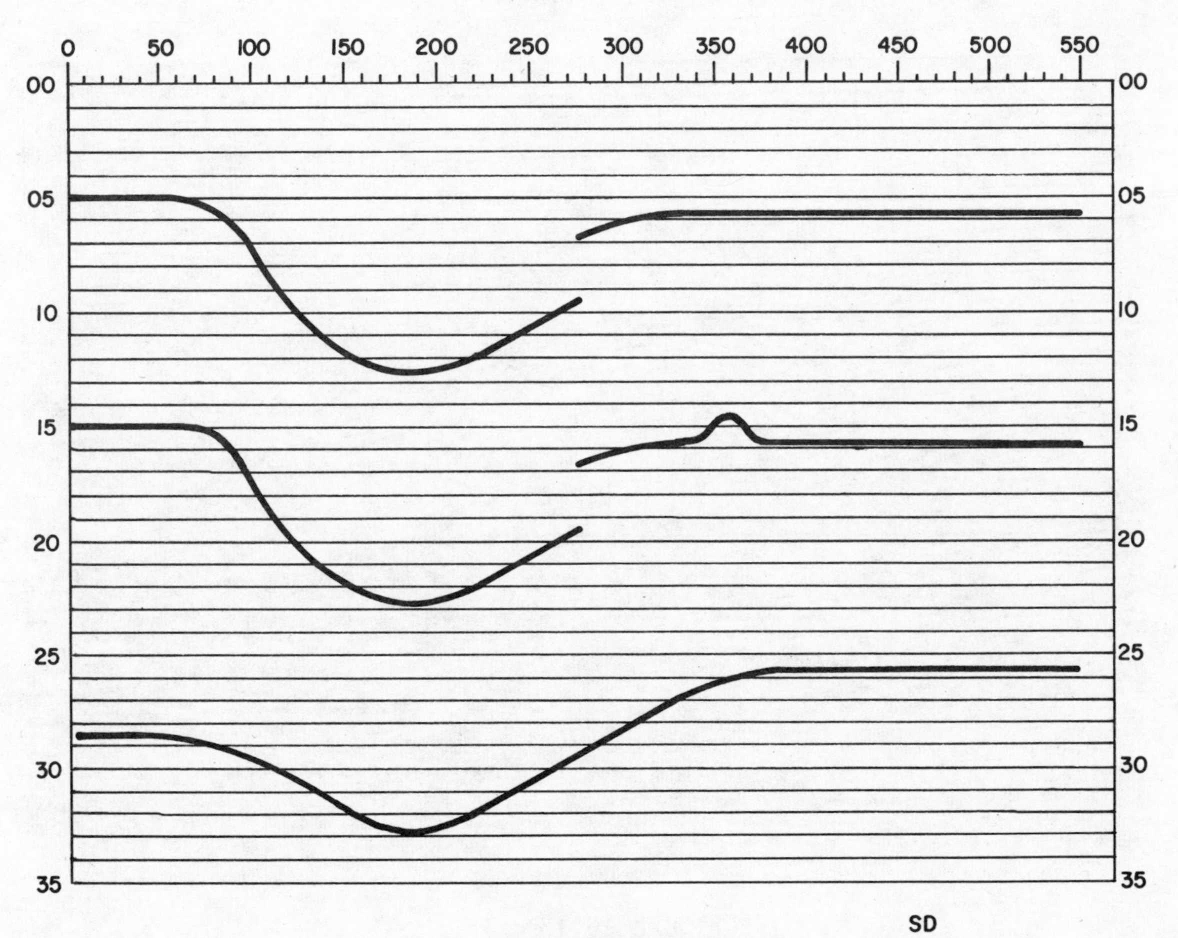

Figure 19.11

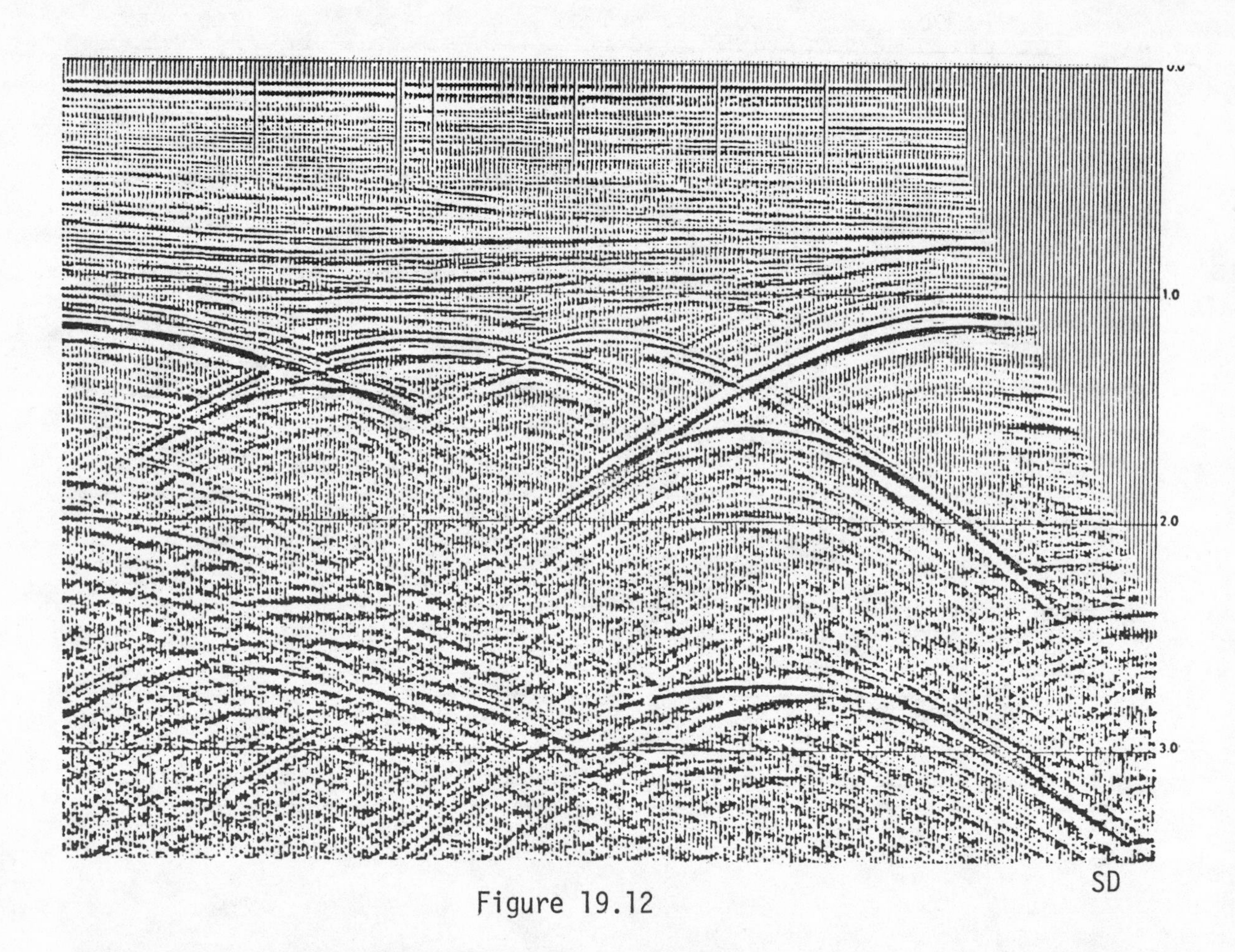

Figure 19.12

SD

Figure 19.13

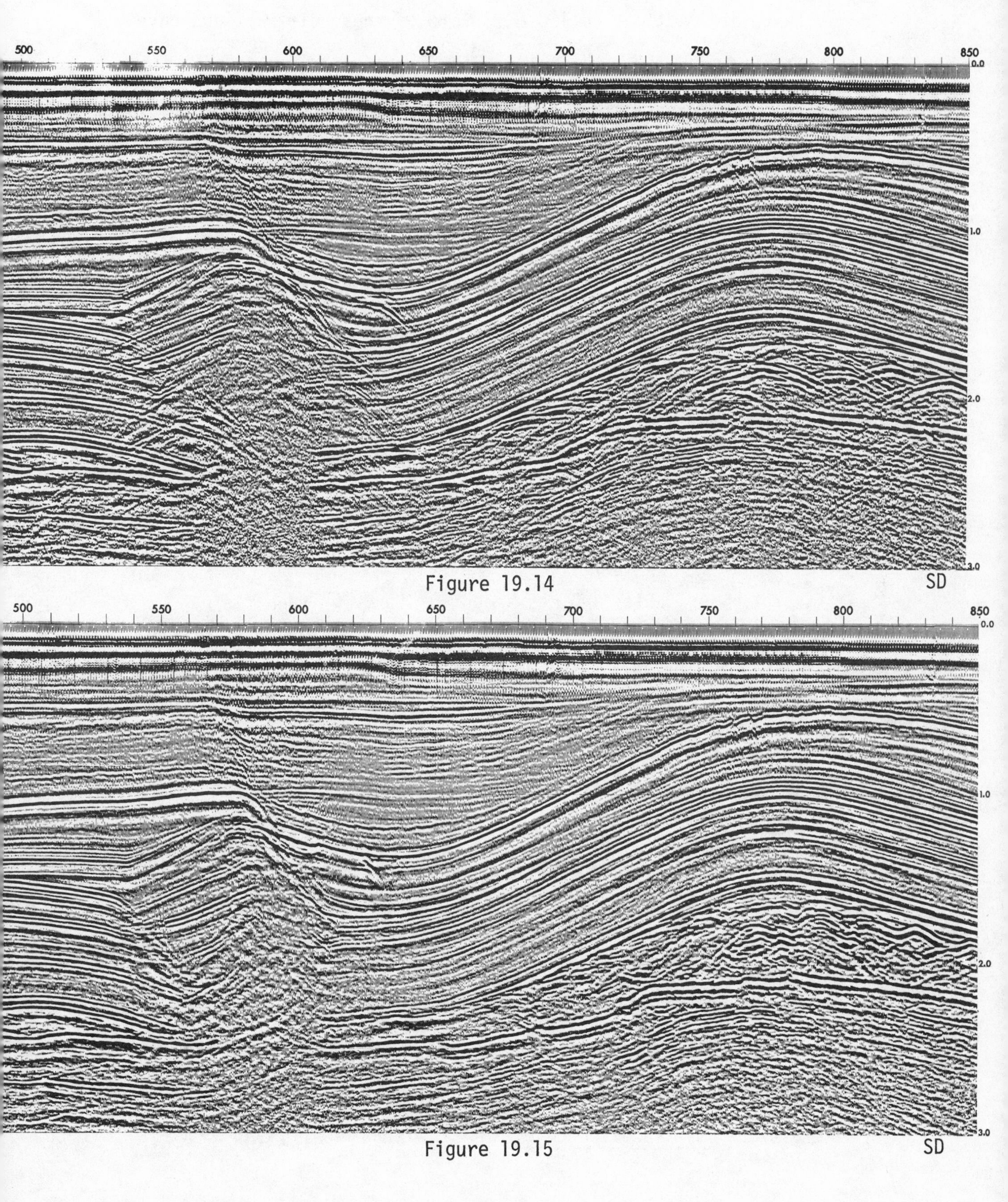

Figure 19.14

Figure 19.15

are caused by huge jostled blocks of anhydrite and other rocks which were affected by the salt movement. Figure 19.15 after migration shows these diffractions collapsed. Some of these diffractions have not collapsed completely, probably because their sources are not in the plane of the section and hence they violate another of the assumptions implicit in migration. However, their shape has been made more intelligible by the migration even though they have not been completely collapsed. Migration is usually a help in making seismic sections easier to interpret even where the implicit assumptions are not satisfied. The base of the salt is broken by normal faults. A faulted anticline underlies the salt layer. There is also a gas accumulation in the anticline above the salt.

Figure 19.16 shows evidence of a number of faults, but it is not clear how many faults are present nor their precise locations. After migrating, Fig. 19.17, a number of minor faults are clear as well as the major faults and the amount of throw on each can be determined quite reliably.

The use of migration is sometimes restricted to areas with appreciable dip on the theory that it will not significantly affect interpretation in areas of flat dip. However, it is the departures from flat dip that are of interest and the sometimes subtle evidences of these need to be located properly to make them sensible. Figure 19.18 shows a section of mainly flat dip but with a few anomalous dips. There is a layer of salt in this section which has been generally disolved away in the region to the right, which has allowed subsequent sediments to sink into the collapse zone. Some closures have resulted and the collapse has also affected subsequent deposition. This section is across the salt solution edge. After migration, Fig. 19.19,

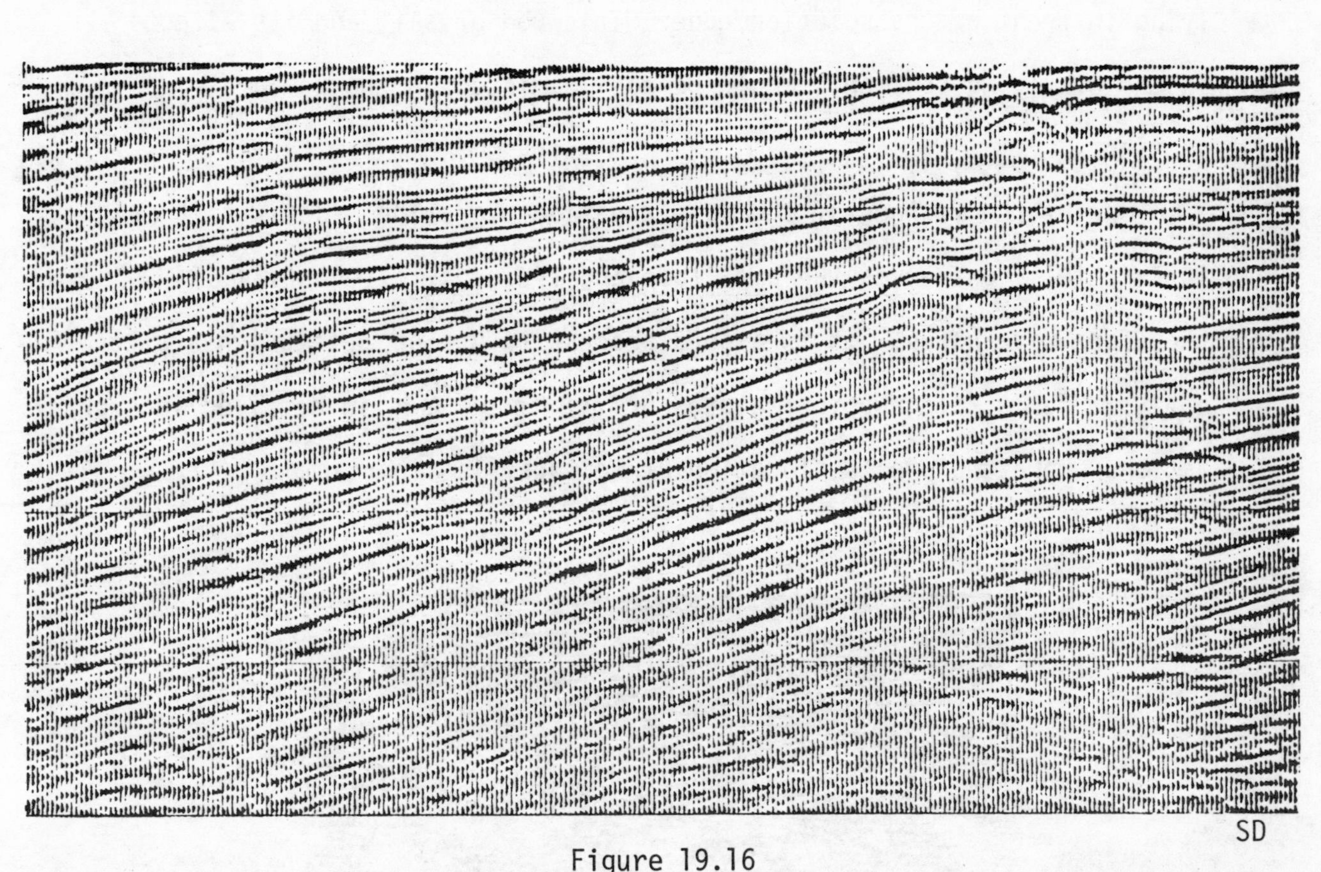

Figure 19.16

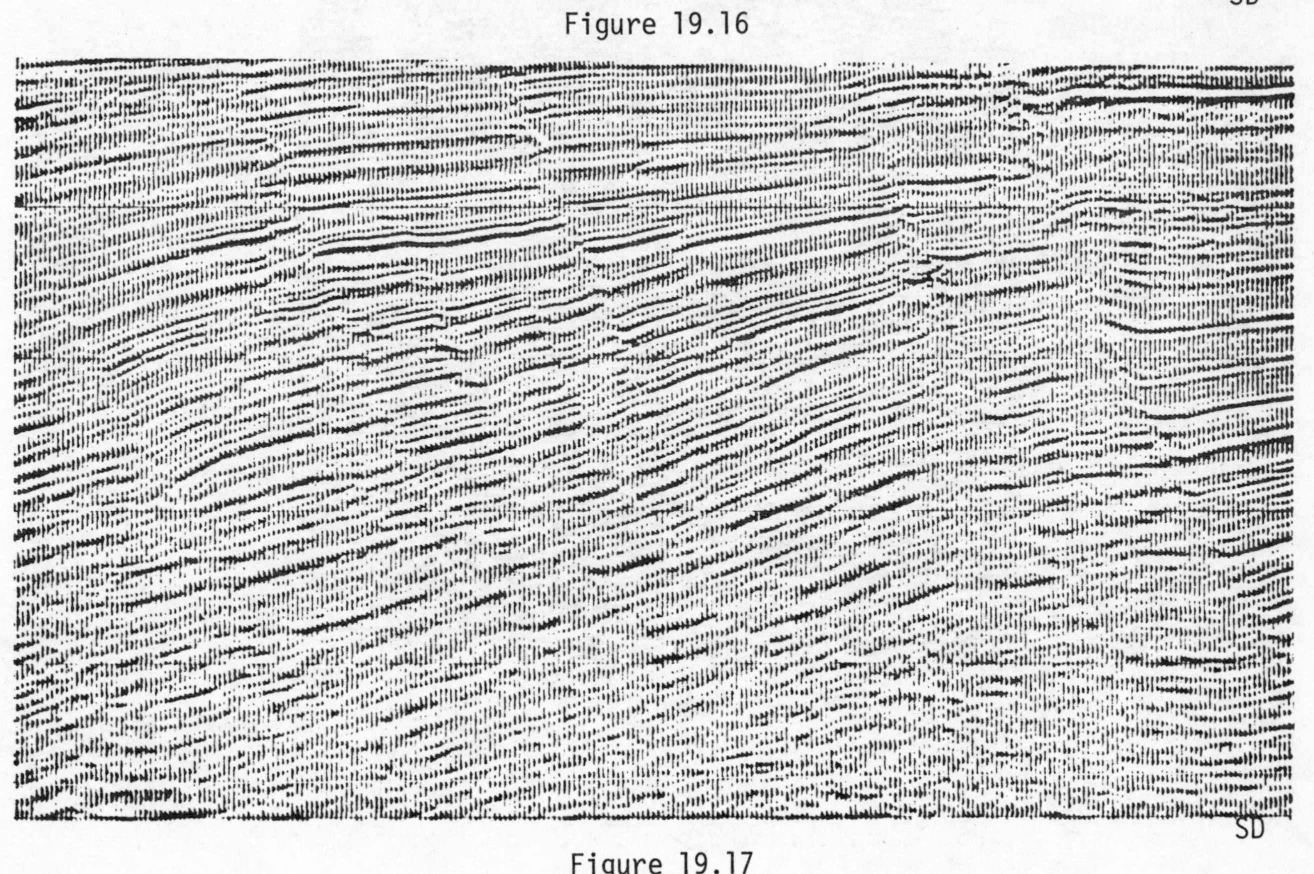

Figure 19.17

the salt edge shows clearly and also a pod of residual salt can be seen lying in front of the solution edge. This pod of salt and its effects in the overlying sediments would be easily missed on the unmigrated section.

One of the virtues of the wave equation migration process is that it faithfully preserves the amplitude and frequency of events and therefore waveshapes. It is a good process to use on data which are to be analyzed for stratigraphic features or are to be used in detailed interpretations where the precise location of features is important.

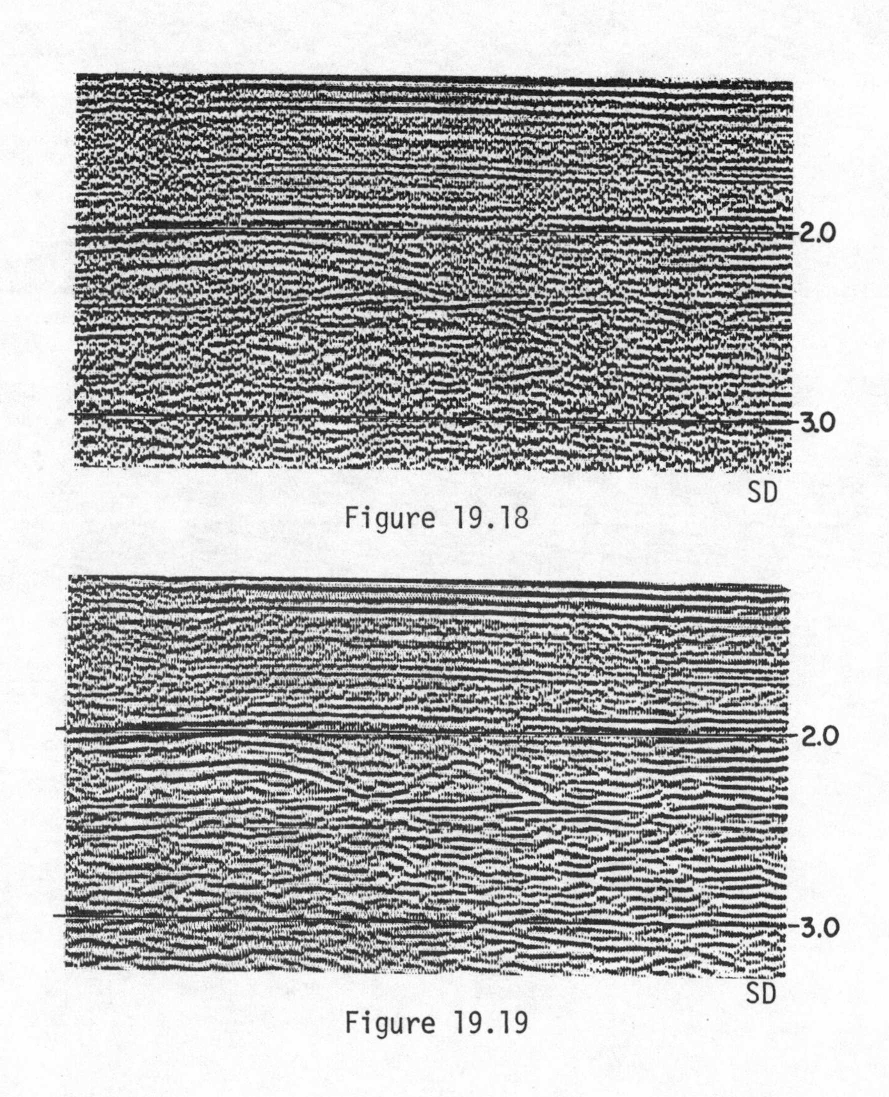

Figure 19.18

Figure 19.19

CHAPTER 20

FAULTING

We shall first look at faulting on a seismic model and second on an actual seismic section, in order to see the criteria by which faults are recognized.

The fault model in Figure 20.1 is a step model of a reflector, taken from Hilterman. Below the model is the seismic section which would result from a line shot perpendicular to the fault. The crosses on the model indicate reflecting points, the projection of observing points on the surface on to the model. The throw on this model is large, about two wavelengths, so the reflection from the bottom of the step has to travel 4 wavelengths further than that from the top

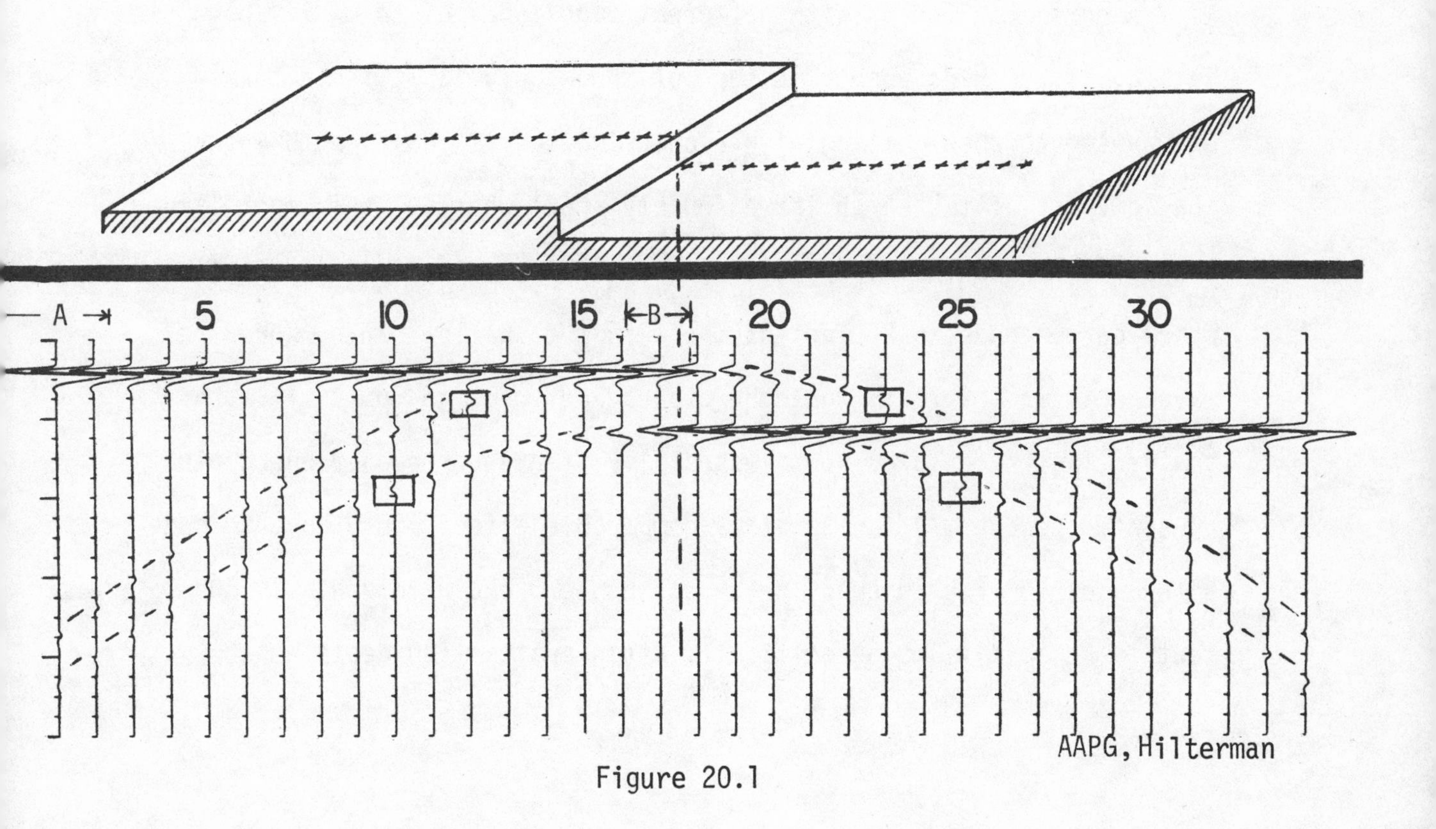

Figure 20.1

surface. Hence the reflections from the upper and lower surfaces terminate fairly cleanly in diffraction curves, as indicated by the dashed lines. The boxes indicate seismic traces an equal distance on either side of the termination point of the fault (the vertical dashed line). Note that the wave shape is the same in boxes which are equal distances from the crest of the diffraction curve except that they are of opposite polarity. Since their amplitude is the same, the amount of energy carried in the two limbs of the diffraction curves is equal. The backward limb which appears underneath the reflection is often not evident, however, because it gets lost in the tail of the reflection. Note also that the amplitude of the reflection event itself decays as it approaches the location of the fault. The amplitude of the reflection just before the fault is reached has only half of the energy which it had when remote from the fault edge, which can be seen by comparing values of the amplitudes at A and B.

A series of faults with different magnitudes of throw is shown in Fig. 11.8. The proper measure of throw on a fault is in terms of wavelength when dealing with a seismic record. Wavelength in the shallow part of the earth is reasonably short whereas it is much larger deeper in the earth. The fault with a throw of a quarter-wavelength, which corresponds to 25 feet of throw in the shallow earth where the wavelength is 100 feet, would be clearly defined, and possibly even smaller faults could be seen, especially if high frequency energy is present. However, a fault with a comparable effect would have to have a throw of 150 feet deeper in the earth where the wavelength is 600 feet. The ability to detect faults deteriorates with depth.

Two seismic lines, shown in Fig. 20.2, intersect at right angles. The adjacent traces of the two sections represent the same location on the surface, so that the data tie at that point. The reflection identified as Σ has fairly distinctive character, consisting of about three cycles which allows that event to be identified at various locations along the lines. It is fairly clear that this reflection

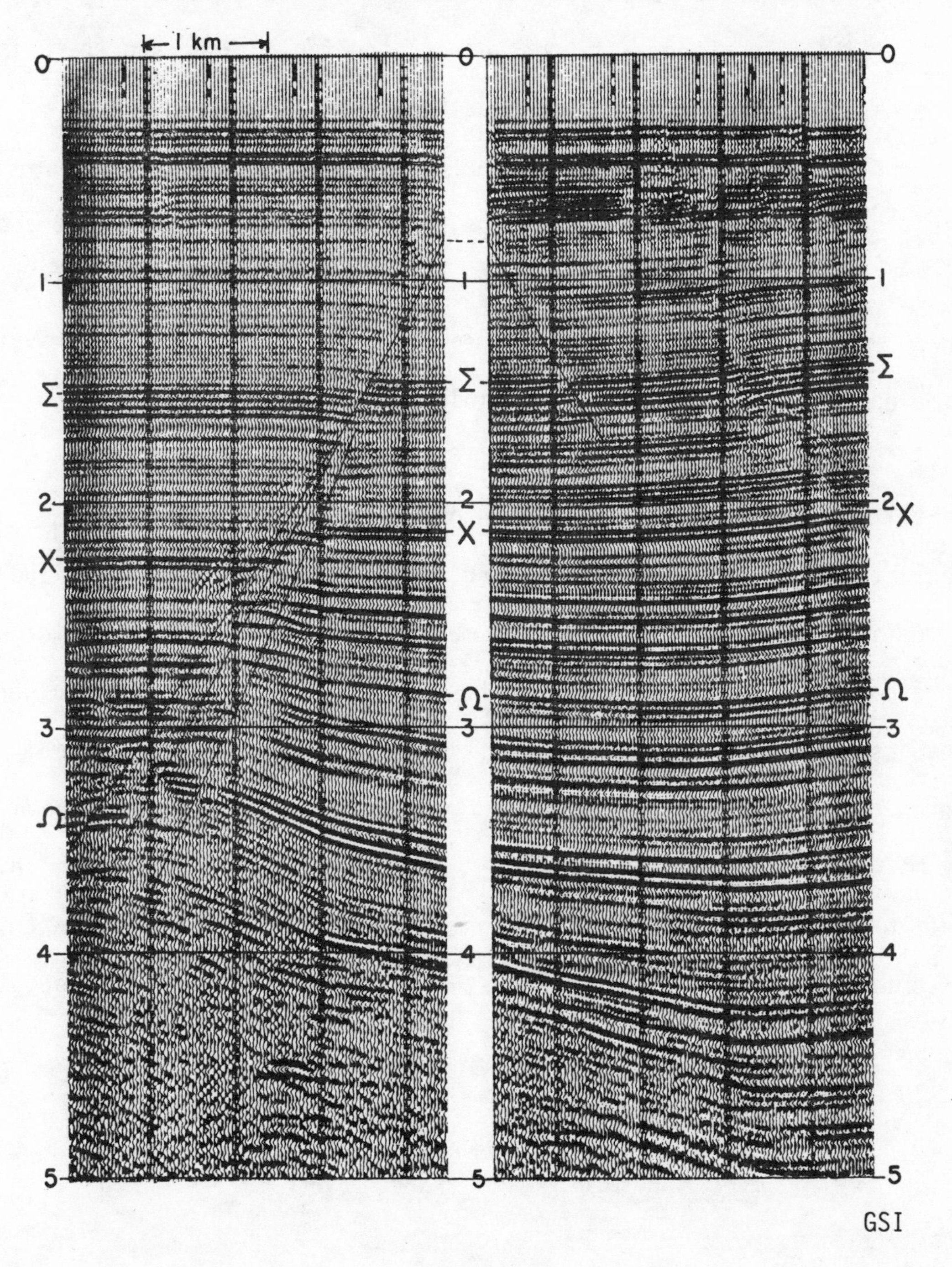

Figure 20.2

terminates and is displaced on the left section, indicating a fairly well defined fault. Farther down along the fault plane are some diffractions and pieces of diffractions, which are also evidences of this fault. In order to honor the diffraction information on the section, the fault has to be split into two faults and the attitude of beds in the narrow slice between the two faults differs slightly. Some of the data change attitudes across the fault plane and there is some distortion of the data underneath the fault plane.

On the shallow Σ horizon, the difference in time across the fault has been converted to depth using the appropriate velocity, and the fault is found to have 300 feet of throw at this location. The throw on the fault is obtained by measuring the difference in arrival time upthrown and downthrown, multiplying by the velocity and then dividing by two because the travel time difference represents two-way travel whereas we are interested in only the one-way throw. If we look at a deeper horizon, the strong event marked X whose amplitude gives it distinctive character, we measure 600 feet of combined throw across the fault and splinter fault, using a velocity which is larger than used for the Σ horizon. We similarly determine 3000 feet of throw on the lower horizon called Ω, which is identified downthrown by more of the seismic line than shown here. The conclusion from this left-hand section, therefore, is that the throw increases with depth, a common situation in the Gulf Coast area where these data were acquired.

This fault extends to the line intersection and on the right section this fault cuts the Σ horizon with only 100 feet of throw. It also appears to decrease its throw with depth and die out before cutting the X horizon. Thus the interpretation of this fault appears to be contradictory, that it is growing with depth on the left-hand section and dying out with depth on the right-hand section. This paradox can be explained by drawing a map as in Fig. 20.3, where the fault intersections at three horizons have been marked. The strike of this fault is northeasterly and the fault is dying out in the northeasterly direction. Lateral variation in fault throw is common and often produces misties if faults are picked erroneously. Traversing a reflection

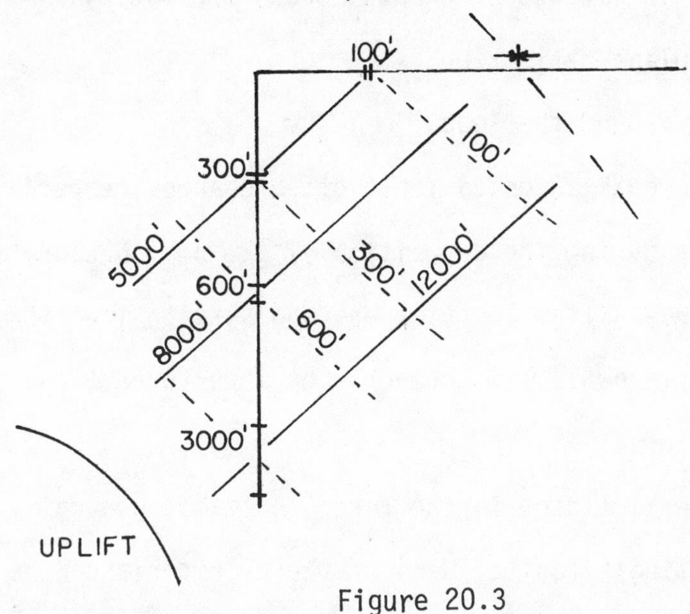

Figure 20.3

around a closed loop is apt to bring one to a different arrival time because the errors in displacement at fault intersections on the loop will not compensate each other unless the fault picks are correct. In the instance shown in Fig. 20.3, the lateral change of throw contributes to the apparent increase in throw with depth on the left-hand section and decrease with depth on the right-hand section. A large domal structure lies just southwest of the south end of this line and the fault being studied is a radial fault off this feature. A characteristic of such faults is that they often die out before they reach the rim syncline, and that is the case in this instance.

On the model and actual seismic sections, we have seen a number of evidences of faulting, as listed below:

A. Abrupt termination of events;

B. Diffractions;

C. Changes in dip, flattening or steepening;

D. Distorted dips as seen through the fault;

E. Cut out or deterioration of data below the fault;

F. Change in pattern of events across the fault;

G. Fault plane reflection;

H. Mistie around the loop.

The pattern of events across a fault often changes, especially if a fault was active during the deposition of the beds because then the beds downthrown may differ in thickness and sorting from the corresponding beds upthrown. This changes the interference and hence the overall data pattern.

All of the criteria listed in the foregoing table generally will not be found with a single fault. Many of these criteria can be caused by situations other than faulting, of course.

Let us now examine the effects of compaction on faulting. We assume a model 2000 foot thickness of shale at the surface of the earth cut by a fault with 50^{0} of dip (Fig. 20.4). Assuming a velocity of about 5000 ft/sec, the displacement on a seismic section will be about 800 msec. The block on a seismic time section is shown at the right. If this block of shale is buried 6300 feet deep, it will be compacted because of the increased weight of the overburden rock; we assume this will cause it to be only 1430 feet thick, which will make the fault dip 40°. Note we have here assumed only burial and compaction; there has been no further movement along the fault plane. The compaction will have increased the velocity so seismic waves will traverse this thickness in less time than if the velocity had been the same as at the surface; we expect therefore a seismic travel time through this block of shale of only 295 msec. On the seismic section it will appear as if the shale has compacted more than it actually has due to the increase of velocity with compaction. Fault surfaces on seismic sections tend to have concave-upward curvature because velocity normally increases with depth. Since faults often have actual concave-upward curvature, the effect of velocity is to accentuate the curvature.

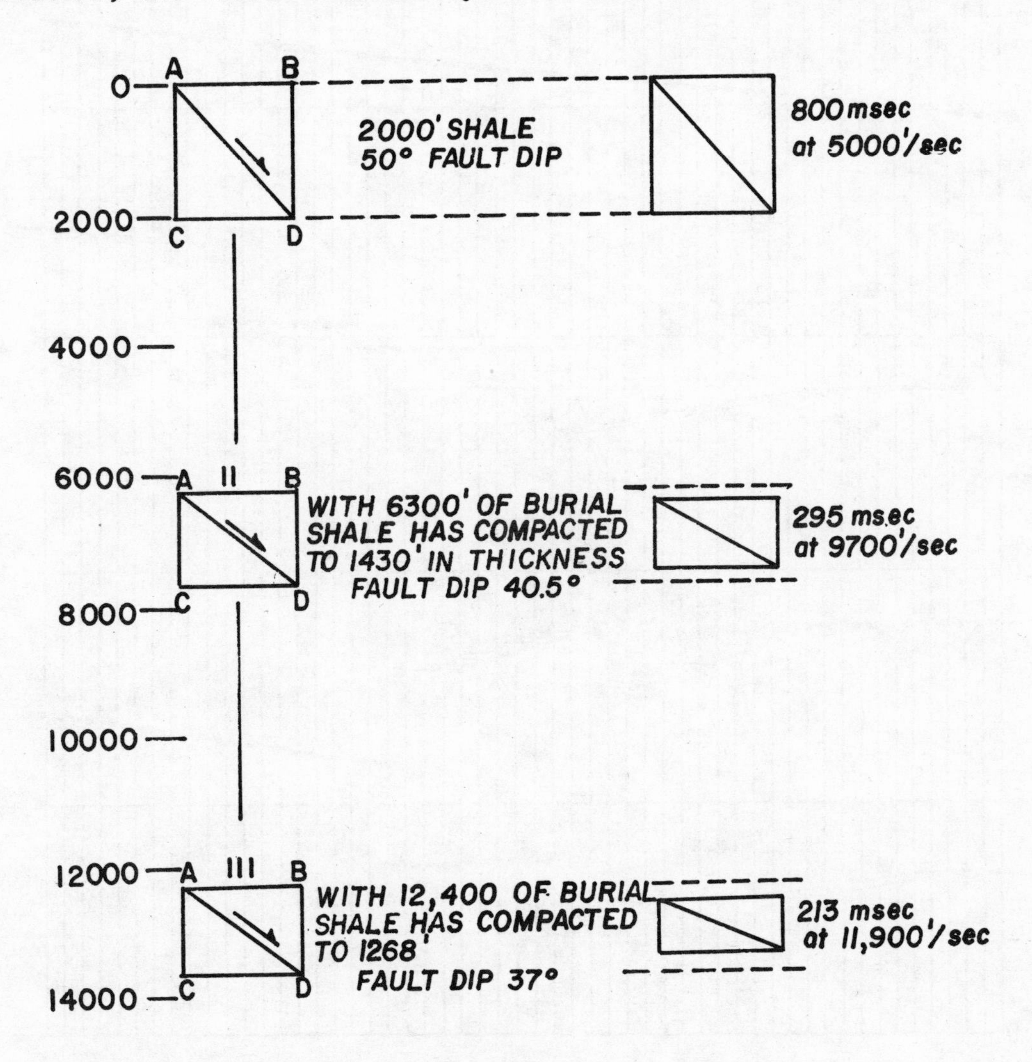

Figure 20.4

The model seismic section in Fig. 20.5 shows several reflection segments and a number of diffractions. Fault terminations line up so that one could interpret a series of reverse faults as shown by the dotted lines. However, the correct interpretation is a series of normal faults as shown by the dashed lines. The segments of data are not necessarily in the correct fault blocks until after the data have been migrated. If data have different attitudes in different fault blocks, they will migrate by different distances. The faults in this model moved at different times and different parts of the section are thicker in different blocks.

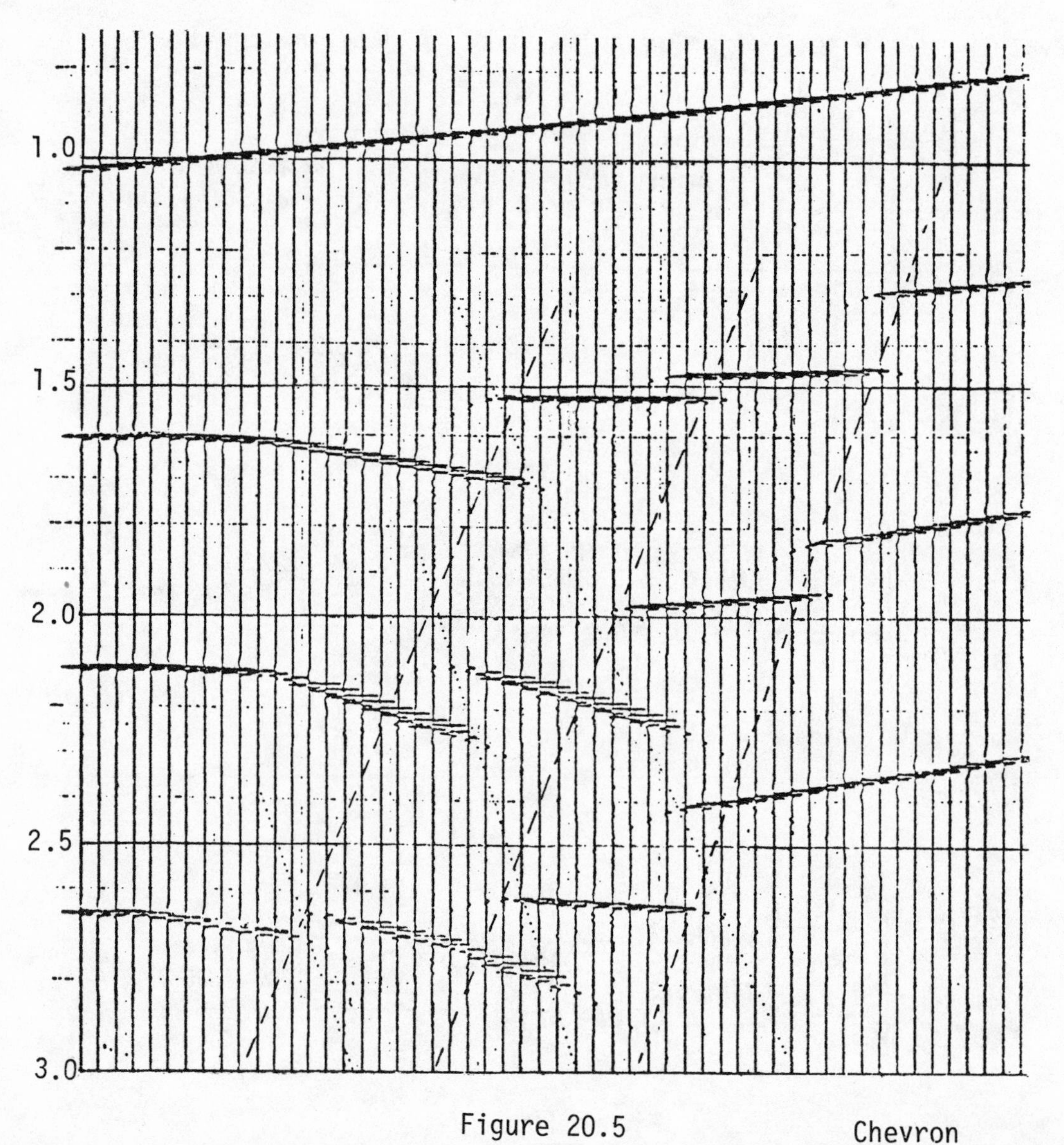

Figure 20.5 Chevron

The fault blocks rotated as they moved so that the reflectors differ in attitude in the various fault blocks. Locations of faults are perhaps best defined by drawing them through the crests of diffraction curves.

The model in Fig. 20.6 illustrates some points about fault plane reflection. The fault is indicated by the terminations of several horizontal reflections and by diffractions. The crests of the diffraction curves locate the fault plane. Note that the fault plane does not lie along the fault plane reflection, unless the latter is migrated properly. Note also that the reflection character of the fault plane reflection changes. The fault opposes different combinations of beds against each other along

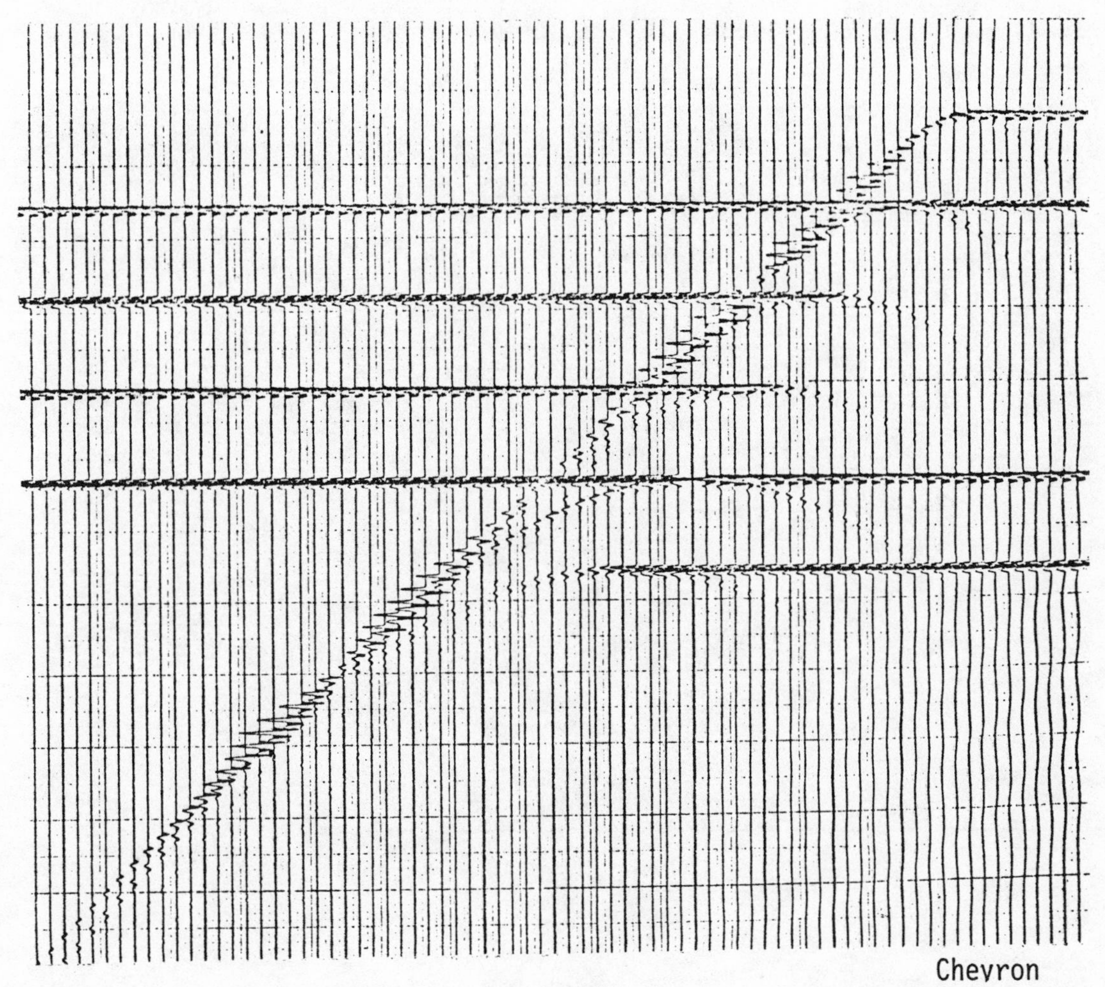

Chevron

Figure 20.6

the fault plane so that the acoustic impedance contrast changes rapidly along the fault plane, producing rapid changes in character of the fault plane reflection, both in amplitude and in polarity.

The seismic section in Fig. 20.7 has considerable vertical exaggeration, that is, it has been compressed horizontally. The sharp clear fault looks almost vertical, but actually the fault plane dips at about 60^0. Vertical exaggeration, which is common on seismic sections, changes the appearance of fault criteria.

A growth fault is shown in Fig. 20.8. The fault is indicated by a number of diffractions. The lower left corner shows an event F which is part of a fault plane reflection and migrates onto the fault plane in its lowest visible portion. Fault plane segments

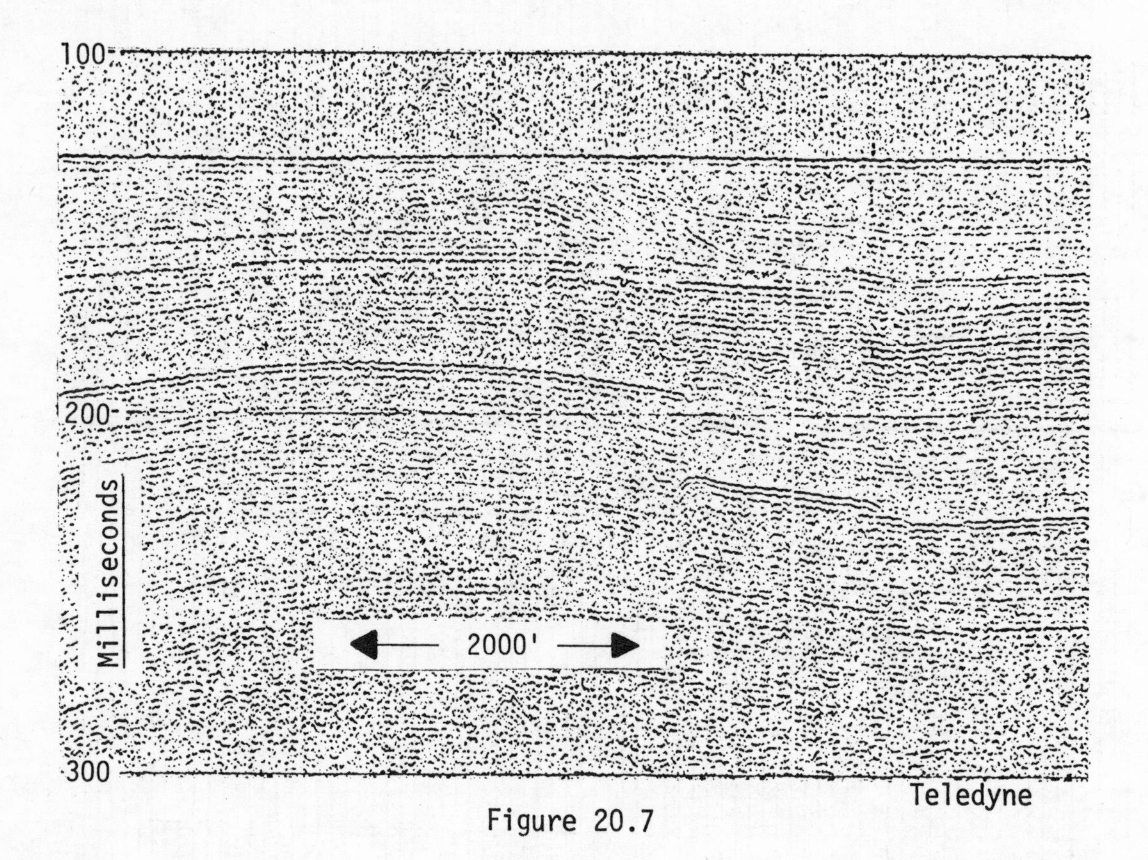

Figure 20.7

are often not recognized as such. This fault involves movement of a massive shale, generally defined by the base of the reflection sequence, and about all that can be correlated easily on the section is this base of reflections. Upthrown and downthrown reflections often cannot be correlated where a fault was active during deposition of the beds because corresponding units are thicker on the downthrown side and hence the individual reflection components interfere differently, producing reflections of different character.

Some other diffraction curves can be observed underneath the principle fault plan in Fig. 20.8. This lineup suggests another fault lying under the first fault. Compensating antithetic faults (A) have also been sketched in Fig. 20.8.

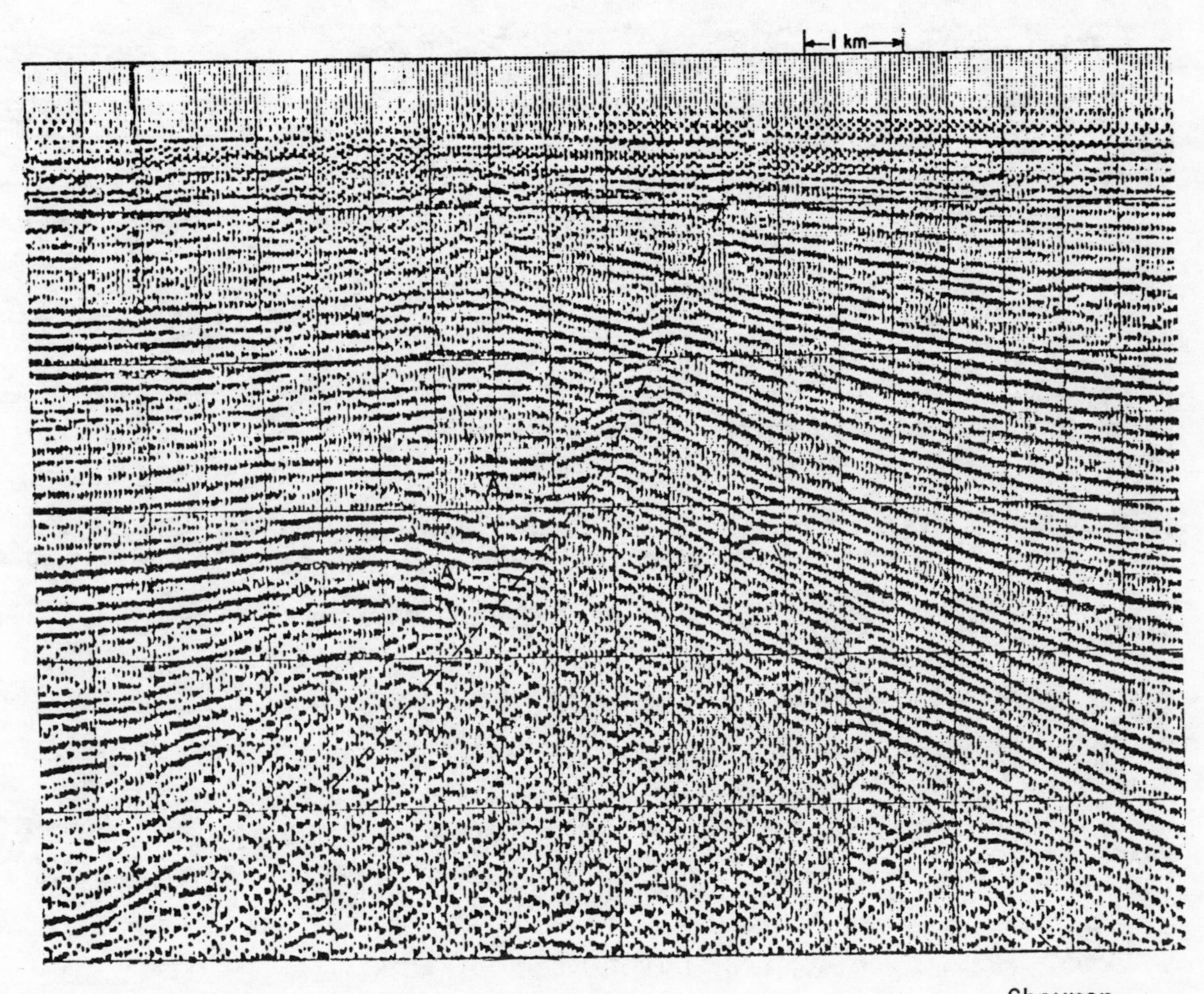

Chevron

Figure 20.8

Figure 20.9 shows two growth faults with the beds thickening rapidly into the faults, indicating that the faults were active at the time these beds were being deposited. There are some distortions and shadow regions under the faults.

A salt dome can be thought of as a circular fault, the upwelling salt being the upthrown side. At the top of Fig. 20.10 is a profile section which gives an enlarged view of shallow reflections. The fault pattern over the salt dome is a graben pattern formed because the sediments were under tension as a result

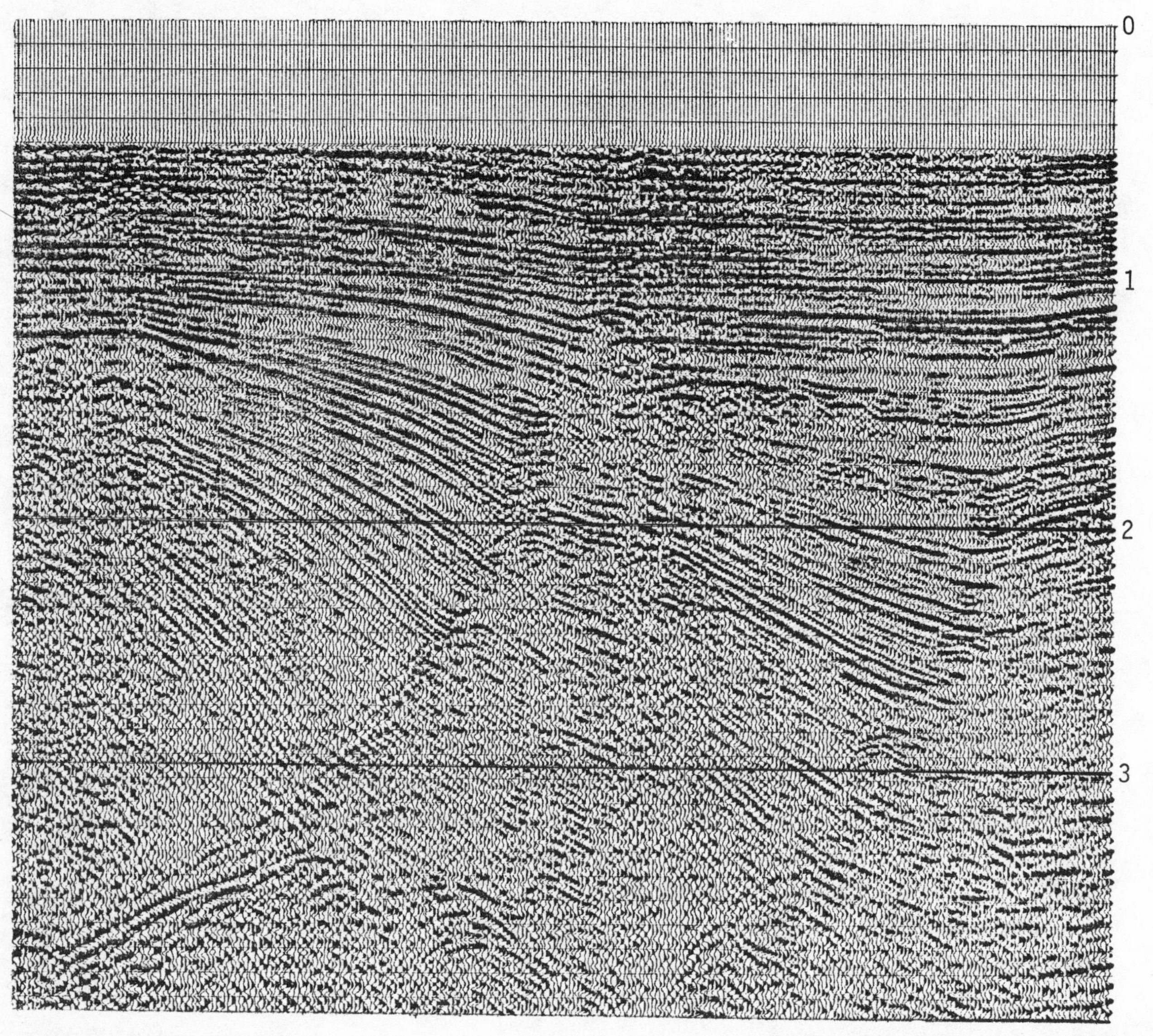

Figure 20.9

Exploration Services Division of Geosource Inc.

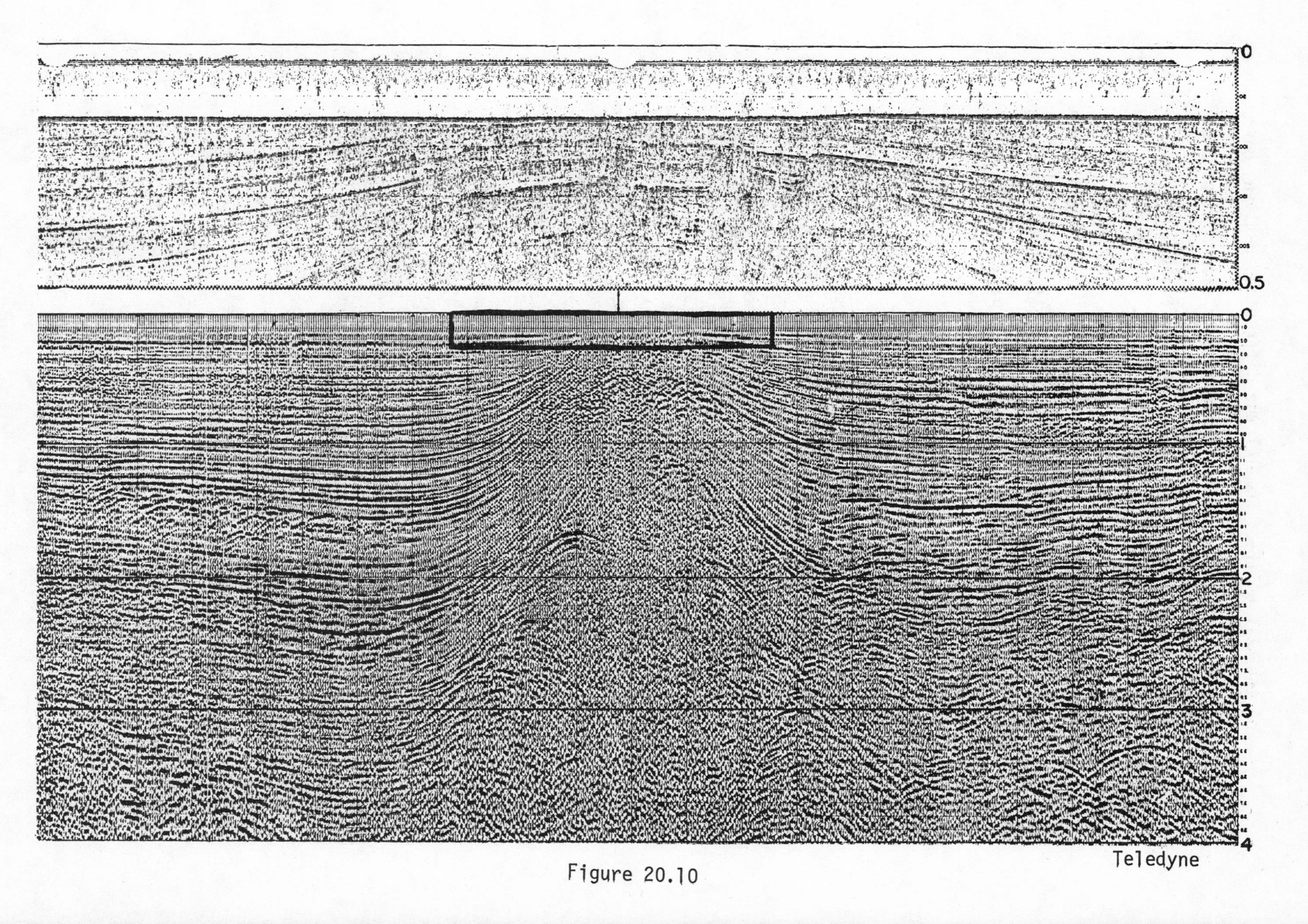

Figure 20.10

of the salt uplift. Some of the diffraction patterns have been marked on Fig. 20.11; the crests of diffractions are often the key in locating the salt dome flank. Migrating the diffractions (and flank reflections) accomplishes the same objective. Defining the flanks of the salt domes precisely is a matter of considerable economic importance because hydrocarbon accumulations are often immediately adjacent to the salt dome.

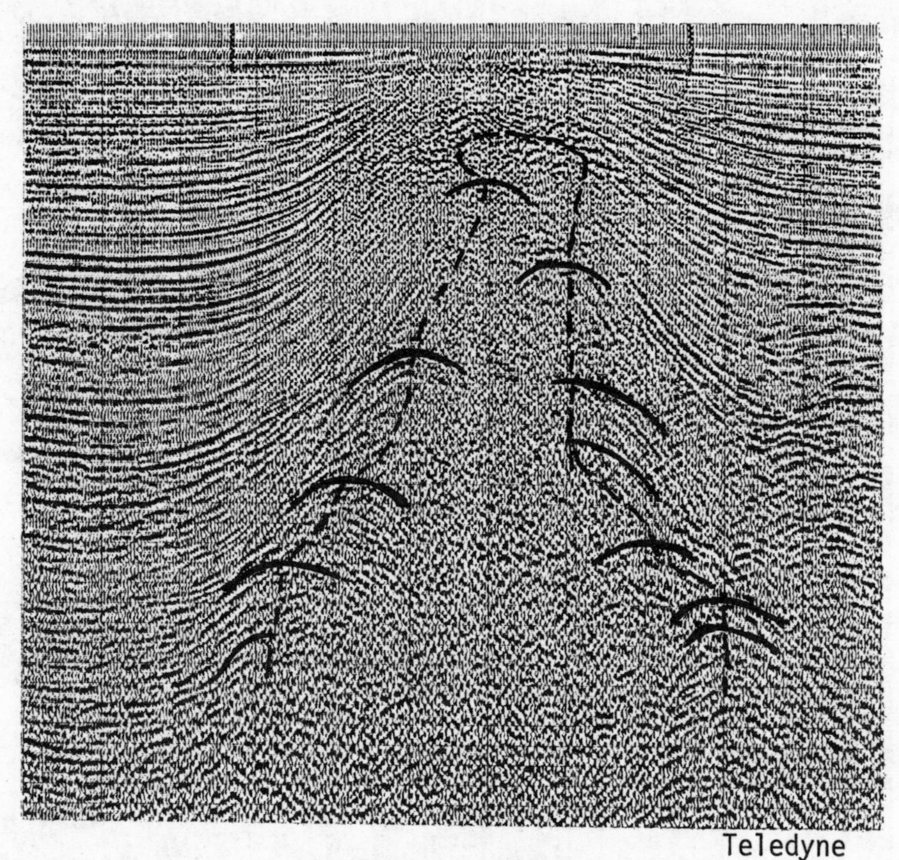

Teledyne

Figure 20.11

CHAPTER 21

THE THREE-DIMENSIONAL PROBLEM

Most seismic interpretation is done as if data are two-dimensional and then conclusions from two-dimensional profile interpretation are combined in maps to give a three-dimensional picture. To better understand three-dimensional effects on seismic data, let us consider a particular reflection event with 2.5 second arrival time on two seismic lines which intersect at right angles. On one of these lines, this event dips westward 45 msec/1000 ft, that is, there is a difference in arrival time of 45 msec between two points on this line separated by 1000 ft. On the seismic line at right angles of this line, the event dips 35 msec/1000 ft northward. The dips we measure are apparent dips seen at the surface rather than actual dips or even components of actual dips. The change in velocity with depth causes ray paths to curve and actual dips are usually larger than apparent dips.

If we resolve these two apparent dips as shown in Fig. 21.1, we obtain a dip of 58 msec/1000 ft. A line in the true dip direction would measure the maximum dip and a perpendicular line is therefore the strike direction. Figure 21.1 shows arrows pointing towards the shot point where observations were made as a reminder that data migrate updip; by convention arrows point downdip.

We shall calculate the <u>depth and location of the reflecting point</u> according to several assumptions: first, ignoring migration and merely converting arrival time to depth; second,

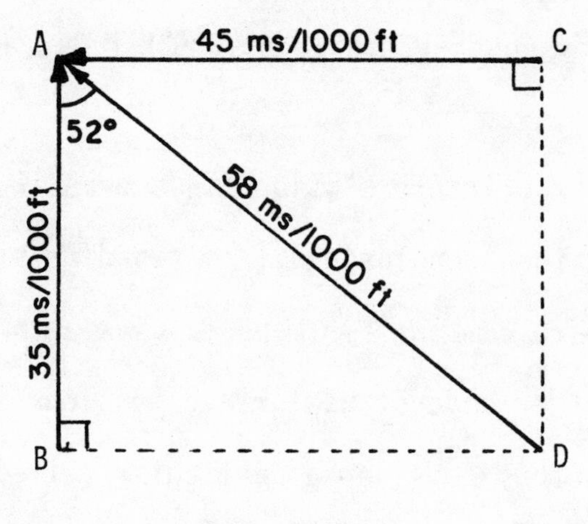

Figure 21.1

assuming that only one of the seismic lines is available and migrating the data as if there is no cross dip, often all we can do in the absence of information about the cross dip; finally, migrating the resolved dip. Figure 21.2 shows a portion of the wavefront chart from which we can read the values resulting from four different assumptions; these results are tabulated below (letters key the respective picks):

Assumption	Depth	Displacement from Shot Point	Dip	
Neglect of Migration	14,900 ft	0	7° * 8° * 10° †	A
Migration of N-S line	14,300 ft	3400 ft S	21° N	B
Migration of E-W line	13,900 ft	4300 ft E	27° W	C
Migration of resolved dip	13,300 ft	5300 ft SE	35° NW	D

* Attitude at datum of apparent dip on the two lines.

† Attitude of resolved dip at datum.

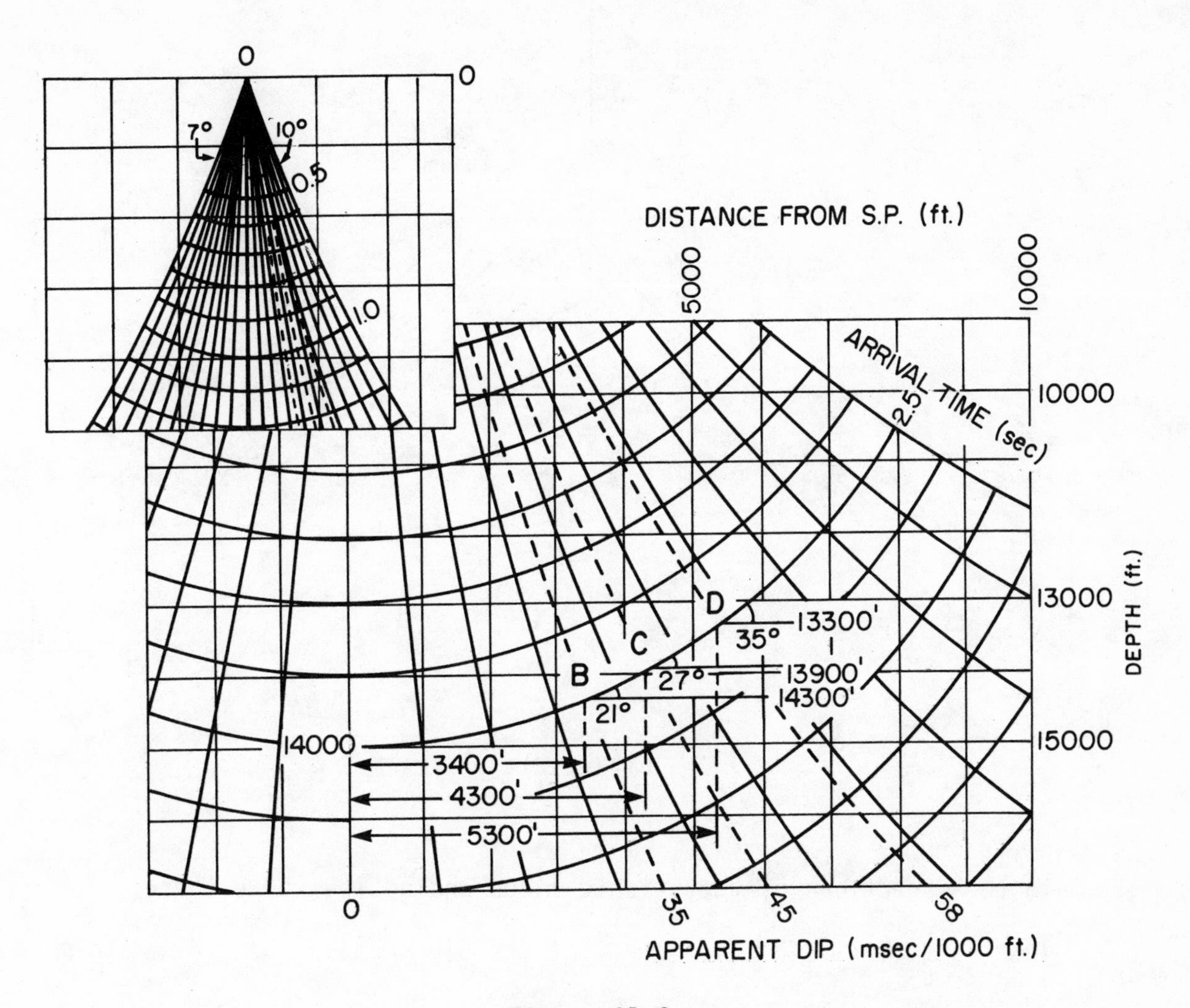

Figure 21.2

These four determinations are also plotted in the isometric diagram shown in Fig. 21.3. The reflecting plane is based on the depth and dip of the resolved dip case. Note that the four determinations are not co-planar, that is they do not lie on the same plane. Note also, however, that the two-dimensionally migrated values give a more nearly correct picture than the unmigrated values even through crossdip was neglected and the migration was not correct.

If the seismic lines are not at right angles, as shown in Fig. 21.4, the correct way of obtaining the resolved dip is by constructing lines at right angles to scaled measures of the dip

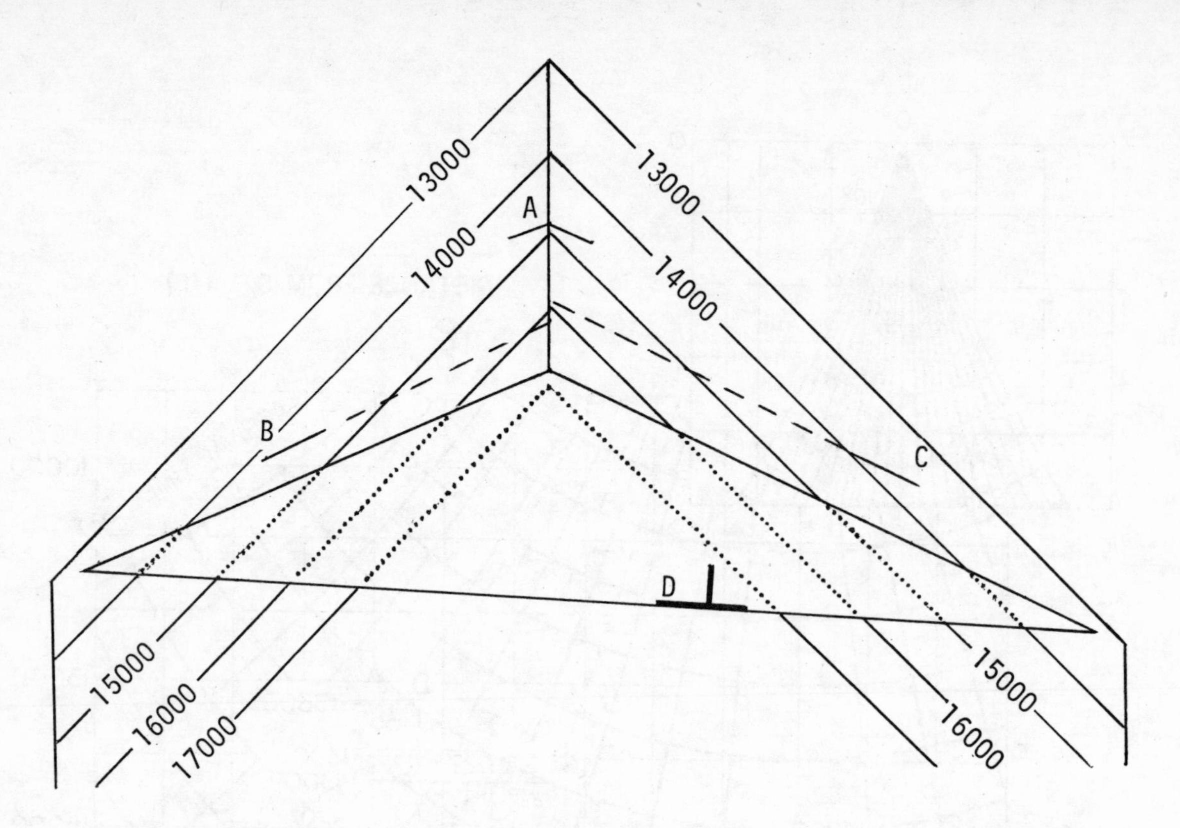

Figure 21.3

in the directions of the seismic lines and noting where these lines intersect. This assures that the dip seen on the two seismic

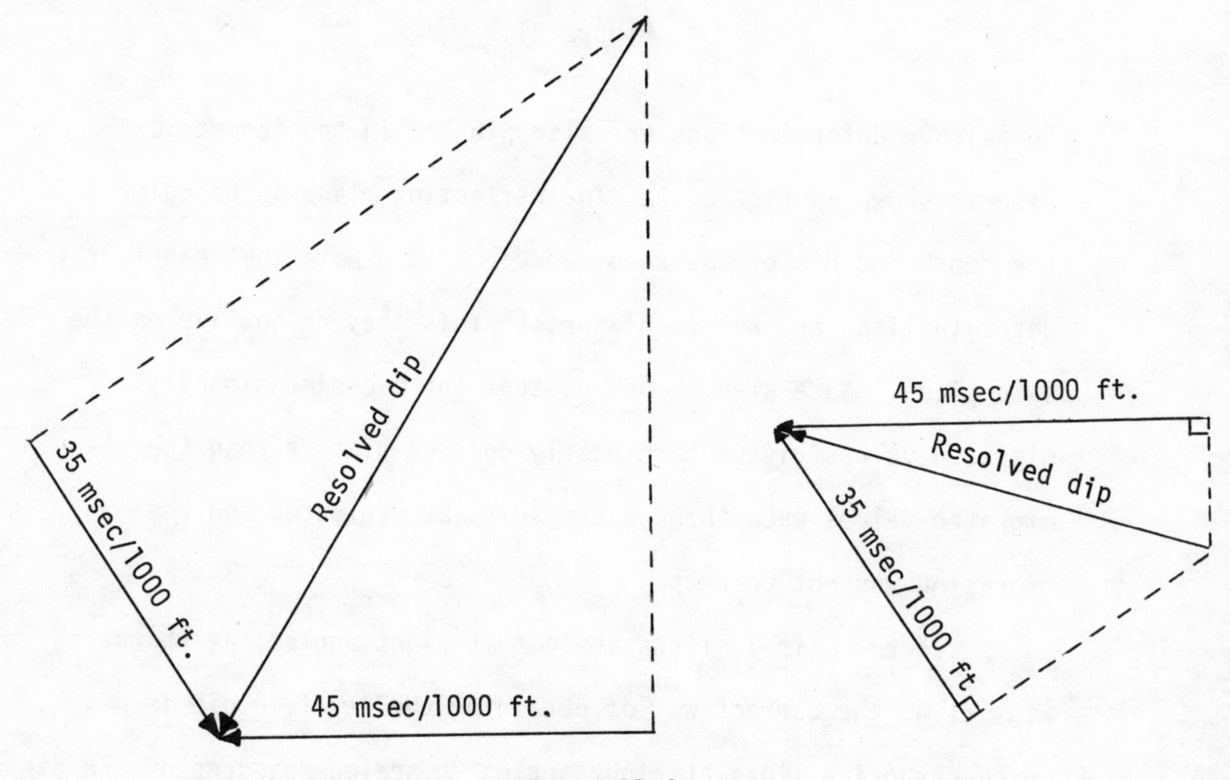

Figure 21.4

lines are dip components.

Now let us consider the effects of dip when tying seismic data to well data. Consider several reflection events on one spread length of a seismic section and dips from a dipmeter survey in a well. If the seismic line is perpendicular to the strike, then the seismic data can be migrated and converted to depth, as shown in Fig. 21.5, and the data tie the well dips.

Sometimes we have the opposite problem, that of locating a seismic line to tie a well. If the line is to be perpendicular to the strike, we can unmigrate the seismic data, as shown in Fig. 21.6, and we then know how far we have to run the seismic line to tie the well data at any given depth. If the seismic line is approach-the well from the up-dip direction, we must extend the line beyond the wellhead to effect the tie. If the seismic line had been coming in along the strike direction, we would have to locate the seismic line down-dip from the well an appropriate distance to tie

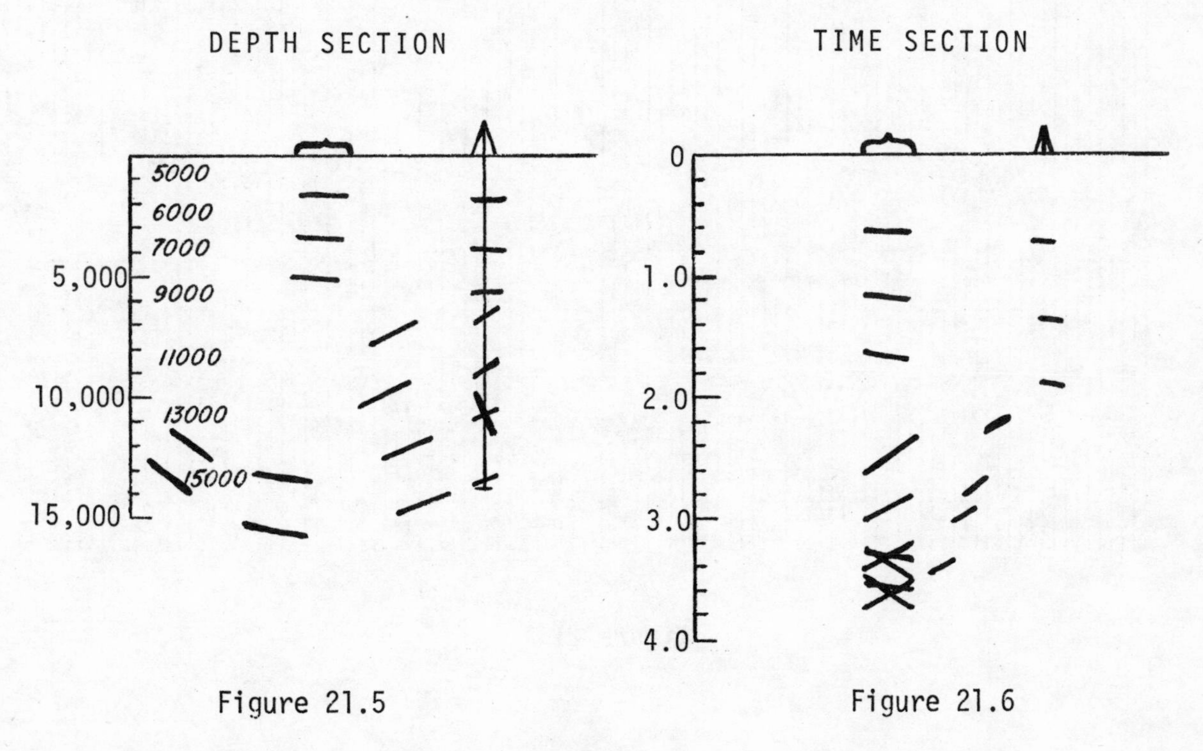

Figure 21.5

Figure 21.6

a particular objective horizon; tying the wellhead would not achieve a tie. Ties of seismic and well data where there is dip are often made incorrectly without realizing that a problem exists. Other problems then follow from the incorrect tie; one has to invent a compensating error, such as incorrectly inserting a fault at some-place where data quality deteriorate, in order to remove a mistie.

Exercise:

Figure 21.7 shows two (model) seismic lines. Figure 21.8 is a base map showing these line locations and Fig. 21.9 shows the onset of the reflection events migrated by conventional methods. The problem is to construct a depth map on the lower of these two horizons. Work the problem before going on to the disucssion.

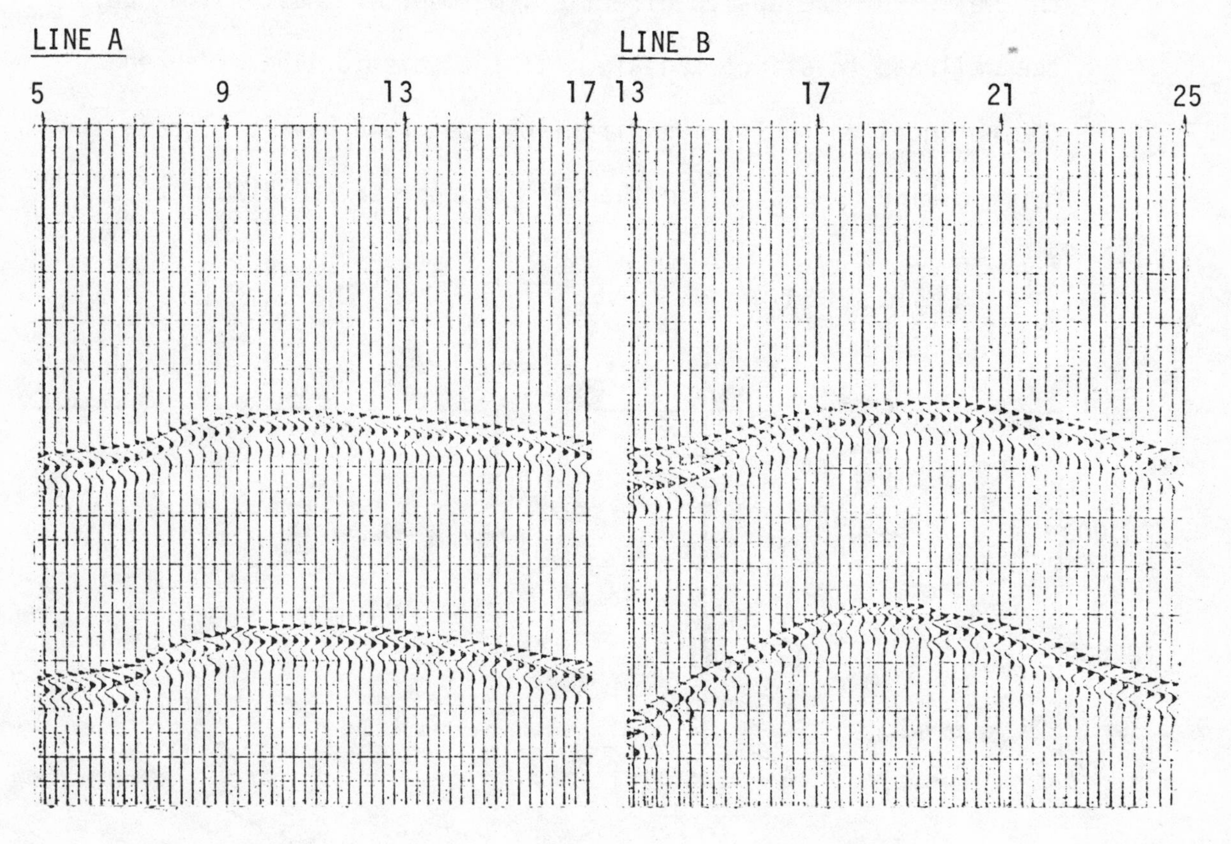

Figure 21.7

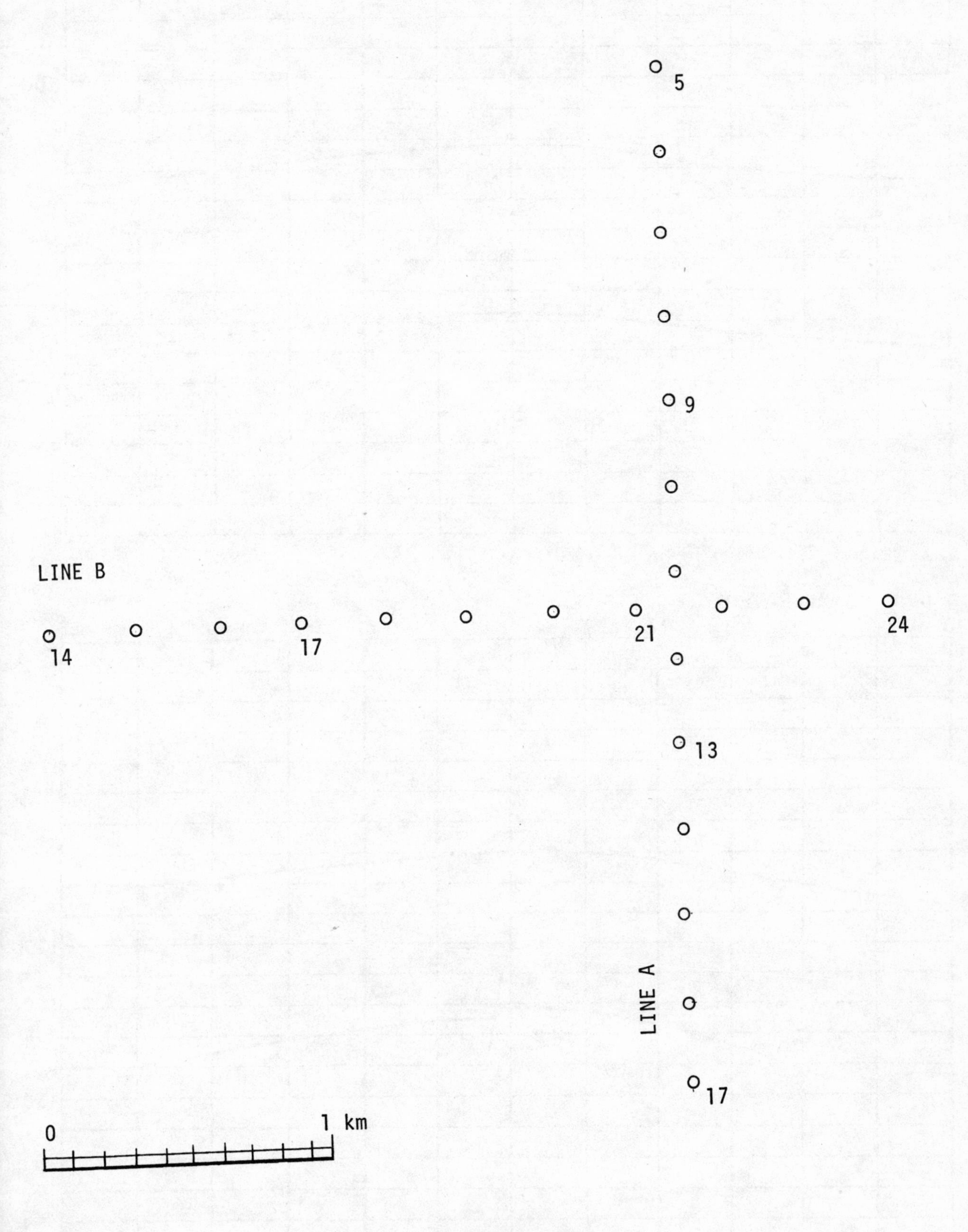

Figure 21.8

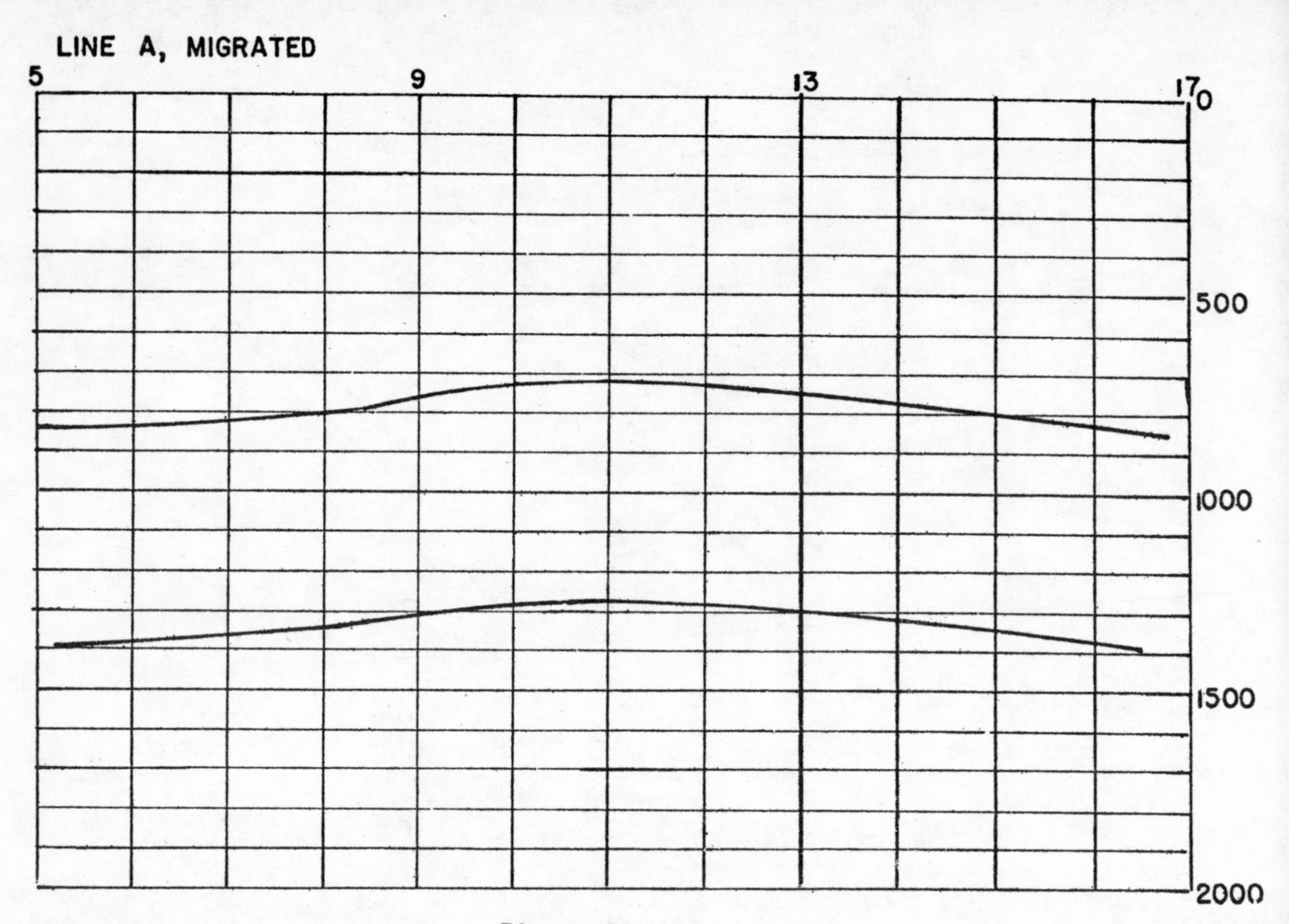

Figure 21.9a

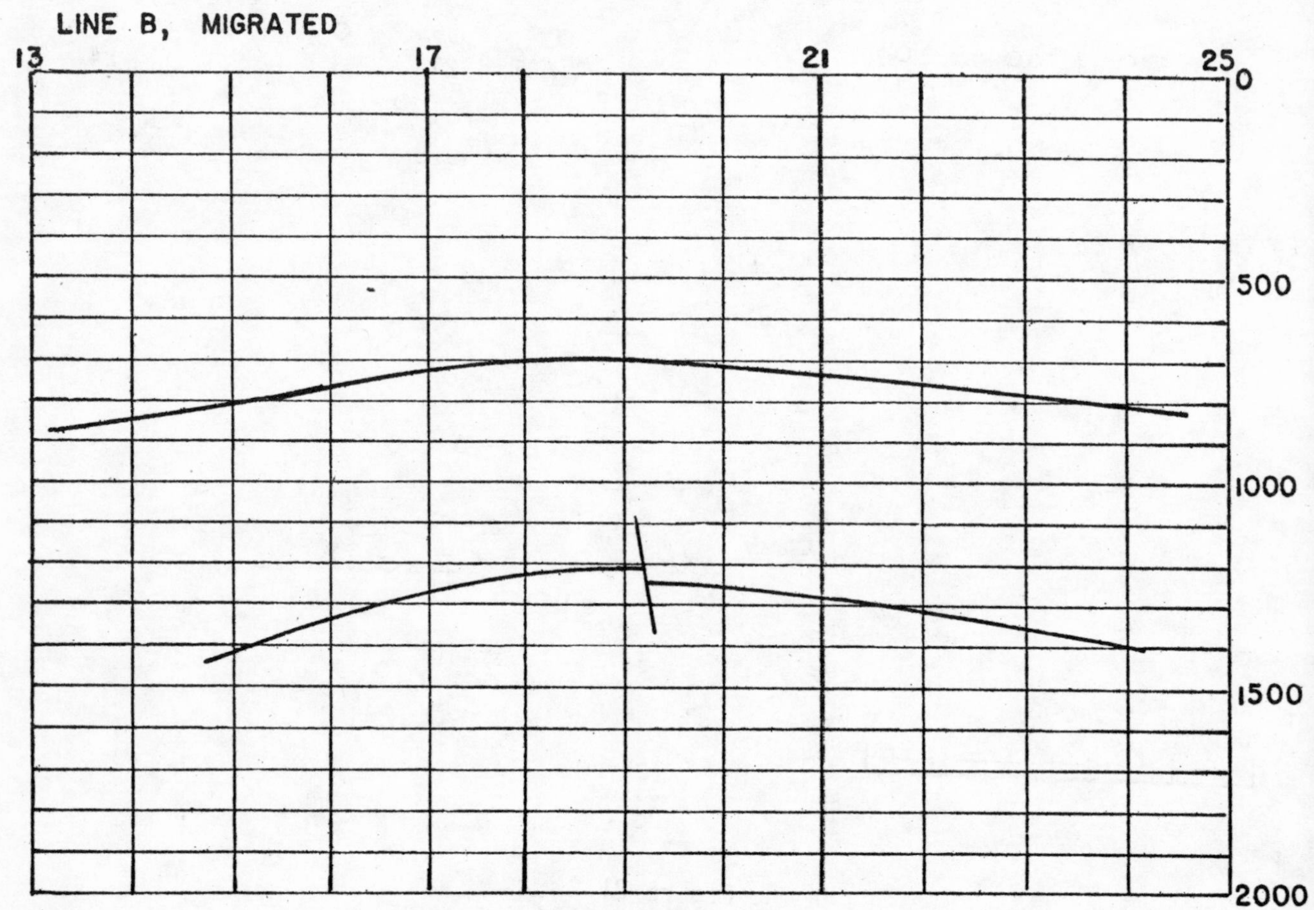

Figure 21.9b

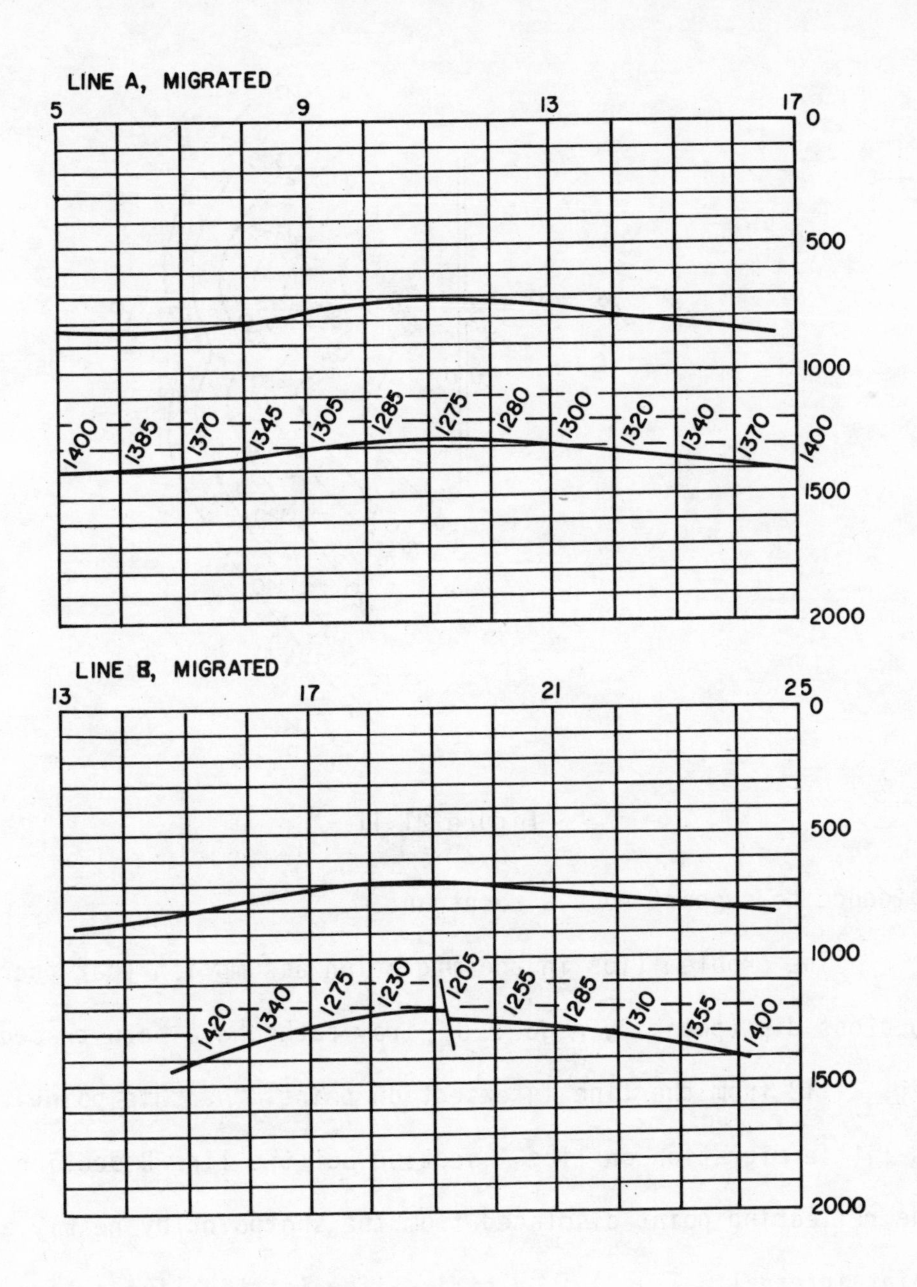

Figure 21.10

Discussion of exercise:

Since our goal is a depth map, we begin by reading depth values from the migrated depth sections, as shown in Fig. 21.10. We then post these data on the base map and contour, as shown in Fig. 21.11. When we contour honoring all the data points, we are forced to map a reentrant along line B cutting into the structure. We may check the seismic lines in Fig. 21.7 to see that the data tie at the line intersection, which they do. At the same time we see no obvious

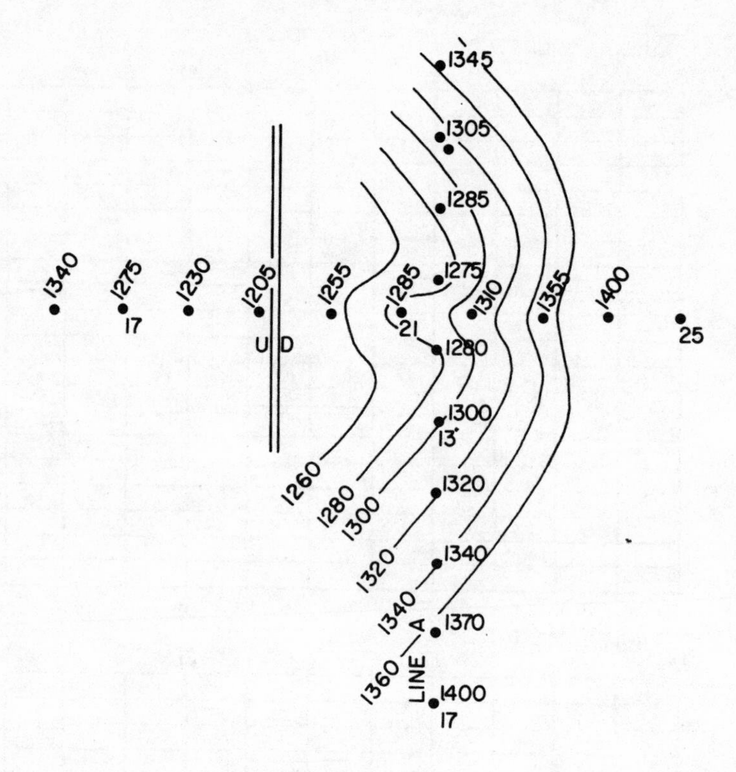

Figure 21.11

evidence to support such a reentrant.

The problem lies in the migration assumption that there is no cross dip (point 2, page 226); ray paths have been traced on Fig. 21.12 from the line intersection point. At this point there is little migration on line A section but the line B section has the reflecting point displaced from the shotpoint by nearly a shotpoint interval. Thus A is a strike line (at the line intersection) and the reflecting points lie up-dip from the shotpoints or west of the line. If we replot line A data at their reflecting points (Fig. 21.13), the data can be contoured smoothly.

The habit of posting data at shotpoints rather than at subsurface reflecting points is widespread but wrong. Note in this example that the data were neither exceptionally steep or exceptionally deep, both of which would accentuate errors. There are, of course, other ways of solving the problem correctly, such as mapping unmigrated data and then migrating the mapped results.

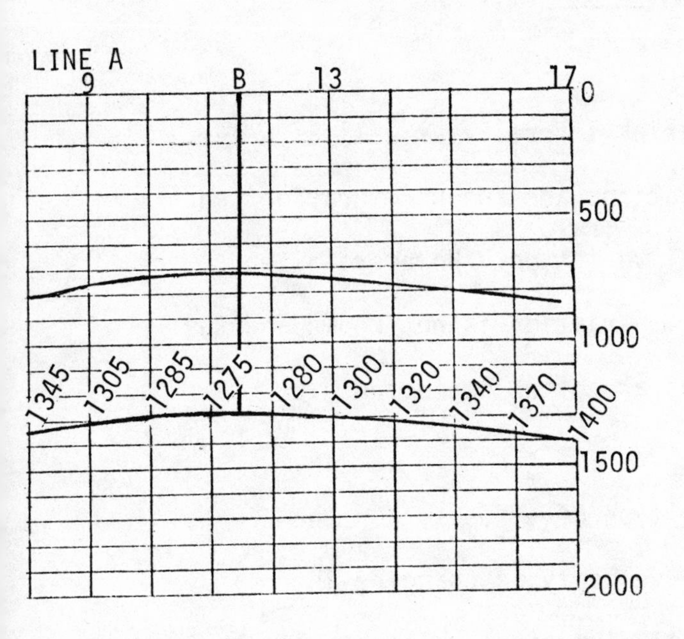

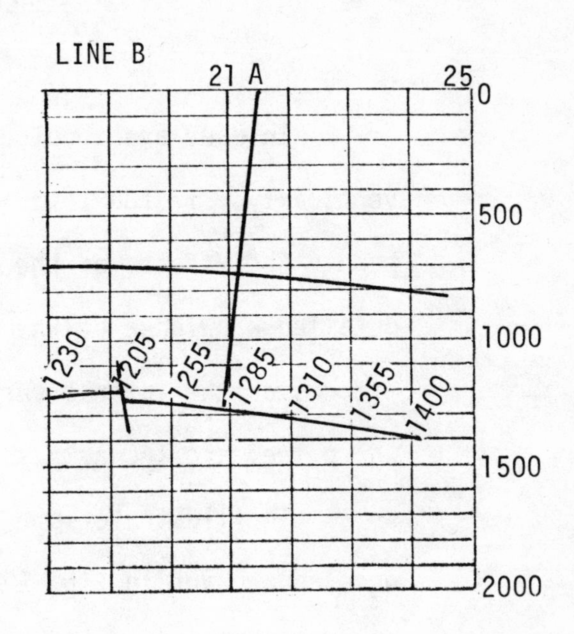

Figure 21.12

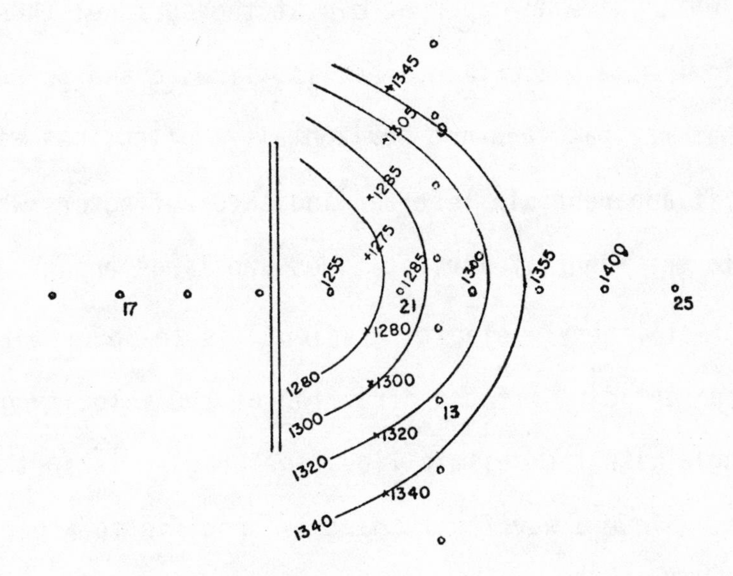

Figure 21.13

CHAPTER 22

VELOCITY COMPLICATIONS

In many areas velocity varies laterally as well as vertically. If the lateral variations are known, migration can handle the effects of the lateral variation. However, often detailed knowledge of the lateral variation is not known and the effects of the lateral variations are often neglected even where they are known.

To illustrate some of the type of effects, for the wavefront chart in Fig. 22.1 the velocity is changing in the horizontal as well as vertical directions. The result is an asymmetric chart. The heavy ray path line is for reflection data which have no apparent dip at the surface; these reflectors do dip in the subsurface, that is, lines drawn perpendicular to that ray path are not horizontal. Reflections with 10 msec/ 1000 ft apparent dip left can indicate reflectors which actually dip to the right if arrival times are large enough. One possible effect of lateral velocity gradients is to reverse the direction of apparent dip. If the direction of the velocity gradient makes an angle with the seismic line, the proglem is further complicated. Since a wavefront chart is specific to a particular velocity distribution, such charts are not practical solutions to the problem of handling lateral velocity gradients since every shotpoint would require a new chart.

The unmigrated section in Fig. 22.2 indicates salt move-

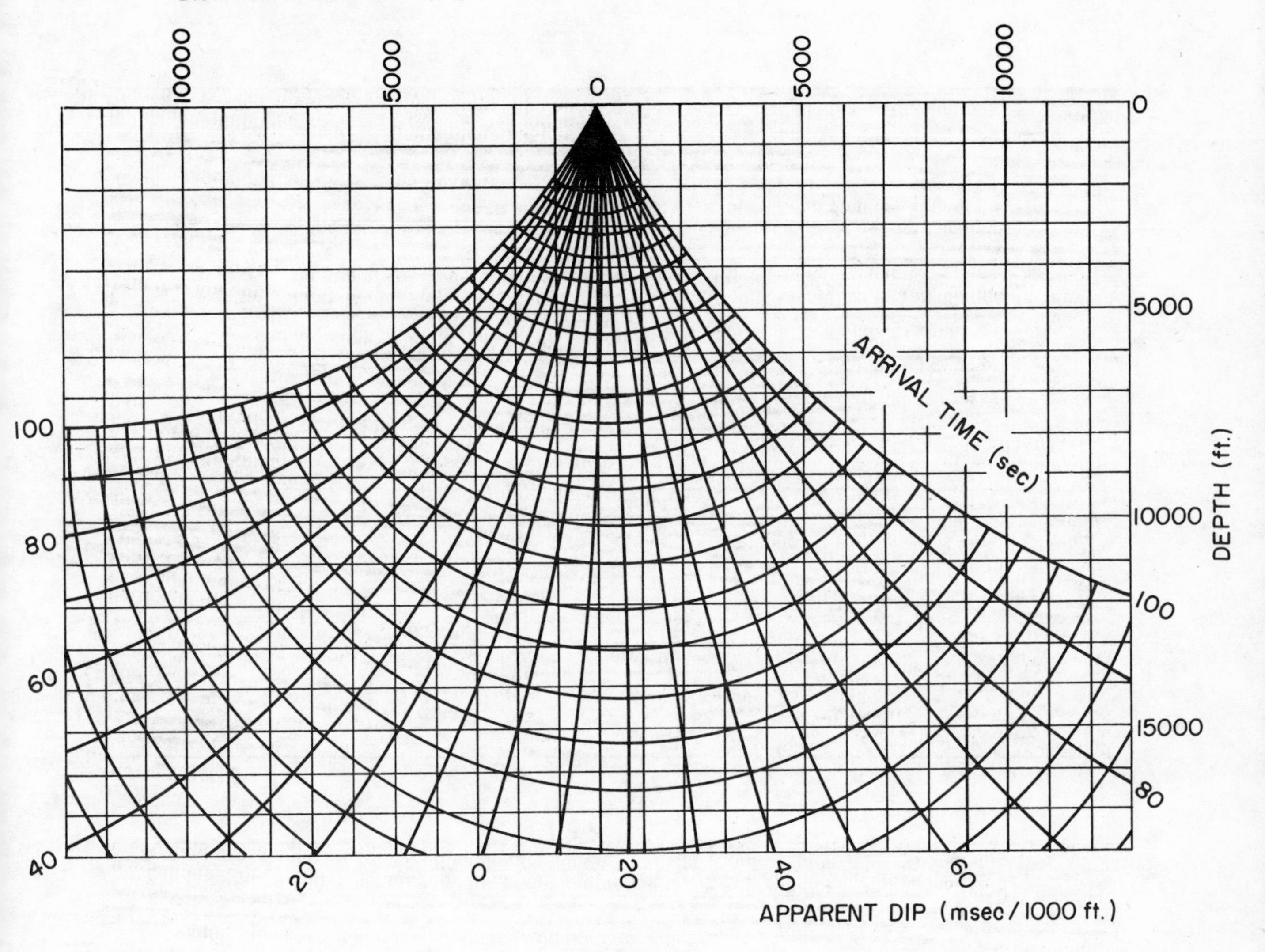

Figure 22.1

ment which has left pillows of salt on the left and right sides of the section. The middle underneath where the salt has been removed may involve some uplift. These same data are migrated in Fig. 22.3. There might be a reversal of dip below the gap in the salt. We need to carry out some calculations to determine if this could be a velocity effect. A straight line is drawn along the base of the salt assuming that the base is uniform. The thickness of the salt is about 400 msec and the salt velocity is 15,000 ft/sec. By multiplying the salt

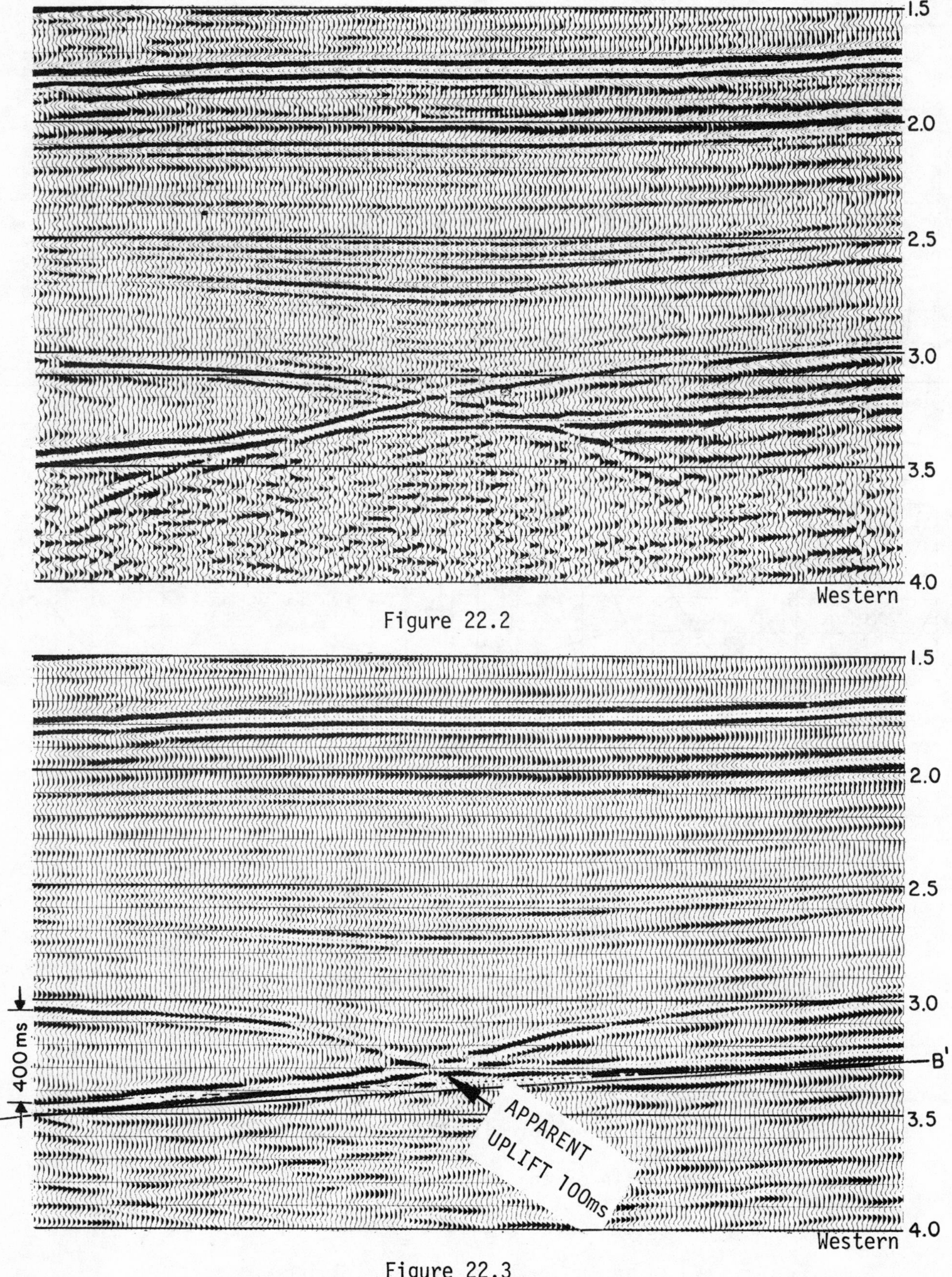

Figure 22.2

Figure 22.3

velocity by half the travel time, we calculate that the salt pillow is approximately 3000 ft. thick. If we fill the region from which the salt has been removed with clastic sediments with a velocity of about 12000 ft/sec., which would be a reasonable velocity, we can calculate how large a velocity anomaly could be produced. If the base salt is flat, there should be 3000 ft. of clastic sediments as the salt removal equivalent, which would have involved 0.5 second travel through them and a depressing of the base of the salt rather than a shallowing of it. Thus if this assumption is correct, there is a significant uplift of larger magnitude than appears.

On the other hand, perhaps the salt-equivalent section is carbonate with a velocity larger than the salt velocity. If the sediments had a velocity of 20,000 ft/sec., the velocity uplift would be of the magnitude that is observed.

Another common problem with marine data is illustrated in Fig. 22.4. The right-hand side of the seismic line is on the Continental Shelf and the section continues across the continental slope into water 3300 ft. deep. There is considerable apparent seaward dip on the reflections. The questions here is, is the apparent dip on the horizon marked HH real or not. The vertical marks on the section locate places for which we shall calculate the depth to this horizon. A velocity distribution is shown in Fig. 22.5. The top dashed curve V_{iA} indicates instantaneous velocity as function of arrival time and the top solid line indicates average velocity $\overline{V}_A$ at location A and

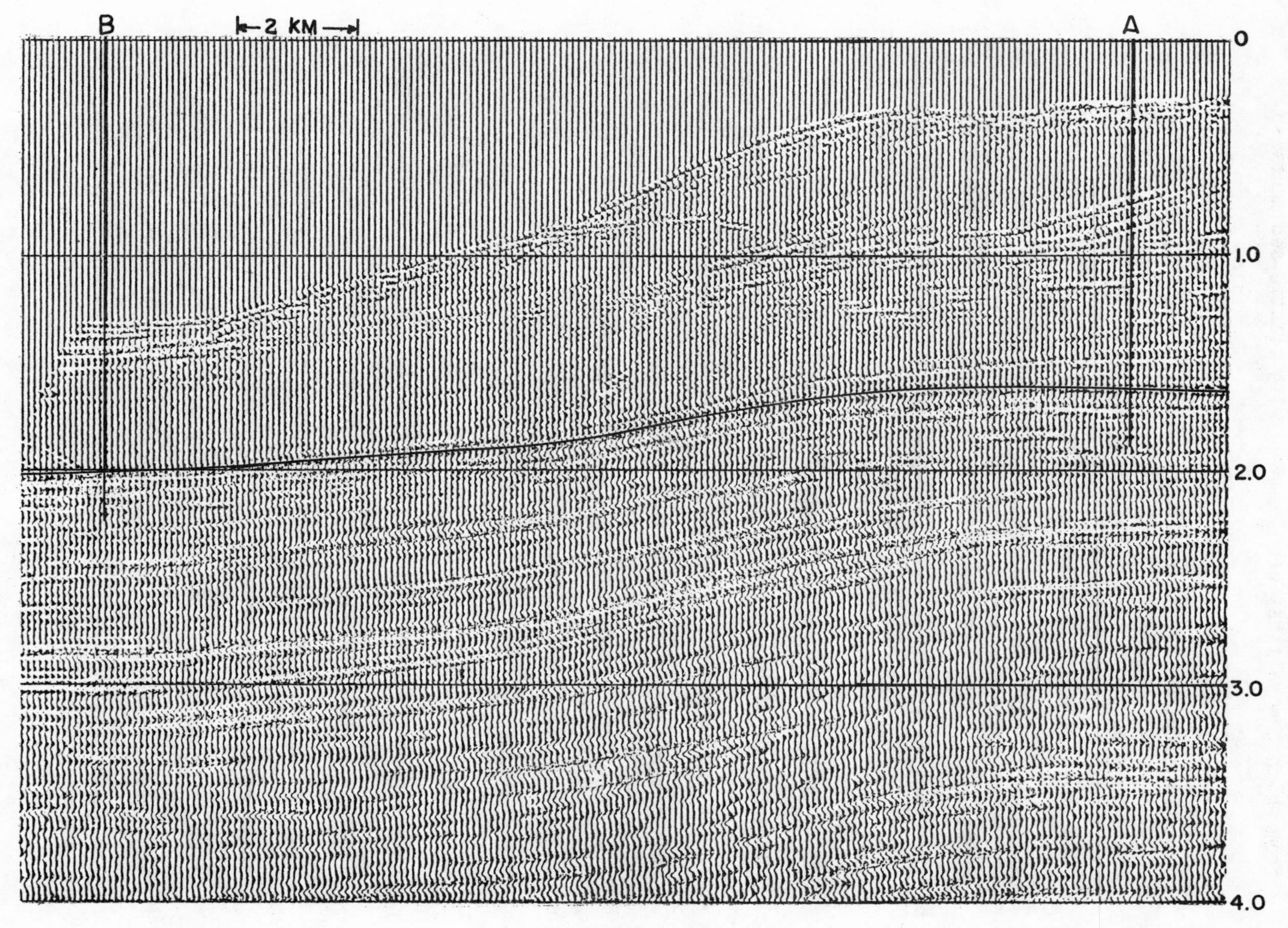

Figure 22.4

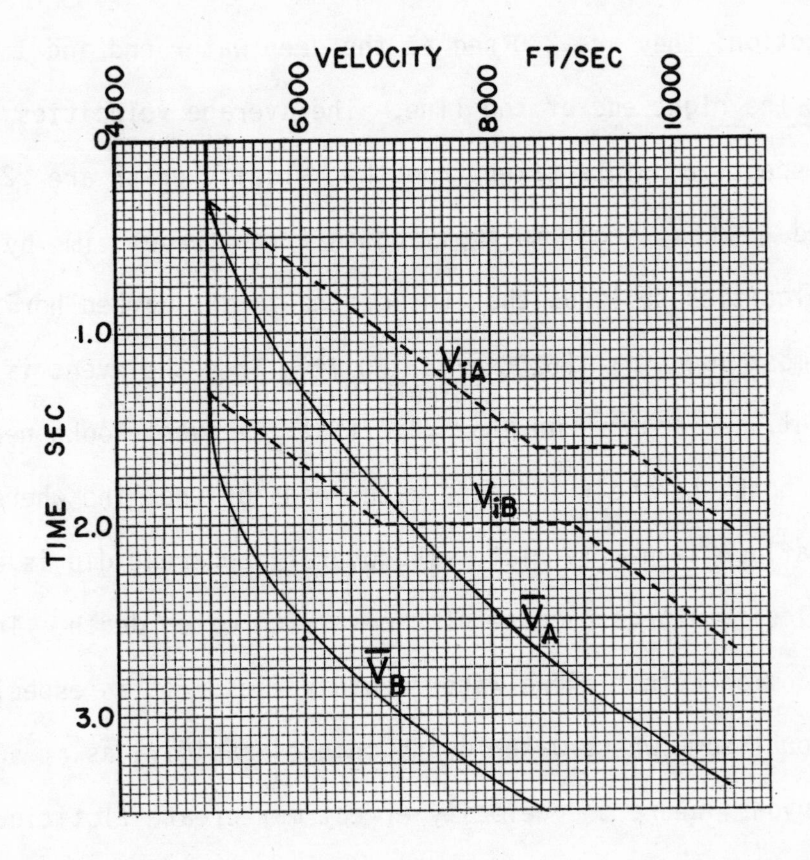

Figure 22.5

the lower two curves indicate the same inferred for location B. The data at location A are known from a survey, those at B are inferred.

Exercise:

1. What is the depth of the indicated horizon at the points A and B?
2. Is the velocity distribution shown in Fig. 22.5 reasonable?

Work this exercise before continuing on to the discussion.

Discussion of exercise:

We pick arrival times at opposite ends of the seismic section; they are 2.0 sec at the deep water end and 1.6 sec at the right end of the line. The average velocities to those respective points taken from the velocity graph are 5200 and 6450 ft/sec. Multiplying half the travel time by these velocities tells us that the depths to the marked horizon are almost identical (5200 and 5160 ft). Thus the event is nearly flat. We observe on Fig. 22.4 that the event looks nearly flat at both the shelf and deep water ends of the line where the seafloor is nearly flat. The overall apparent dip is a velocity effect because of the varying water depth. This is a common effect where water depth varies. It is especially troublesome where water depth changes rapidly, as at submarine canyons, where the velocity effect may create fictitious structures.

The dashed curve on Fig. 22.5 for the right end of Fig. 22.4 shows the constant velocity of water down to 0.3 sec, which is the travel time in 750 ft of water. Below this the curve shows a uniform linear increase of velocity until the arrival time of the strong event is reached, where the velocity suddenly increases (perhaps because of a change from unconsolidated to consolidated sediments). Below this the velocity again increases with depth.

At the deep water end of the line, the constant water velocity extends to 1.3 sec, corresponding to a water depth of 3250 ft. Below the water bottom the velocity increases at the same rate as at the shelf end of the line. Remembering that the effective stress on a rock is the difference between the overburden and fluid pressures and that adding water on top increases both these pressures by the same amount, the sediments just below the water bottom should feel the same differential pressure, regardless of the amount of water above, and therefore should have the same velocity. Thus the velocity curves appear reasonable.

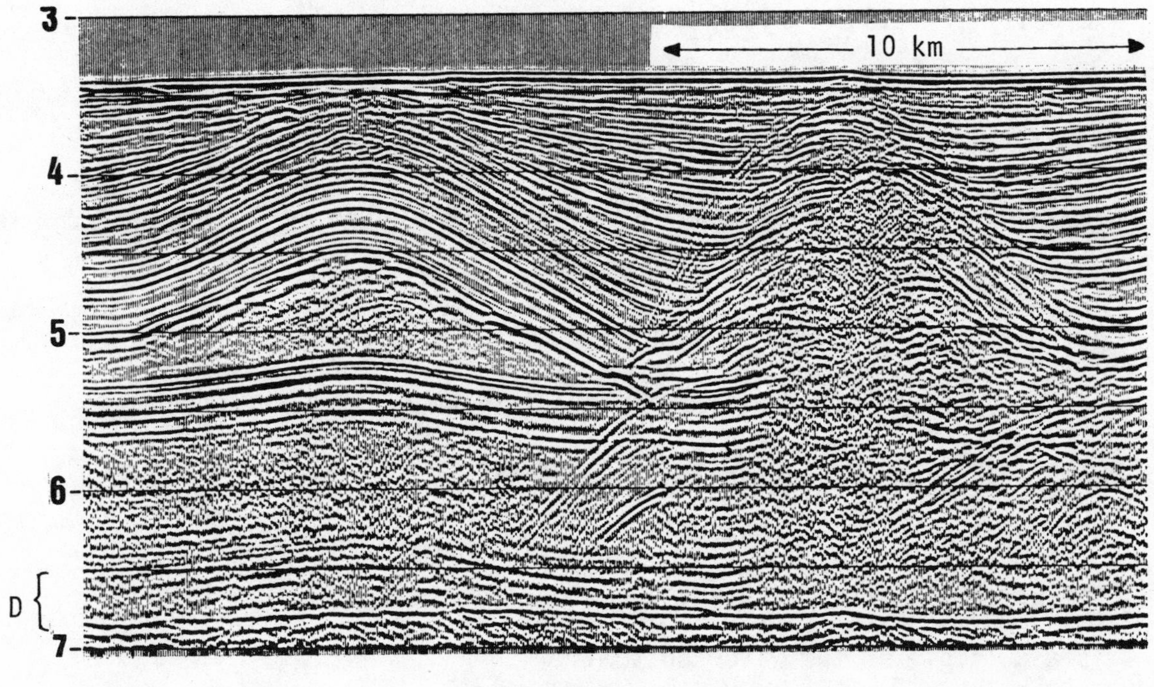

Figure 22.6

CGG
(CGG non-exclusive Vaporchoc offshore survey - Mediterranean Sea)

Figure 22.6 shows two uplifts in very deep water which we interpret to be caused by salt diapirism. The salt pillow on the left has not pierced through the sediments. The event which probably represents the base of the salt appears to have some uplift under the salt pillow, which we suspect is a velocity uplift. To check that the amount of uplift is of reasonable magnitude to be caused by velocity, we calculate the thickness of the salt as 4500 ft (0.6 sec salt thickness at 15,000 ft/sec). Assuming that the salt has thinned to zero between the salt features, we calculate the velocity of the same thickness of other sediments which replaces the salt, assuming that the traveltime through them is the 0.6 sec plus the 0.180 sec of suspected velocity uplift. This yields a velocity of 11,500 ft/sec, which is reasonable for clastic sediments.

The salt on the right did pierce through some of the sediments, which has resulted in some disruption of the data. We can sketch in the salt zone through the crests of the diffractions (Fig. 22.7).

Below this salt pillow on the left we see a divergence of reflections (D on Fig. 22.6). However the arrival time of the event at 6.8 sec is double the 3.4 sec arrival time of the water bottom reflection, and so we conclude that this event is a multiple of the water bottom.

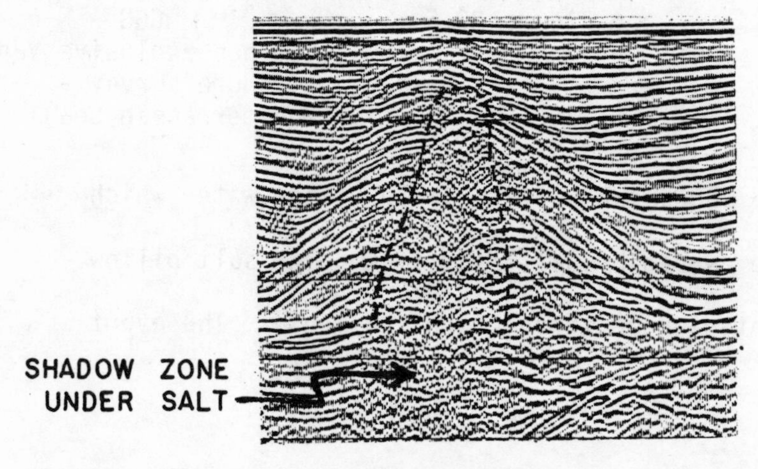

Figure 22.7

CHAPTER 23

STRATIGRAPHIC INTERPRETATION

The improvement in data quality which has resulted from data processing allows us to see details in seismic records which formerly were not clear. The types of information which we derive from seismic data are summarized in the following table:

arrival time	→	depth
differences with location	→	dip
differences with offset	→	velocity
differences in amplitude	→	reflectivity
angular relationships	→	geologic history
patterns	→	depositional situation
combinations of above	→	gross lithology, stratigraphy and fluid content

Lithologic — Sedimentary — Historic

	Velocity-density contrast	Gross lithology	Rock nature	Lateral lithologic variation	Fluid content	Bed thickness/spacing	Bedding continuity	Depositional process	Sediment source direction	Depositional environment	Subsidence/growth/erosion	Uplift from maximum burial
Reflection amplitude	✓			✓	✓	✓						
Dominant frequency		✓			✓	✓						
Reflection continuity							✓	✓		✓		
Abundance of reflections		✓										
Reflection configuration			✓					✓	✓	✓	✓	
Interval variations			✓						✓		✓	
Interval velocity		✓		✓	✓							✓

Figure 23.1

Another way of looking at stratigraphic inferences from seismic data is illustrated in Fig. 23.1. From the types of seismic measurements listed in the left-hand column, we can infer things about geologic features such as those listed across the top.

Let us look at some examples of stratigraphic inferences from seismic data. A progradational pattern P across the middle of Fig. 23.2 indicates an ancient delta. The delta built out from the right side of the section toward the left. Just above this an unconformity U can be seen. Note the suggestion of a broad gentle channel just below this unconformity at the left part of this section.

Figure 23.2

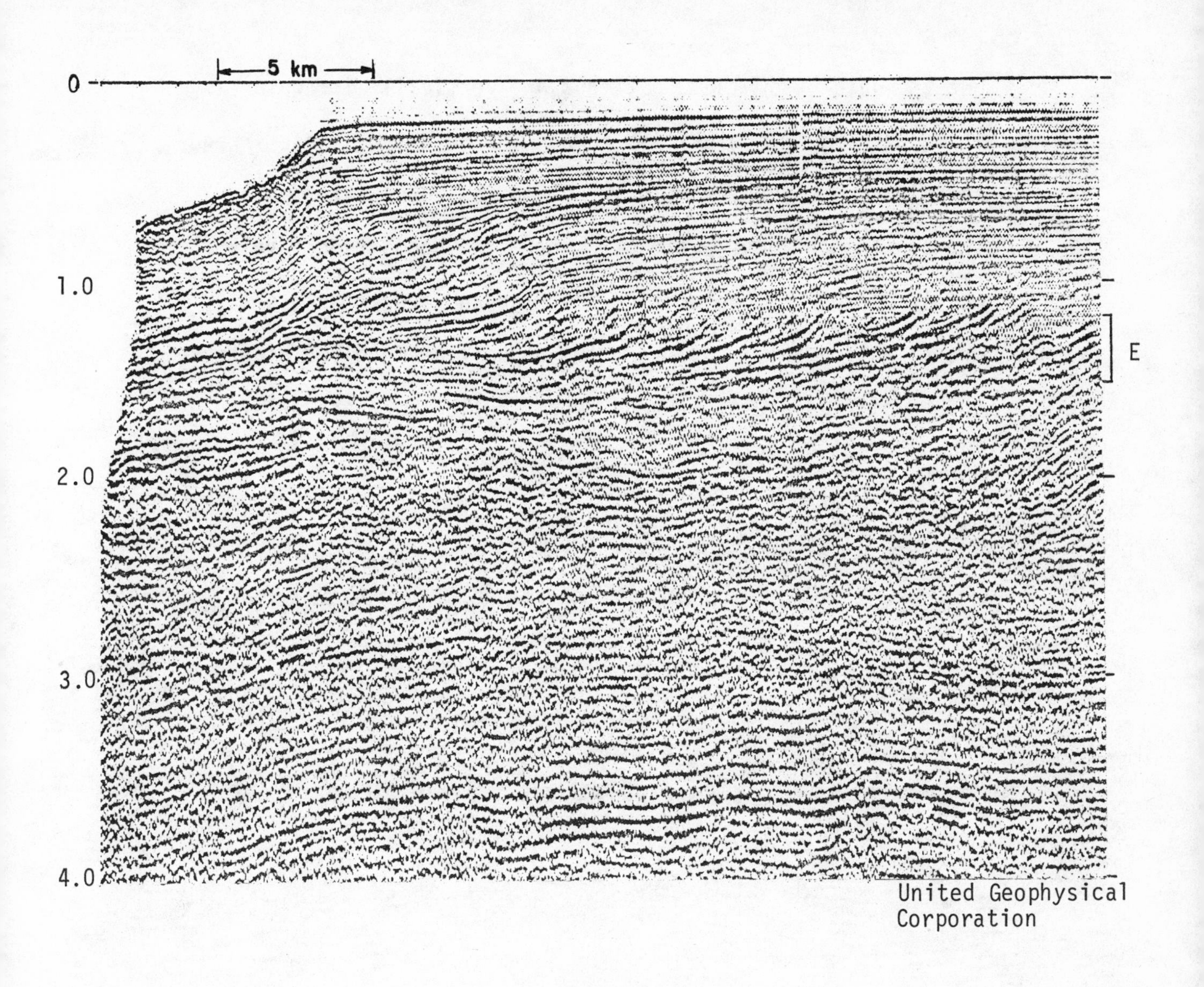

Figure 23.3

The top of Fig. 23.3 shows the ocean bottom reflection, and in the upper left-hand corner is the continental edge with the continental slope to the left. Similar continental edge patterns can be seen crossing the section from the upper left corner toward the right side around E; these clearly indicate earlier locations of the edge of the continental shelf. Some periods involved mainly outbuilding, other upbuilding, and some of the patterns are truncated by errosional cycles.

The seismic data above these continental edge patterns are flat with no irregularities to indicate statics variations, but the data below these patterns are vertically irregular as if weathering irregularities remain in the data. These irregularities are caused by velocity variations associated with the continental edge patterns. Where the deposition involved mainly carbonates, velocities are high and deeper arrivals are slightly early, but where the deposition involved clastics, the velocities are slower and deeper reflections are delayed.

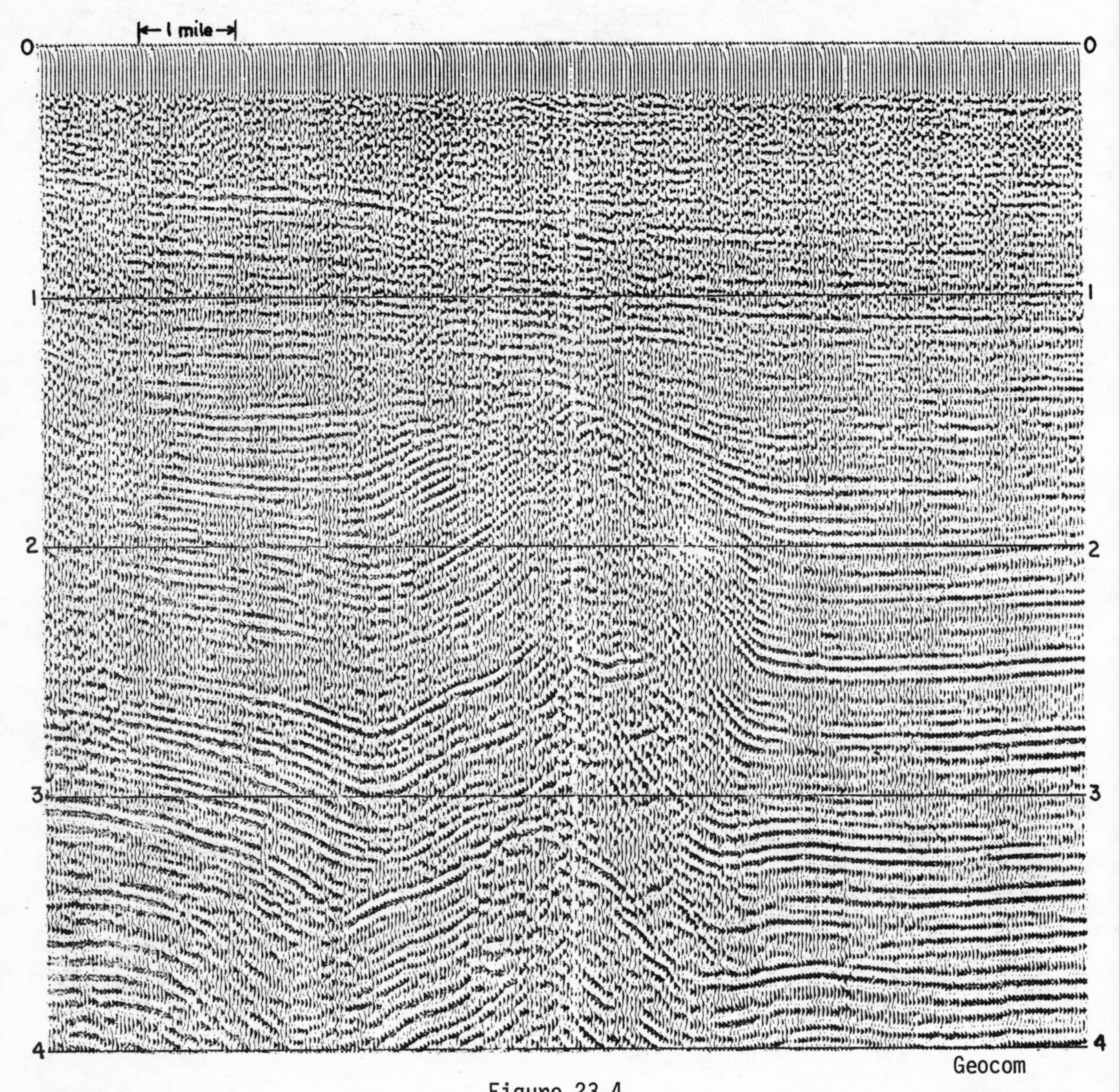

Figure 23.4

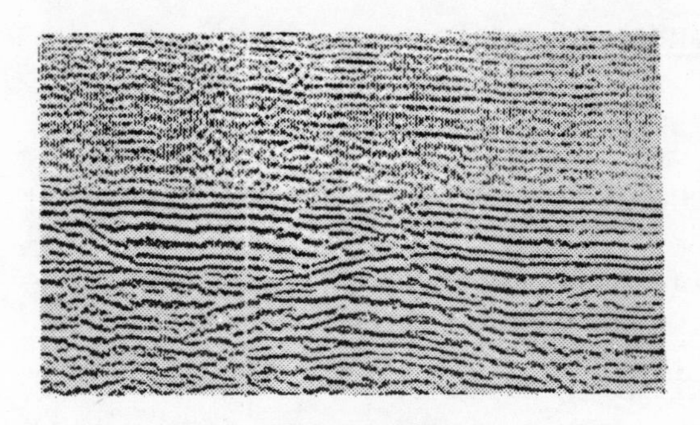

Figure 23.5a

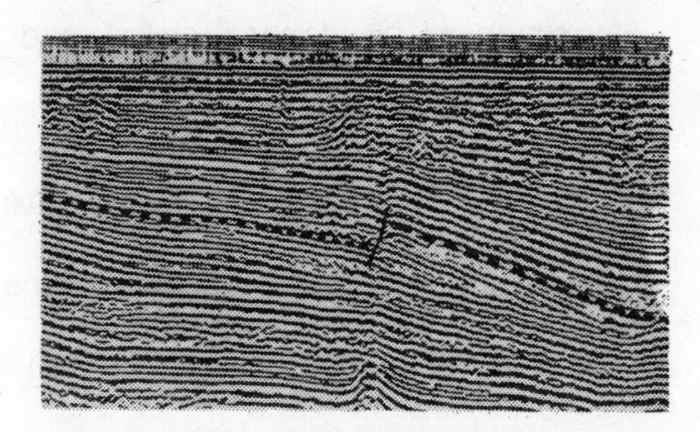

Figure 23.5b

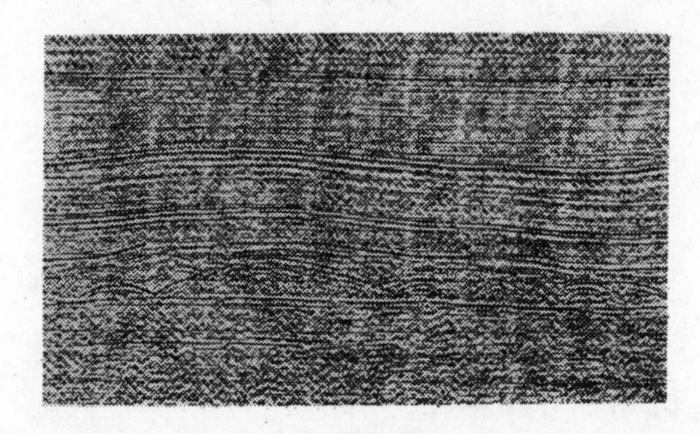

Figure 23.5c

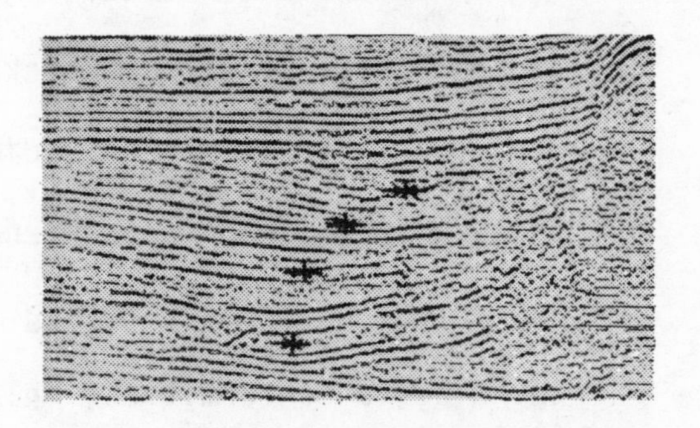

Figure 23.5d

Thus we can infer something about lithology of the continental edge patterns from these velocity anomalies.

Figure 23.4 shows tight compressional folding. The section has been migrated, apparently rather well. Some of the lithic units are bounded by reflections which are essentially parallel through the folding pattern, indicating competent rock units. Other sectional units show thickening and thinning as a result of the compressive stresses; these units involve rocks which flow when subject to stress, such as the shales.

Several other patterns which imply stratigraphic features are shown in Fig. 23.5. Figure 23.5a shows a marked change in character from a massive clastic section to a section of thick layered carbonates. The shale to a thick limestone gives rise to a strong, low-frequency reflection.

Figure 23.5b shows an interesting pattern indicating a

reversal in direction of the source of sediments, which can be seen despite an extremely ringy character. Above the dotted line all of the units thin to the left and below it, to the right.

Figure 23.5c shows the variable thickness and diffractions commonly associated with salt flowage. Similar patterns can be seen in Fig. 19.14.

Figure 23.5d shows migration of a rim syncline with depth. Local synclinal thickness indicates that the salt was moving at the time those sediments were being deposited. Many seismic stratigraphic inferences result from working out the history of an area by noting regions of local or regional thinning, unconformities, onlap patterns, etc.

Sangree proposed a classification based on the relative attitude of seismic events. He called these seismic facies; his classification scheme is shown in Fig. 23.6. Such a classification system is useful in systematizing our analysis of stratigraphic features in seismic data.

Seismic facies classification also extends to the overall form of seismic sequence units, some of which are shown in Fig. 23.7.

Sangree also shows (Fig. 23.8) an example of changes in seismic patterns where the lithology grades from non-marine on the left, through a zone of interfingering sands and shales into marine shale deposition. Note that a seismic pick would follow a contemporaneous time surface through the changes of pattern despite the changing facies.

Angularities between seismic reflection may be of several types, as shown in Fig. 23.9. Such angularities may mark seismic

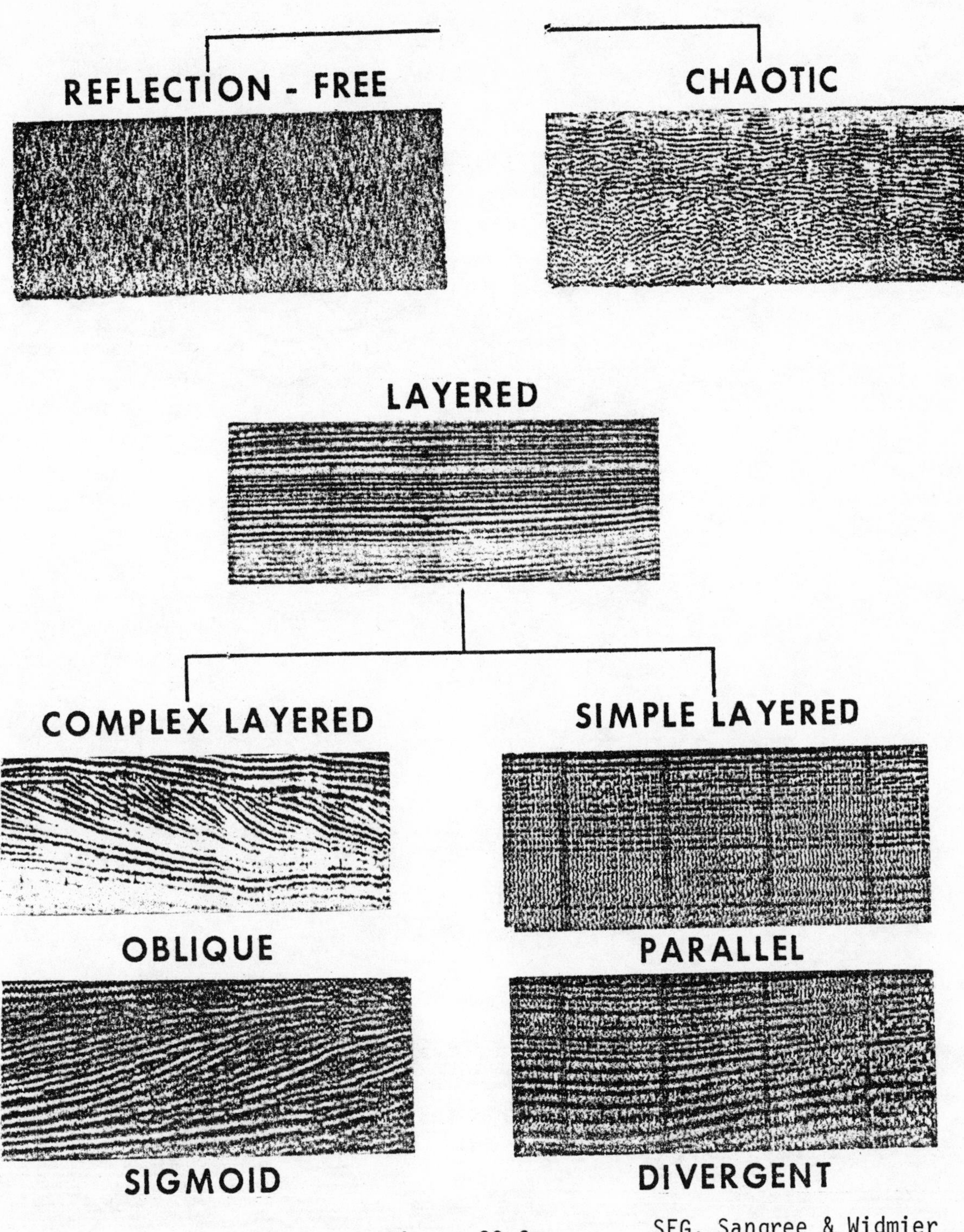

Figure 23.6

SEG, Sangree & Widmier

sequence boundaries of regional, even worldwide extent. Defining seismic sequence boundaries is the key to one type of seismic stratigraphic interpretation. Seismic sequence boundaries are not everywhere marked by angularities in the seismic data.

EXTERNAL FORM OF SEISMIC FACIES UNITS

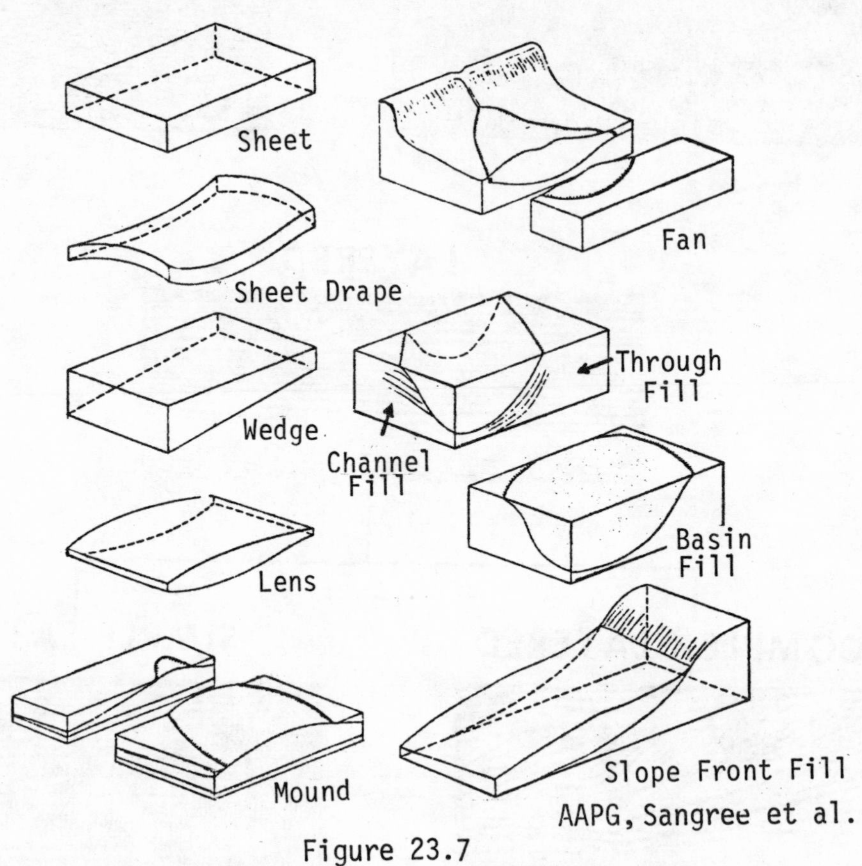

Figure 23.7

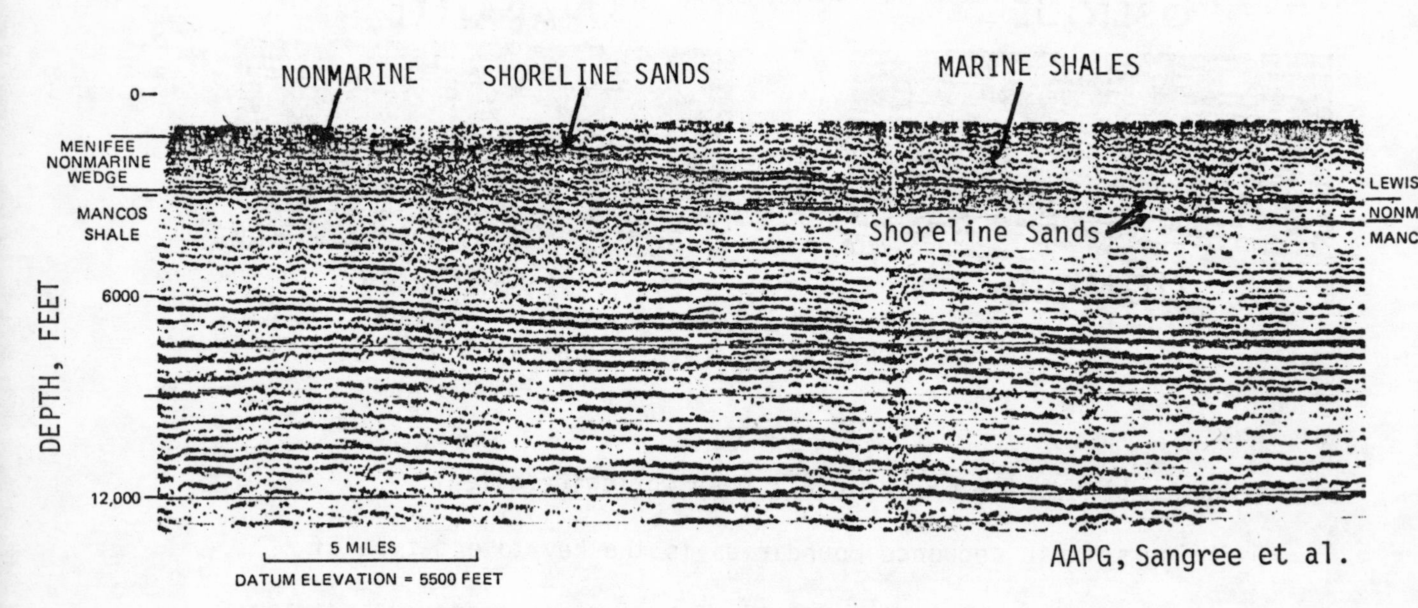

Figure 23.8

ABOVE A SEISMIC SEQUENCE BOUNDARY

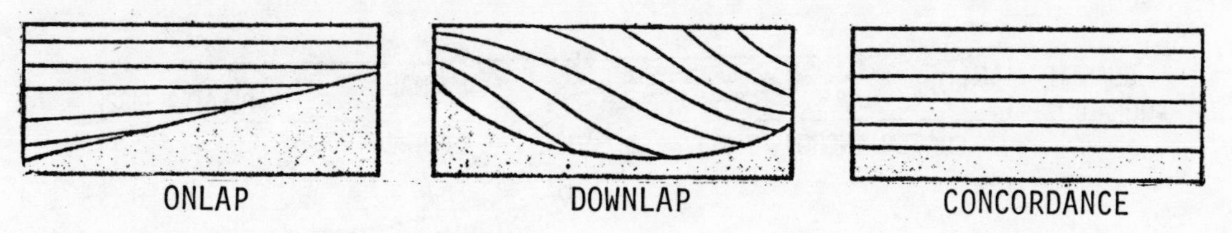

BELOW A BOUNDARY

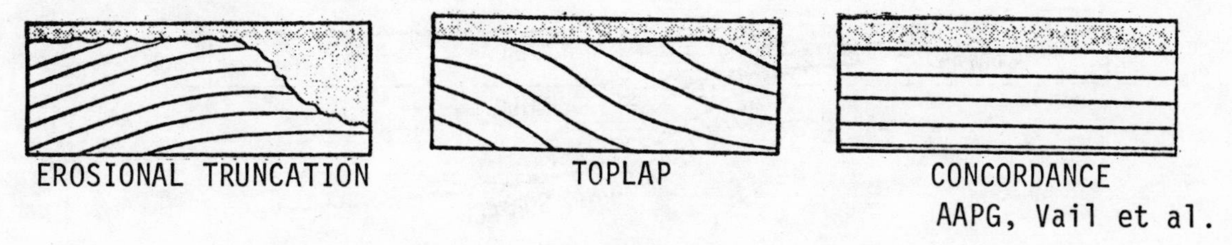

AAPG, Vail et al.

Figure 23.9

"Concordance over concordance" is a type boundary which usually has to be determined elsewhere, where beds pinch out either from below or from above the boundary, and then the boundary pick is simply carried in the same manner as other types of mapping horizons are carried. Seismic sequence boundaries are often strong reflections which may also aid in their identification. The patterns shown in Fig. 23.9 may occur in various combinations. Distinctions between "truncation" and "toplap" and between "onlap" and "downlap" are not always made, and when made are not always made in exactly the same way. Vail et al. define them in terms of present-day geometry.

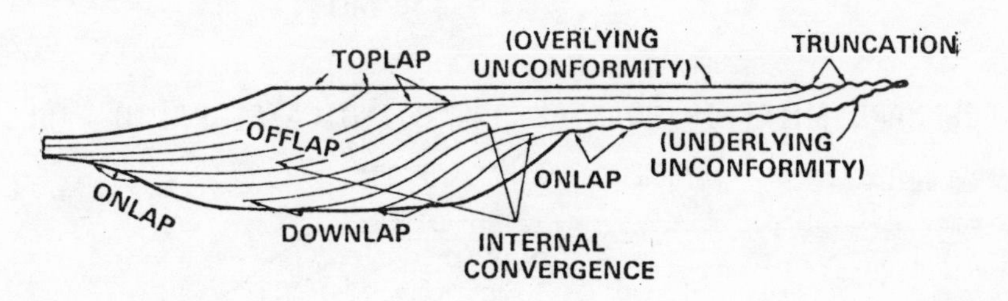

Vail et al.

Figure 23.10

LOW TERRIGENOUS INFLUX - TRANSGRESSION

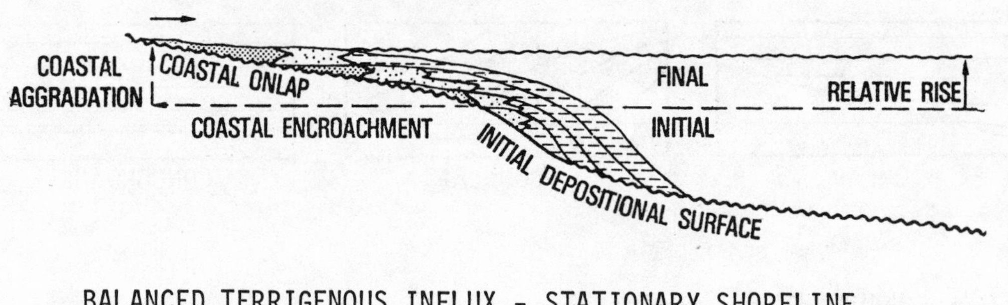

BALANCED TERRIGENOUS INFLUX - STATIONARY SHORELINE

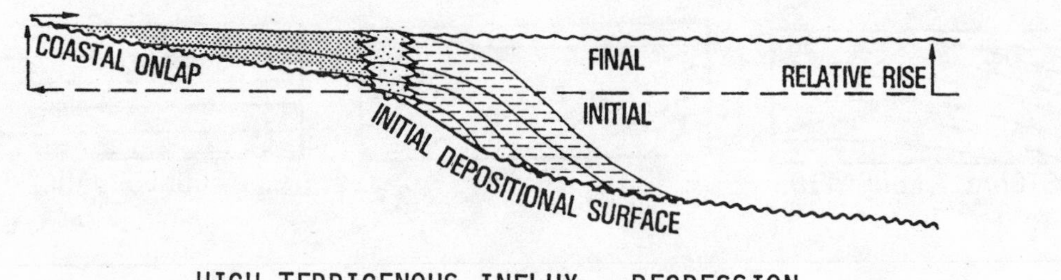

HIGH TERRIGENOUS INFLUX - REGRESSION

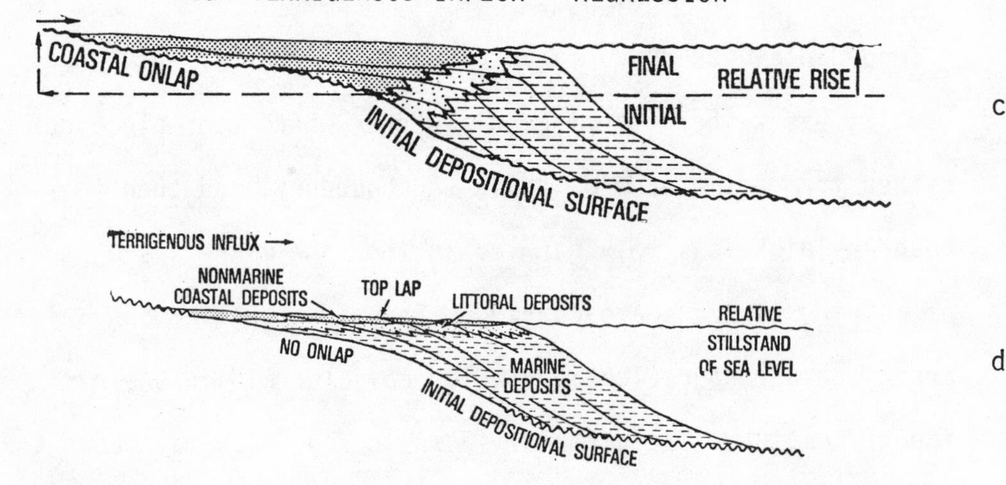

DOWNWARD SHIFT IN CLINOFORM PATTERN INDICATES GRADUAL FALL

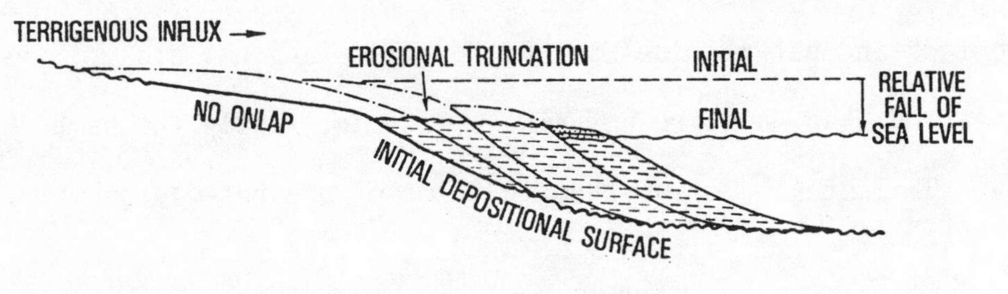

DOWNWARD SHIFT IN COASTAL ONLAP INDICATES RAPID FALL

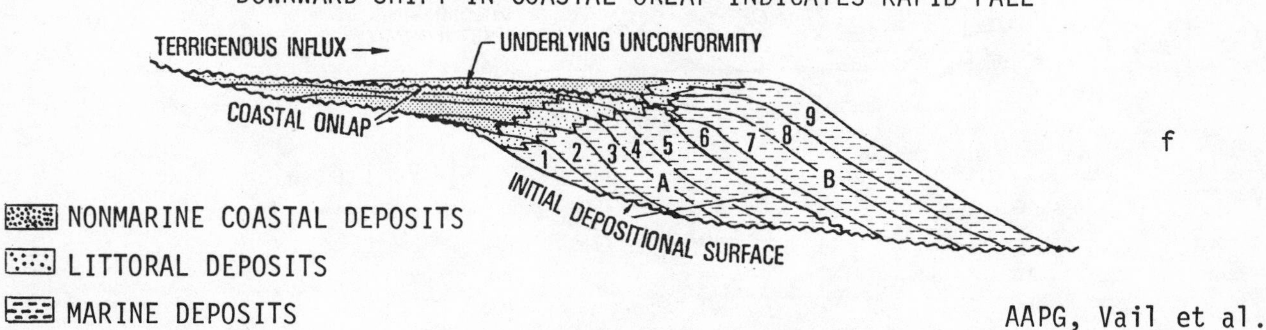

AAPG, Vail et al.

Figure 23.11

but some apply genetic concepts to their definitions.

A typical seismic sequence is shown in Fig. 23.10. A seismic sequence boundary, representing a hiatus in the depositional pattern, indicates a rise or fall of relative sea level, eustatic variations, which result in deposition shifting at a different location. The models in Fig. 23.11 indicate depositional patterns expected for a rise, stillstand and fall in sea level. If a low rate of sediments influx accompanies a rise of sea level, as in Fig. 23.11a, a transgression may result, but if the influx rate is high, a regression, as in Fig. 23.11c. If there were a gradual fall of sea level, we might expect a pattern as shown in Fig. 23.11e. Actually we rarely see this pattern, which suggests that gradual fall of sea level is uncommon. The pattern usually seen is the rapid fall of sea level pattern shown in Fig. 23.11f. It appears that rise and fall of sea level is not symmetrical throughout geological history, periods of gradual rise in sea level being followed by sudden falls of sea level.

Many of the sea level changes seem to be worldwide and these can be summarized on global cycle charts such as Fig. 23.12. The "supercycles" such as indicated on this chart can be further broken down into lesser cycles. If the local seismic sequences can be related to the global chart, it may be possible to date the sedimentary units based entirely on seismic data.

One key often useful in getting local sequences in register with the global chart hinges on identifying the unit associated with periods of exceptionally low sea level. When sea level is high, the seas lap on over the continental shelves and sedimentary units are broadly distributed, there being many sediment sources. On the other hand, when there is lowstand of sea level, the sea retreats off the continental shelf, sediment sources then have a long distance to traverse before reaching the sea and tend to become entrenched in a few canyons across the shelves resulting in local sources. The picture is illustrated in Fig. 23.13. We may be able to tell from seismic data whether we are looking at a local pod of sediments or a fairly widespread unit, and the local units indicate the periods of extremely low sea level. There are relatively few of these periods so we may be able to make a unique identification to the global chart and then relate other units also to the master charts.

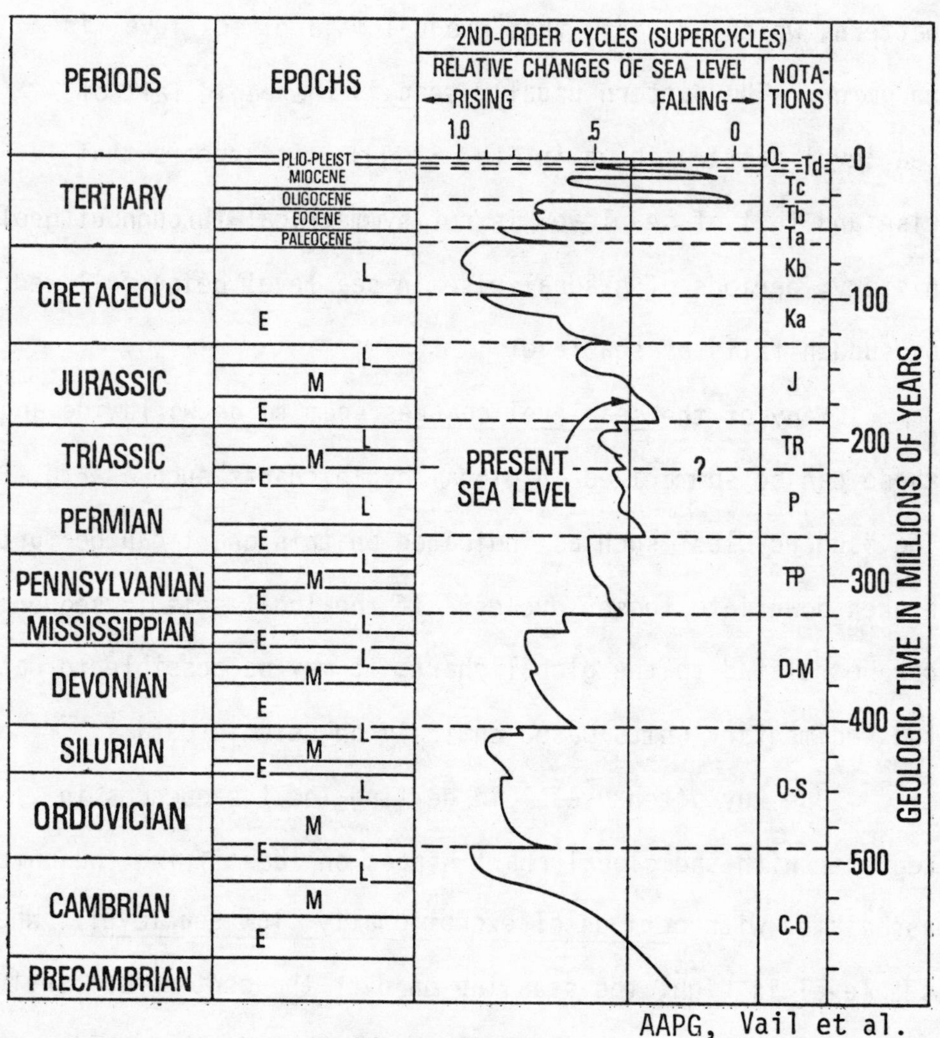

Figure 23.12

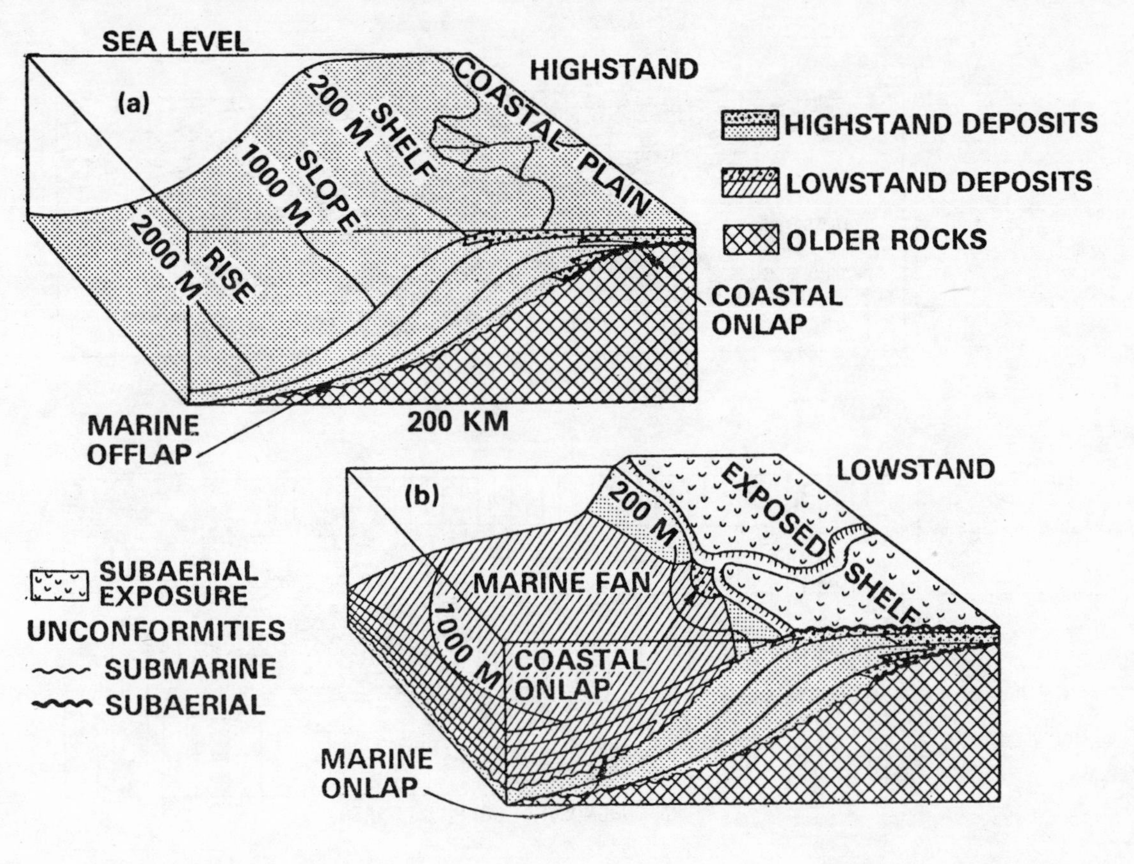

AAPG, Vail et al.

Figure 23.13

The procedure for going from a seismic section to the eustatic chart is via an intermediate chronostratigraphic chart as shown in Fig. 23.14. The stratigraphic section shown at the top of the figure results directly from seismic sequence analysis of seismic sections. The chronostratigraphic chart merely shows the geographical distribution of the seismic sequences. The vertical scale of the chronostratigraphic chart is in periods of time and the horizontal scale is in terms of location. Compare the unit labeled B on both diagrams; it has a very limited lateral extent and sea level was probably low during its deposition. From such reasoning an eustatic level chart can be constructed, as shown at the bottom of Fig. 23.14.

In order to apply these principles to actual seismic data,

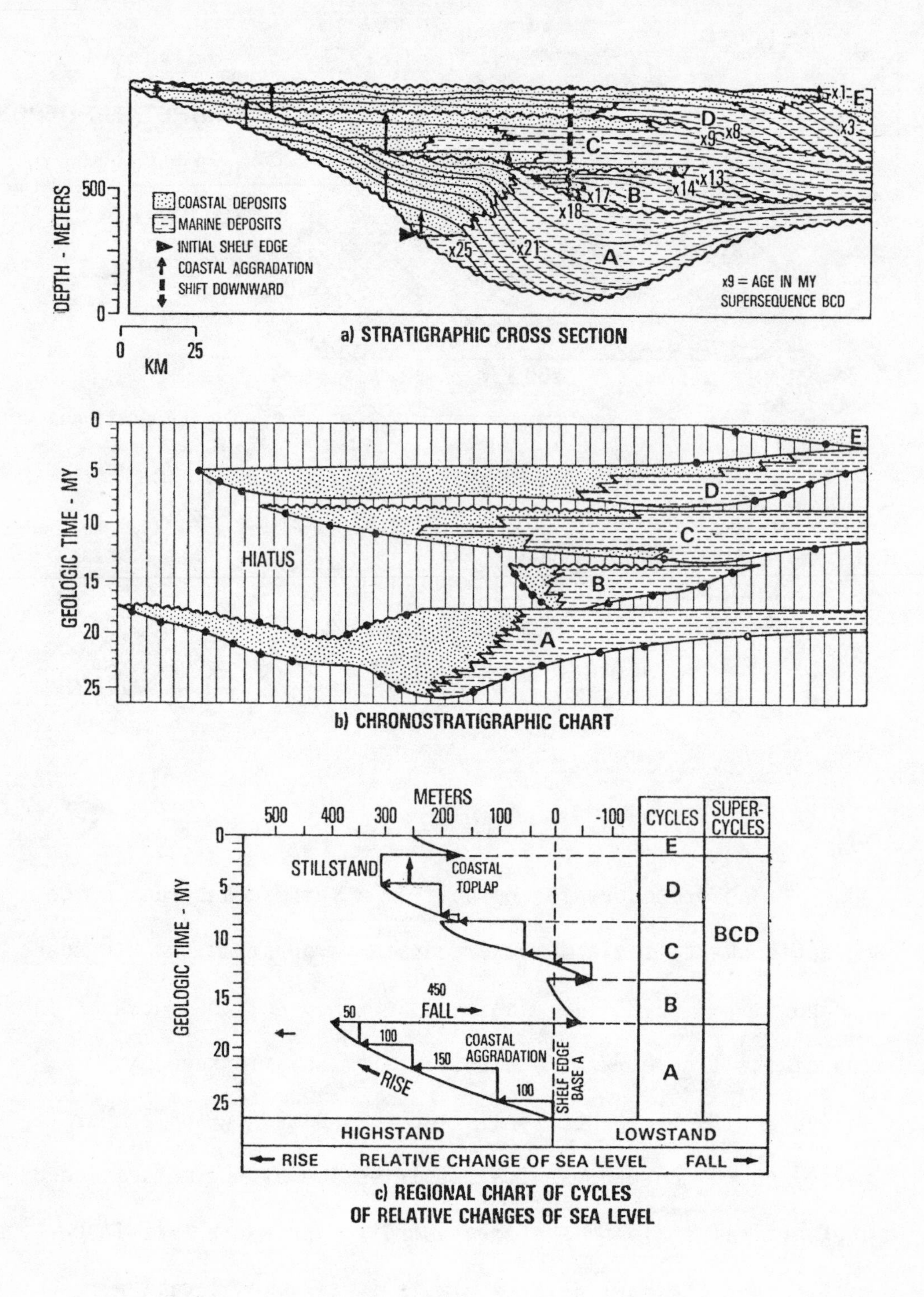

AAPG, Vail et al.

Figure 23.14

we need a section of regional length such as the one in Fig. 23.15.

Exercise:

Define the seismic sequence boundaries in Fig. 23.15.

Figure 23.15a

AAPG, Vail et al.

(Section continues on Page 290 with some overlap.)

(Section continues from Page 289 with some overlap.)

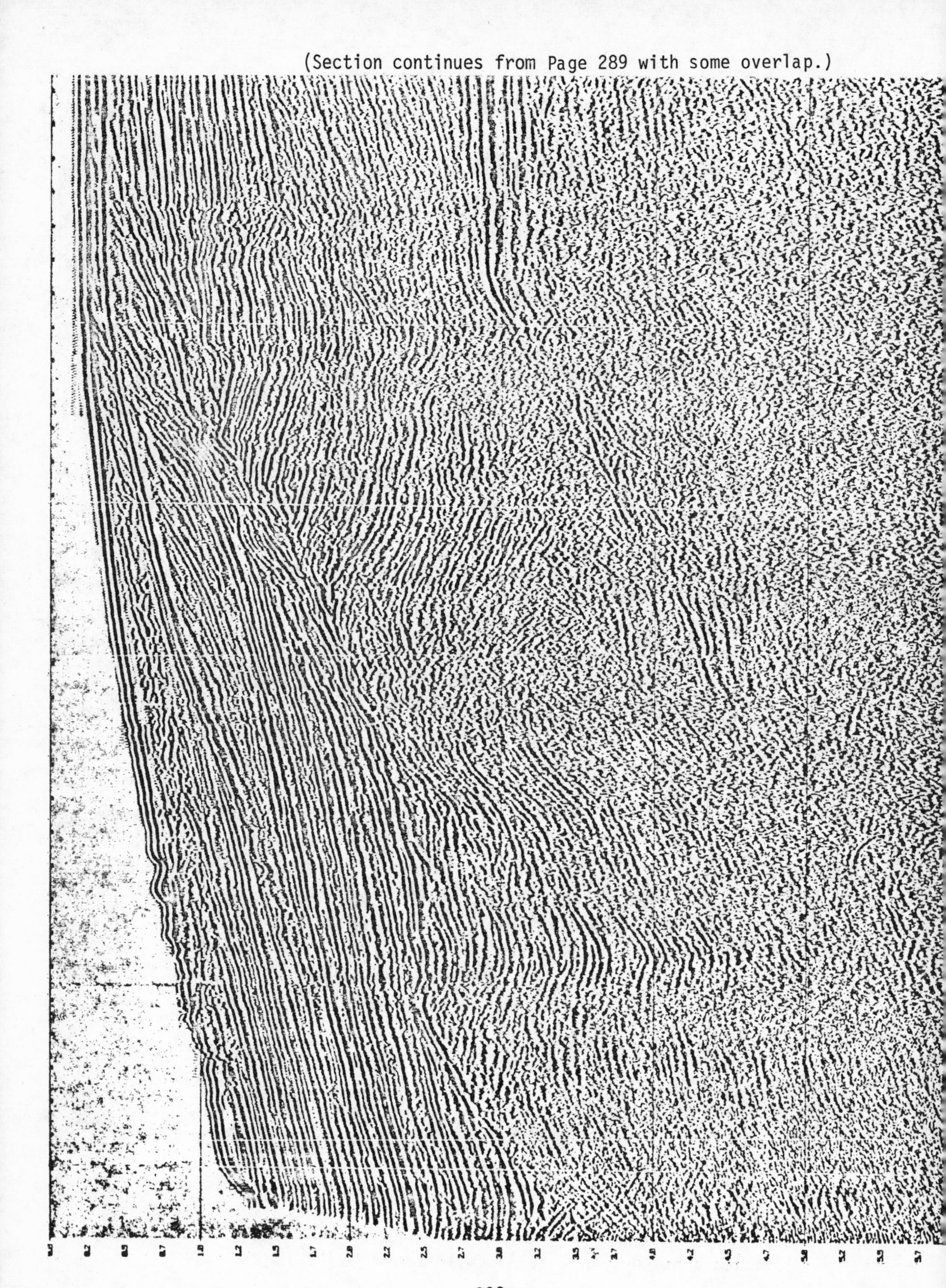

Discussion of exercise:

Some of the angularities between seismic events are marked on Fig. 23.16 by arrows. These define the sequence boundaries,

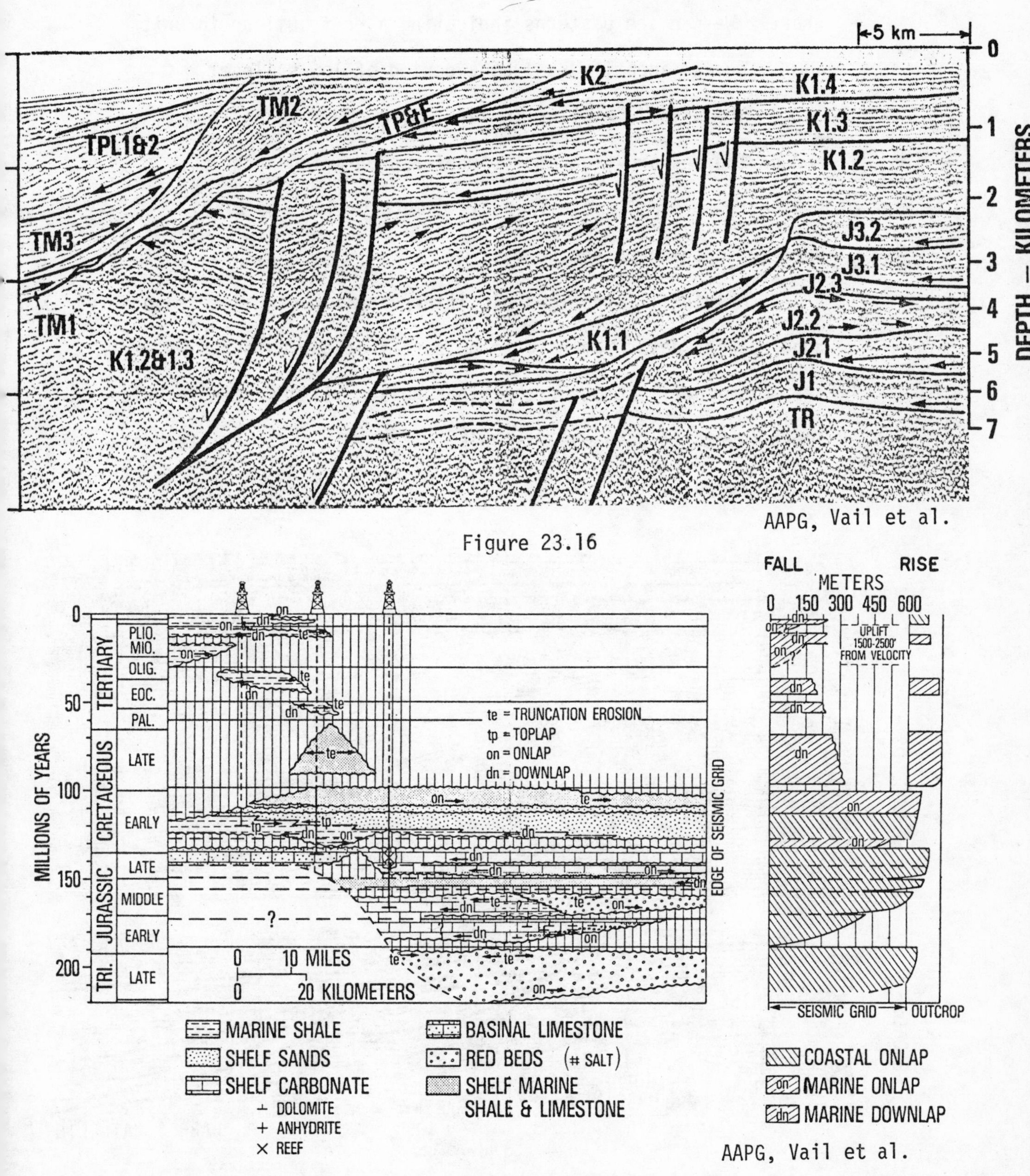

AAPG, Vail et al.

Figure 23.17

also often characterized by appreciable change in acoustic impedance and so show up as strong reflections. Figure 23.17 shows the corresponding chronostratigraphic chart and eustatic chart. We can see patterns indicating a reef buildup in unit J3.1. We can identify K1-1 as a period of low sealevel.

The seismic section in fig. 23.18 indicates a reef mass and on the seaward side of the reef a prograding wedge of sediments is deposited. There is a more-or-less parallel pattern in the back reef area.

Meaningful stratigraphic interpretation requires reasonably good quality data. We have to be sure that the patterns we see are seismic reflections rather than noise. It has been technical improvements in data quality which makes this sort of interpretation more possible today.

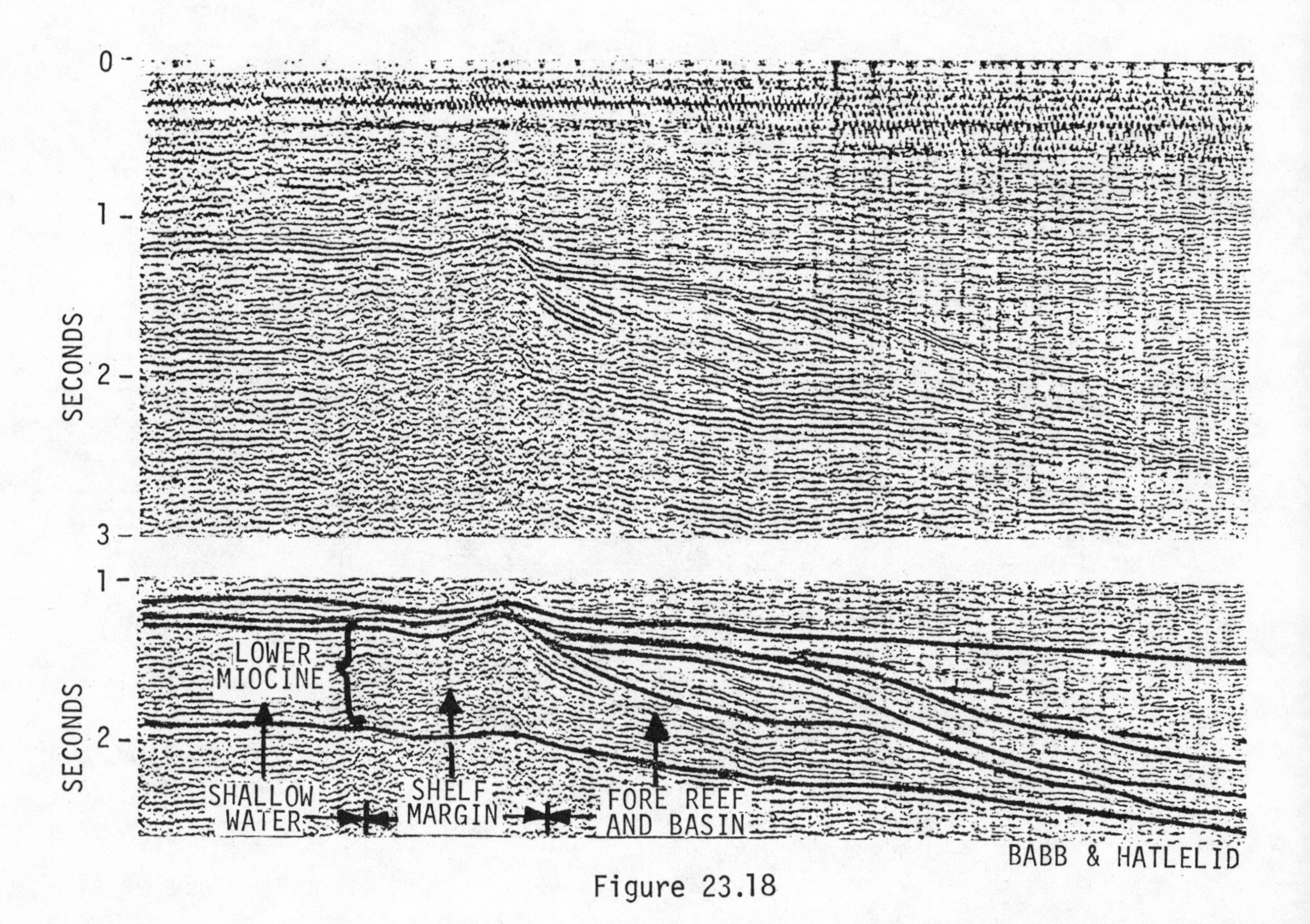

Figure 23.18

CHAPTER 24

BRIGHT SPOT

Direct hydrocarbon detection from seismic evidences is the subject of this unit. This subject is also often called bright spot. Bright spot got its name because it was evidenced by an increase in amplitude, but other evidences are often included as part of the same subject. Bright spot is also often thought of as a subset of seismic stratigraphic interpretation.

At the outset let us disclaim any reliable one-to-one correspondence between seismic evidences and hydrocarbon accumulations, although under some circumstances such evidences do exist.

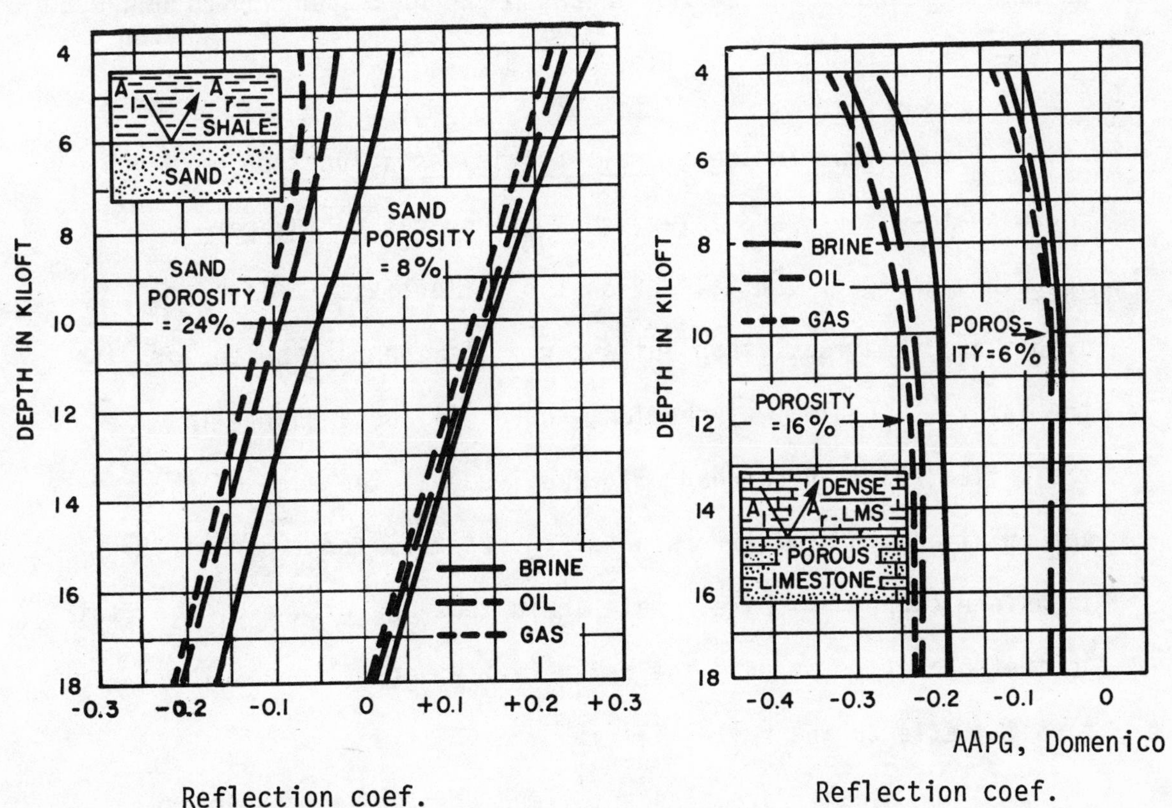

Reflection coef.

Figure 24.1

Reflection coef.

Figure 24.2

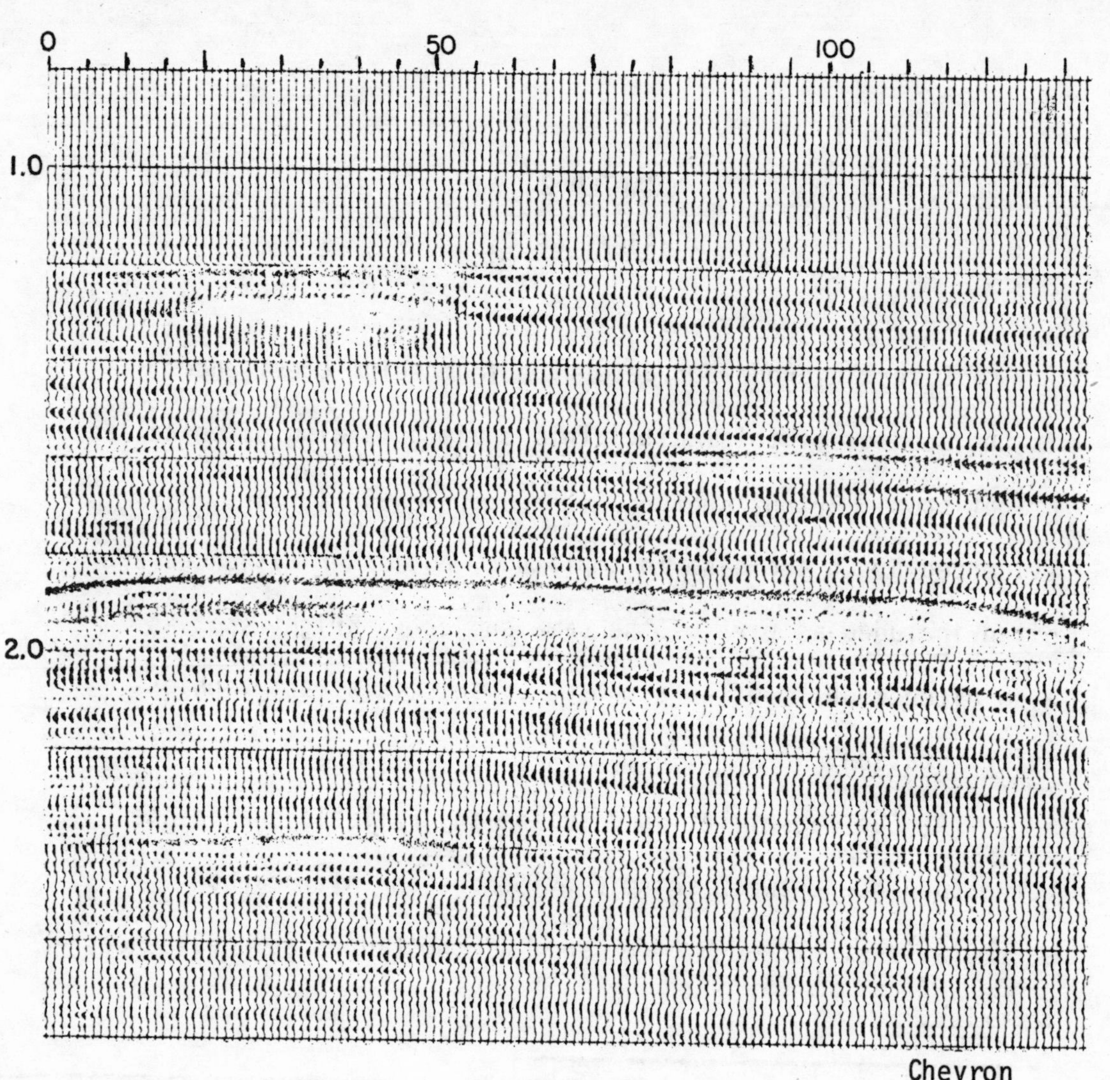

Figure 24.3

Figure 24.1 indicates reflectivity as a function of depth for a shale-porous sand interface. Oil or gas in the pore space has a significant effect on the reflectivity. If oil or gas are in a local trap but water fills the pore space for the laterally equivalent section, there will be a change in the reflectivity and a change in amplitude at the edge of the accumulation. A similar graph in Fig. 24.2 is for a porous limestone capped by shale. Note again that the nature of the fluid in the pore space affects the reflectivity, and also how the porosity affects the reflectivity.

Other effects are also sometimes associated with hydro-

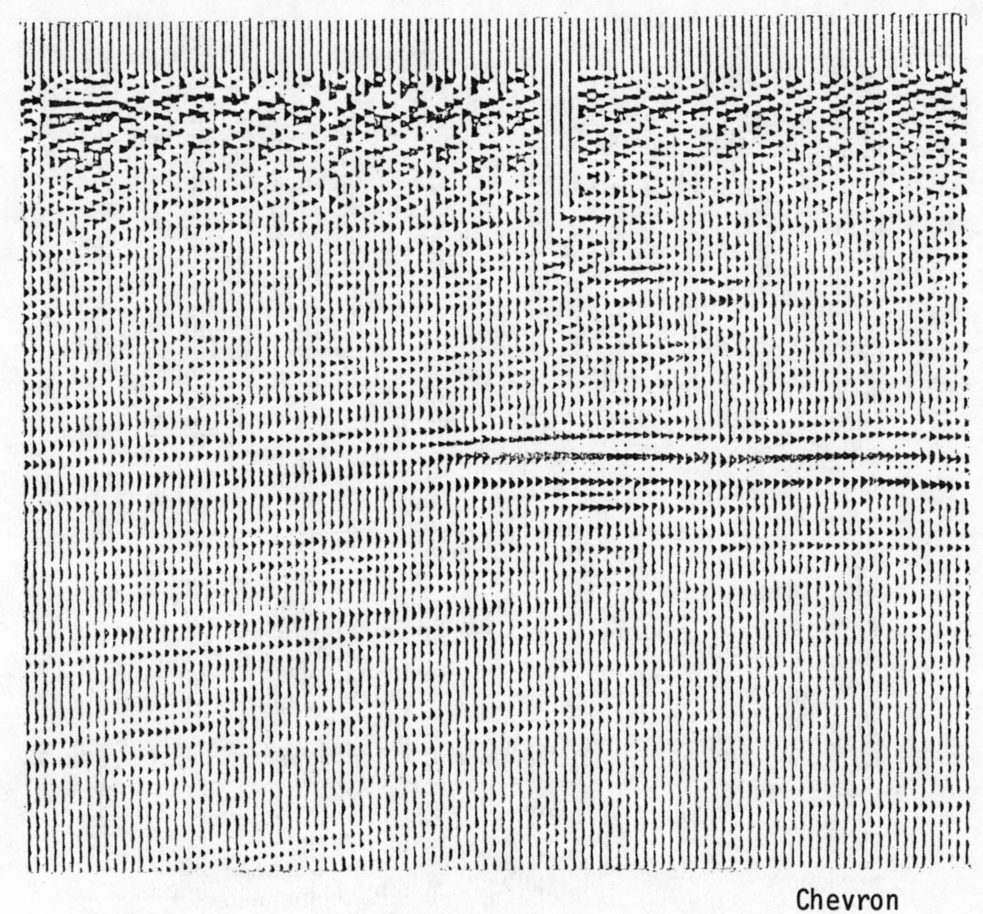

Figure 24.4

carbon accumulations. Where gas fills the pore space, velocity is less than if the pore space were filled with water and deeper reflected energy which has to pass through the accumulation will be delayed slightly. This might result in a <u>sag</u> in reflections <u>under a gas sand</u>. The amount of sag is relatively small, 6 msec for a 100 ft. thick gas sand of velocity 6500 ft./sec. vs. 8000 ft./sec., but still might be measurable on a seismic section if the sand were sufficiently thick. The seismic section in Fig. 24.3 shows a gas reservoir in the top left (from traces 18 to 52 at 1.2 sec.) and another gas reservoir slightly lower and toward the right (from about traces 65 to 117 at 1.5 sec.). A deeper or strong reflection at about 1.85 sec. appears to have slight sag because of delays introduced by the gas sands above it.

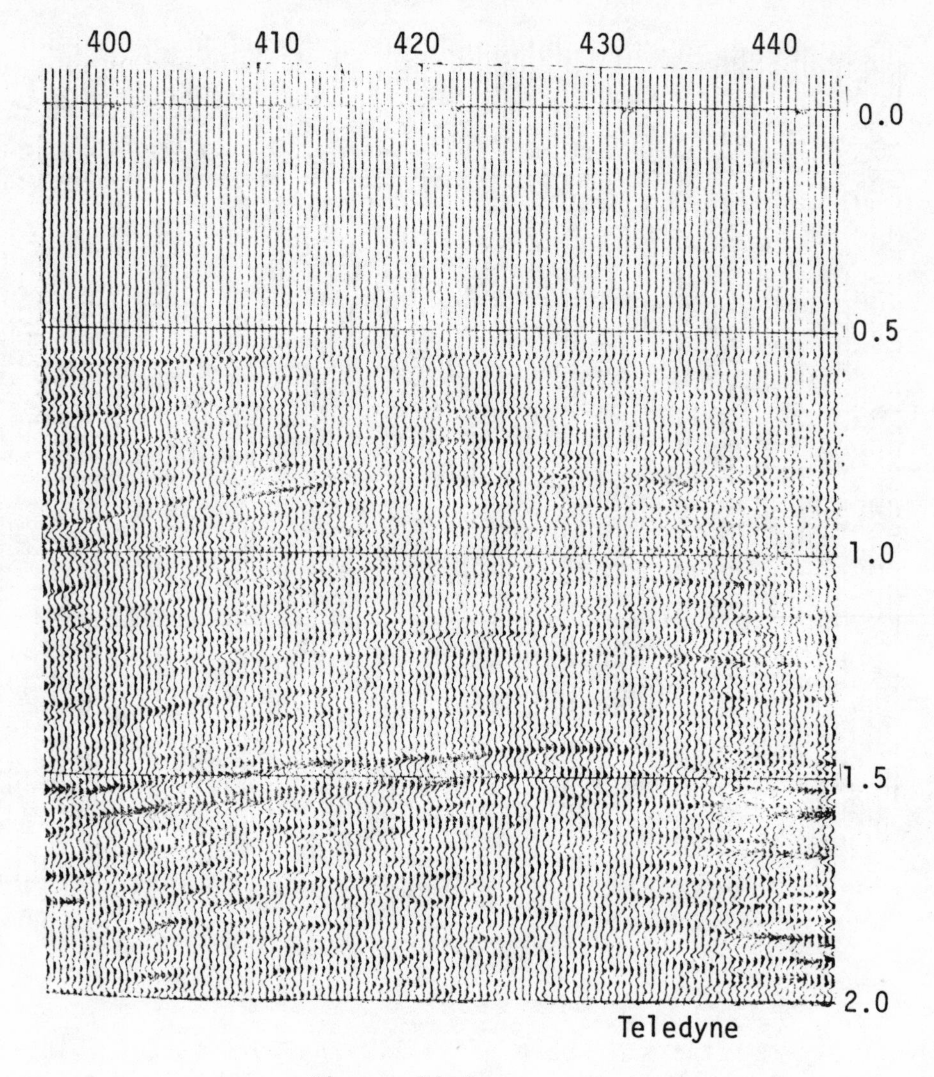

Figure 24.5

A shadow zone is sometimes seen under a gas accumulation, as in Fig. 24.4. We could expect a maximum of 10% loss of energy in the transmitted wave because of the energy reflected from a very contrasting gas sand. Increased absorption of energy in hydrocarbons is another possible contributing explanation to the observed shadow zones under gas sands. Improper amplitude handling in processing (normalizing) may also result in indications of a shadow zone.

Strong reflectors such as gas sands are also generators of multiples. A multiple of the upper gas sand can be seen in Fig. 24.3 at about 2.4 sec. It can be useful to process so as to emphasize multiples in looking for a hydrocarbon accumulation, though this does not produce a section useful for ordinary interpretation.

Phasing at the edges of a possible accumulation may also evidence it.

The presence of gas in limestone may lower the reflection coefficient in contrast to a capping of shale, resulting in a weakening of data or dim spot as evidence for hydrocarbon accumulation (Fig. 24.5). A well at S.P. 430 found gas in limestone at 1.4 sec.

Usually our reflections are from some degree of structure, but if the reflections are from a fluid contact, a flat spot may result and be the tip-off to an accumulation of hydrocarbons. A calculation of such a flat spot is shown in Fig. 10.12 and the section in Fig. 19.14 and 19.15 has such a flat spot in the crest of the anticlinal curvature at about 1.25 sec. The flat reflection is from a gas-water contact.

We have seen in Fig. 14.12 that a very small amount of gas creates a very large effect of the reflectivity. The presence of only a small amount of gas can deceive us by producing a significant anomaly without evidencing a commercial reservoir.

While reflection strength is the principal criteria for hydrocarbon accumulation, variation in frequency content is often associated with it. A low-frequency shadow often lies underneath a hydrocarbon accumulation. Color displays can be a great help in seeing such features.

The graph in Fig. 24.6 from Neidell illustrates the variation of amplitude with thickness of a wedge. The interference pattern of reflections from the top and bottom result in maximum amplitude at a quarter wavelength thickness, the "tuning thickness." The amplitude for thicknesses less than the quarter-wavelength point may be useful in helping determine the thickness of thin reservoirs. The reflection from a thin bed 5-10% of a wavelength thick is diminished only about half in amplitude compared with the reflection from a massive bed.

Bright spot technology sometimes allows us to make good estimates about where hydrocarbon accumulate and quantitative work in estimating the geographic limits of a reservoir is probably possible. In fact I believe there is a bright future for geophysics in the area of development geology. However, considerable work still needs to be done before we understand the phenomena clearly and meanwhile we must be aware of the limitations of our present understanding.

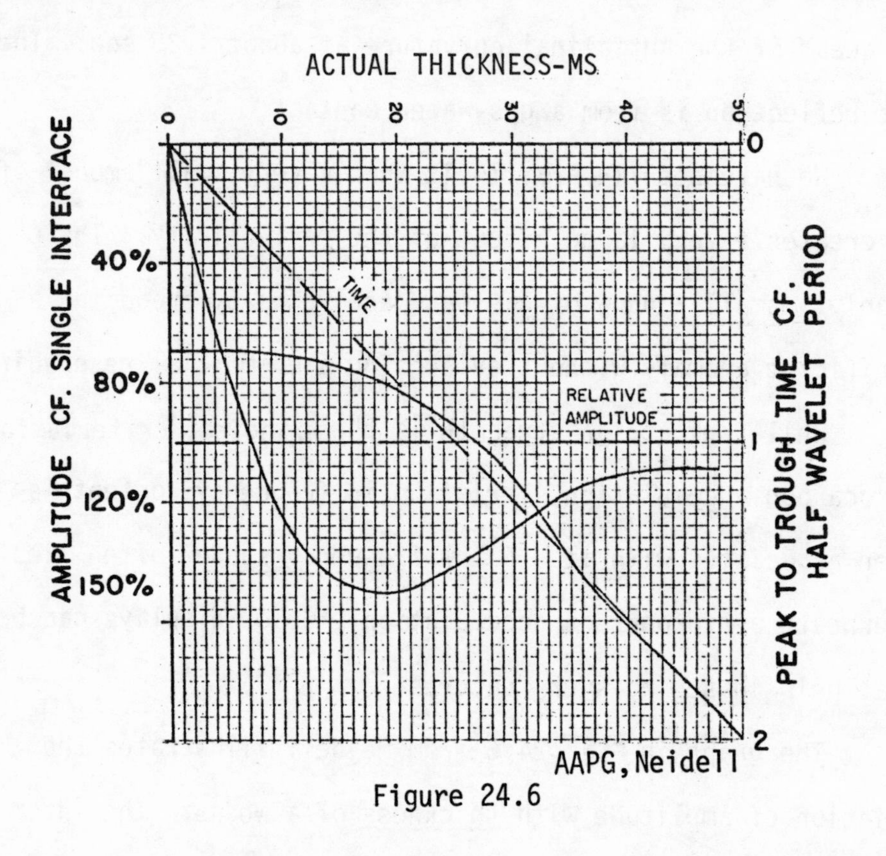

Figure 24.6

CHAPTER 25

COMPLEX TRACE ANALYSIS

Complex trace analysis and color displays are two examples of recent technology that help in seeing features in seismic data which otherwise might be missed.

Transformations of data from one form to another are common in Signal Analysis. Approaching data from different points of view often results in new insight and the discovery of relationships not otherwise evident. While rearrangement of data does not create new information, the effect is the same if it results in the discovery of significance which was not otherwise evident.

The transformation of seismic data from the time domain to the frequency domain is the most common example of data rearrangement to aid in the solution of certain problems. The Fourier transform, which accomplishes this, allows us to look at average properties of a reasonably large portion of a trace, but it does not permit examination of local variations. Analysis of seismic data as an analytic signal, complex trace analysis, is a transform technique which retains local significance. Like Fourier techniques, it provides new insight and is useful in the solution of certain problems.

Complex trace analysis effects a natural separation of amplitude and phase information, two of the quantities, called <u>attributes</u>, which are measured in complex trace analysis. The amplitude attribute is also called reflection strength. The phase information is both an attribute in its own right and the basis for instantaneous frequency measurement. Amplitude and phase information are also recombined in the additional attributes, averaged weighted frequency and apparent polarity.

Signal analysis also involves a communications problem: how to convey to an interpreter the information content of the data, a sense of the reliability of measurements, and a picture as to how information elements relate to each other. Thus data <u>display</u> is an inherent part of the problem. Seismic data are conventionally displayed in variable area, variable density, wiggle, or a combination of these forms. Display scale and vertical to horizontal scale ratio are variables whose judicious choice can facilitate the discovery of data significance. Display parameters also include trace overlap, bias, and color. Color is especially effective in communicating the results of complex trace analysis. N.A. Anstey and M. T. Taner pioneered the use of color for displaying seismic attributes.

Complex trace analysis treats a seismic trace $f(t)$ as the real part of an analytic signal or complex trace, $F(t) = f(t) + jg(t)$. The quadrature, conjugate or imaginary component $g(t)$ can be uniquely determined from the seismic trace by standard mathematical techniques. The conventional seismic trace is usually proportional to the velocity of motion of particles or the kinetic energy density is proportional to the square of the seismic trace amplitude. If we add to this the potential energy we obtain density, a quantity which is presumably proportional to the seismic energy of reflected waves. This is the objective of concentrating attention on the amplitude of the analytic signal.

Complex trace analysis results in calculations of a number of attributes.

Reflection strength, the magnitude of F(t), is the amplitude of the total seismic energy at any instant of time. Seismic sequence boundaries and sharply changed depositional conditions often give rise to strong reflection strength.

Attention in attribute analysis is usually directed at changes which occur along the bedding. Gradual lateral changes may indicate variations in the thickness of beds, which change interference patterns. Local sharp increases or decreases may indicate hydrocarbon accumulations where conditions for trapping are favorable.

Instantaneous phase is a measure of the direction of the vector F(t). It is an angle related to the proportion of the seismic energy that is kinetic at any moment; it carries the arrival-time information. The phase display emphasizes coherence and hence structural features, such as faults and changes in dip as at unconformities. Interfering events with different apparent velocity tend to show clearly even where they have different strengths. Phase aids in seeing pinchouts, progradational patterns, and flat spots.

Instantaneous frequency is the time derivative of the instantaneous phase. Frequency values often characterize continuous or semi-continuous reflections and so provide a useful correlation tool. Variations in the interference of nearby reflections, as at pinchouts and the edges of hydrocarbon-water surfaces, tend to change the instantaneous frequency. Hydrocarbon accumulations, including liquid ones, often seem to be associated with a lowering of frequency. Fracture development in brittle rocks is also sometimes associated with a lowering of frequency.

Average weighted frequency, a smoothing of the product of instantaneous frequency and reflection strength, emphasizes the frequency of the stronger events and smooths irregularities because of noise. The frequency values so obtained agree roughly with "dominant frequency values" determined by measuring peak-to-peak times or other similar phase points. Like instantaneous frequency displays, average weighted frequency displays sometimes are excellent for correlation. Liquid hydrocarbon accumulations are sometimes especially evidenced by characteristic low frequencies.

Apparent polarity, the sign of a trace at reflection strength maxima, is another attribute often calculated. The positive or negative sense of apparent polarity refers to the sign of the associated reflection coefficient where the reflection indicates a single interface and where the absolute polarity can be established. Inasmuch as most reflections are interference composites, polarity often lacks a clear correlation with reflection coefficient and hence the qualifying adjective, "apparent".

The effectiveness of complex trace analysis (and color display) depends on data quality. Complex trace analysis allows looking at data in new ways which help in seeing features of geologic and hydrocarbon accumulation significance, but does not invent information not in the original data. The various attributes tell more as a set than they do individually. Features are often anomalous in systematic ways on various displays and so reinforce confidence. Differences in the way features appear in different displays also convey geologic information.

Most stratigraphic interpretation begins with the structural setting. Attribute patterns aid in correlation, and vertical offset of patterns helps establish the throw across faults. However. it is the variation along the bedding that is of principal interest with all of the displays. Where data quality and processing are sufficiently good that data represent primary reflected energy in correct proportion with little noise, a lateral change in attributes indicates a lateral change in geology. The nature of the change in geology may not be apparent, however, without ties to well control and other data. The more additional data that can be brought together with geologic reasoning in an interpretation, the more significance that can be attributed to these measurements. Those familiar with geologic situations find significance which others miss.

CONCLUSIONS

If there is any common conclusion to be drawn from the foregoing diversity of subjects, I think it is this: to achieve the most probable, most useful geologic information, we must utilize all our resources and have all our efforts working toward our common goal. Specifically,

(1) we must find common solution to all our data, gravity, magnetic, seismic, well, other geologic data; a conflict with any of these indicates something is wrong;

(2) we must communicate our information and our objectives so that data acquisition, processing and interpretation decisions are consistent with our objectives.

SOURCES OF ILLUSTRATIONS

Illustrations were provided by the following companies, who are hereby thanked:

Amoco Production Company (Pan American)
Compagnie Generale de Geophysique (CGG)
Chevron Oil Company
Geophysical Service Inc. (GSI)
Petty-Ray Geophysical, Inc.
Seiscom Delta Inc. (SD)
Seismograph Service Corporation
Teknica Resource Development Ltd.
Teledyne Exploration Company
United Geophysical Corporation
West Australian Petroleum Co. Ltd. (Wapet)
Western Geophysical Company of America

Illustrations were also taken from published literature, as follows:

Angona, F.A., Two dimensional modeling and its application to seismic fault problems: Geophysics V.25, p. 468ff, 1960.

Domenico, S.N., Lithology and velocity: notes for AAPG-SEG Stratigraphic interpretation of seismic data school, 1977.

Faust, L.Y., Seismic velocity as a function of depth and geologic time: Geophysics, V.16, p. 192ff, 1951.

Garotta, R. and D. Michon, Continuous analysis of the velocity function: Geophysical Prospecting V.15, p. 584ff,

Grant, F. and G. West, Interpretation theory in applied geophysics: McGraw Hill Book Co., 1965.

Gardner, G.H.F. L.W. Gardner and A.R. Gregory, Formation velocity and density - the diagnostic basics for stratigraphic traps: Geophysics V.39, p. 770ff, 1974.

Gregory, A.R., Some aspects of rock physics from laboratory and log data that are important to seismic interpretation: notes for AAPG-SEG Stratigraphic interpretation of seismic data school, 1977.

Hagedoorn, J.G., A process of seismic reflection interpretation: Geophysical Prospecting, V.2, p. 85ff, 1954.

Hilterman, F.J., Lithologic determination from seismic velocity data: notes for AAPG-SEG Stratigraphic interpretation of seismic data school, 1976.

Jankowsky, W., Empirical investigation of some factors affecting elastic wave velocities in carbonate rocks: Geophysical Prospecting, V.18, p. 103ff, 1970.

Lindseth, Roy, Seislogs: notes for AAPG-SEG Stratigraphic interpretation of seismic data school, 1976.

Neidell, N.S., and E. Poggiagliolmi, Stratigraphic modeling and interpretation - geophysical principles and techniques, in C.E. Payton, ed., Seismic stratigraphy - applications to hydrocarbon exploration: Am. Assn. of Petroleum Geologists, 1977.

Nettleton, L.L., Gravity and magnetics in oil prospecting: McGraw Hill Book Co., 1976.

O'Doherty, R.F., and N.A. Anstey, Reflections on amplitudes: Geophysical Prospecting, V.19, p. 430ff, 1971.

Richards, T.C., Wide angle reflections and their application to finding limestone structures in the foothills of Western Canada: Geophysics, V.25, p. 285ff, 1960.

Sangree, J.B., and J.M. Widmier, Seismic interpretation of clastic depositional facies: in Vail etal (see below).

Vail, P.R., R.M. Mitchum, R.G. Todd, J.M. Widmier, S. Thompson, J.B. Sangree, J.N. Bubb and W.G. Hatlelid, Seismic stratigraphy and global changes of sea level: in C.E. Payton, ed., Seismic stratigraphy - applications to hydrocarbon exploration: Am. Assn. Petroleum Geologists, 1977.

INDEX